Molecular Imaging i...

...e Practical Approach Series

SERIES EDITORS

D. RICKWOOD
Department of Biology, University of Essex
Wivenhoe Park, Colchester, Essex CO4 3SQ, UK

B. D. HAMES
Department of Biochemistry and Molecular Biology,
University of Leeds, Leeds LS2 9JT, UK

Affinity Chromatography
Anaerobic Microbiology
Animal Cell Culture (2nd Edition)
Animal Virus Pathogenesis
Antibodies I and II
Behavioural Neuroscience
Biochemical Toxicology
Biological Data Analysis
Biological Membranes
Biomechanics—Materials
Biomechanics—Structures and Systems
Biosensors
Carbohydrate Analysis
Cell–Cell Interactions
(The) Cell Cycle
Cell Growth and Division
Cellular Calcium
Cellular Interactions in Development
Cellular Neurobiology
Centrifugation (2nd Edition)
Clinical Immunology
Computers in Microbiology
Crystallization of Nucleic Acids and Proteins
Cytokines
The Cytoskeleton
Diagnostic Molecular Pathology I and II
Directed Mutagenesis
DNA Cloning I, II, and III
Drosophila
Electron Microscopy in Biology
Electron Microscopy in Molecular Biology
Electrophysiology
Enzyme Assays
Essential Molecular Biology I and II
Essential Developmental Biology
Experimental Neuroanatomy
Fermentation
Flow Cytometry
Gas Chromatography
Gel Electrophoresis of Nucleic Acids (2nd Edition)

Gel Electrophoresis of Proteins (2nd Edition)
Gene Targeting
Gene Transcription
Genome Analysis
Glycobiology
Growth Factors
Haemopoiesis
Histocompatibility Testing
HPLC of Macromolecules
HPLC of Small Molecules
Human Cytogenetics I and II (2nd Edition)
Human Genetic Disease Analysis
Immobilised Cells and Enzymes
Immunocytochemistry
In Situ Hybridization
Iodinated Density Gradient Media
Light Microscopy in Biology
Lipid Analysis
Lipid Modification of Proteins
Lipoprotein Analysis
Liposomes
Lymphocytes
Mammalian Cell Biotechnology
Mammalian Development
Medical Bacteriology
Medical Mycology
Microcomputers in Biochemistry
Microcomputers in Biology
Microcomputers in Physiology
Mitochondria
Molecular Genetic Analysis of Populations
Molecular Imaging in Neuroscience
Molecular Neurobiology
Molecular Plant Pathology I and II
Molecular Virology
Monitoring Neuronal Activity
Mutagenicity Testing
Neural Transplantation
Neurochemistry
Neuronal Cell Lines
NMR of Biological Macromolecules
Nucleic Acid and Protein Sequence Analysis
Nucleic Acid Hybridisation
Nucleic Acids Sequencing
Oligonucleotides and Analogues
Oligonucleotide Synthesis
PCR
Peptide Hormone Action
Peptide Hormone Secretion
Photosynthesis: Energy Transduction
Plant Cell Culture
Plant Molecular Biology
Plasmids (2nd Edition)
Pollination Ecology
Postimplantation Mammalian Embryos
Preparative Centrifugation
Prostaglandins and Related Substances
Protein Architecture
Protein Engineering
Protein Function

Protein Phosphorylation
Protein Purification Applications
Protein Purification Methods
Protein Sequencing
Protein Structure
Protein Targeting
Proteolytic Enzymes
Radioisotopes in Biology
Receptor Biochemistry
Receptor–Effector Coupling
Receptor–Ligand Interactions
Ribosomes and Protein Synthesis
RNA Processing I and II
Signal Transduction
Solid Phase Peptide Synthesis
Spectrophotometry and Spectrofluorimetry
Steroid Hormones
Teratocarcinomas and Embryonic Stem Cells
Transcription Factors
Transcription and Translation
Tumour Immunobiology
Virology
Yeast

Molecular Imaging in Neuroscience

A Practical Approach

Edited by

N. A. SHARIF

Alcon Laboratories Inc., Fort Worth, Texas

Oxford New York Tokyo

Oxford University Press, Walton Street, Oxford OX2 6DP
Oxford New York Toronto
Delhi Bombay Calcutta Madras Karachi
Kuala Lumpur Singapore Hong Kong Tokyo
Nairobi Dar es Salaam Cape Town
Melbourne Auckland Madrid
and associated companies in
Berlin Ibadan

Oxford is a trade mark of Oxford University Press

A Practical Approach is a registered trade mark
of the Chancellor, Masters, and Scholars of the University of Oxford
trading as Oxford University Press

Published in the United States
by Oxford University Press Inc., New York

A catalogue record for this book is available from the British Library

Library of Congress Cataloging in Publication Data
Molecular imaging in neuroscience: a practical approach / edited by
N. A. Sharif. – 1st ed.
(Practical approach series)
Includes bibliographical references and index.
1. Molecular neurobiology – Methodology. 2. In situ hybridization –
Methodology. 3. Flow cytometry – Methodology. 4. Fluorescence
microscopy – Methodology. 5. Neurotransmitter receptors – Research –
Methodology. I. Sharif, N. A. (Najam Arfeen) II. Series.
QP356.2.M634 1993 612.8'0724–dc20 93–17576
ISBN 0–19–963381–9 (hbk.) ISBN 0–19–963380–0 (pbk.)

Typeset by Footnote Graphics, Warminster, Wilts
Printed in Great Britain by Information Press Ltd, Eynsham, Oxon

Preface

A few years ago I edited a book entitled *Brain Imaging: Techniques and Applications*. This covered the discovery, refinement and application of a few selected imaging techniques in a reviews-type format. The present book has been compiled to emphasize the practical aspects of certain technologies that have matured enough to be of use to many disciplines, in particular to neuroscience.

Imaging technologies are among the most rapidly growing tools of modern life-science research. This is certainly due to the enormous increase in the availability and use of powerful sophisticated computers and other dedicated machines to further the study of structure and function of the brain. Research in this arena has also been fuelled by the appreciation that a multidisciplinary approach, employing many different imaging techniques to tackle a given problem, is a most effective and productive way to conduct neurobiological research today. While pictures themselves are powerful media for recording important information, the ability to analyse and quantify these images by computer-assisted scanning methods can greatly enhance their value. Hence, the need to be quantitative.

This book details laboratory protocols for imaging neurotransmitter and receptor mRNAs using radioactive and non-radioactive *in situ* hybridization histochemistry. In addition, methods are provided for visualizing and quantifying transmitter and drug receptors, enzymes, and second messenger transducing elements using quantitative autoradiography, sequence-specific antibodies and immunocytochemistry. As a natural extension of the above procedures, the book describes protocols for imaging biological activity (for example, Ca^{2+} mobilization; pH changes) in living cells, and for evaluating the pharmacology of the cell surface receptors and their signalling machinery, using real-time computer-assisted video microscopy, flow cytometry, and spectrofluorometry.

It is particularly pertinent to launch a book on *Molecular Imaging in Neuroscience* in this 'The decade of the brain'. It is clear, however, that the task of covering all aspects of imaging technology is a difficult if not impossible one given the vastness and the dynamic and ever changing emphasis of the field. With these caveats in mind, I have attempted to assimilate chapters that deal with some of the most common and practically viable areas of imaging research as applied in the laboratory. Inevitably, therefore, there are likely to be some omissions and some overlap in the material presented in the different chapters. Nevertheless, this book should provide a useful place to start thinking about modern imaging technology and its uses in neurobiology and other fields.

I extend my gratitude to the contributing authors and the publisher for making this book a reality. Hopefully the readers will accept this offering in the spirit of its overall objectives! The book should prove useful for beginners and established researchers wishing to exploit these contemporary imaging techniques for neuropharmacological, neuroanatomical, and pharmaceutical research.

Fort Worth N.A.S.
March 1993

Acknowledgements

I would like to thank very much all the contributors for their commitment to this book and for writing fine, informative chapters by the imposed deadlines! The generous support and enthusiasm of the staff of IRL at Oxford University Press is gratefully acknowledged.

I wish to thank my scientific mentors, especially Dr Peter Roberts, Dr David Burt, Professor Tim Hawthorne, Dr Ray Hill, Dr John Hughes, Dr Roger Whiting, and Dr Bill Smith for their friendship, guidance, and inspiration over the years.

Last, but not least, it is with the greatest of pleasure that I acknowledge the continued and generous support and encouragement of my wife (Amrita) and my children (Imran and Nadhia), and my parents, brother, and sisters, to all of whom I dedicate this book.

Contents

Contributors

FRANK BALDINO, Jr.
Cephalon Inc., 145 Brandywine Parkway, W. Chester, PA 19380, USA.

M. A. CAREW
Department of Neurobiology, AFRC Institute of Physiology, Babraham, Cambridge CB2 4AT, UK.

MARTA E. COUCE
Department of Anatomy and Cell Biology, East Carolina University School of Medicine, Greenville, NC 27858-4354, USA.

I. DAVISON
Department of Neurobiology, AFRC Institute of Physiology, Babraham, Cambridge CB2 4AT, UK.

JOHN F. DUNNE
Institute of Molecular Immunology (S3-6), Syntex Discovery Research, 3401 Hillview Avenue, Palo Alto, CA 94304, USA.

S. B. DUNNETT
Department of Experimental Psychology, University of Cambridge, Downing Street, Cambridge CB2 3EB, UK.

R. M. EGLEN
Institute of Pharmacology, Syntex Discovery Research, 3401 Hillview Avenue, Palo Alto, CA 94304, USA.

M. HORTON
Haemopoiesis Research Group, ICRF, St Bartholomew's Hospital, London, UK.

J. HOYLAND
Department of Neurobiology, AFRC Institute of Physiology, Babraham, Cambridge CB2 4AT, UK.

J. JONASSEN
Department of Neurobiology, AFRC Institute of Physiology, Babraham, Cambridge CB2 4AT, UK.

ALLAN I. LEVEY
Department of Neurology, Emory University School of Medicine, Woodruff Memorial Building, Suite 6000, P.O. Drawer V, Atlanta, GA 30322, USA.

J. M. LEWIS
Scripps Research Institute, Blake Rm. 209, CVB 4, 10666 N. Torrey Pine Road, La Jolla, CA 92037, USA.

MICHAEL E. LEWIS
Symphony Pharmaceuticals, Inc., 76 Great Valley Parkway, Malvern, PA 19355, USA.

P. M. LLEDO
Institut Alfred Fessard, CNRS, Gif-sur-Yvette, France.

JACQUELINE F. McGINTY
Department of Anatomy and Cell Biology, East Carolina University School of Medicine, Greenville, NC 27858-4354, USA.

W. T. MASON
Department of Neurobiology, AFRC Institute of Physiology, Babraham, Cambridge CB2 4AT, UK.

JOHN T. RANSOM
Institute of Molecular Immunology (S3-6), Syntex Discovery Research, 3401 Hillview Avenue, Palo Alto, CA 94304, USA.

ELAINE ROBBINS
Cephalon Inc., 145 Brandywine Parkway, W. Chester, PA 19380, USA.

G. SHANKAR
Institut-Alfred Fessard, CNRS, Gif-sur-Yvette, France.

N. A. SHARIF
Present address: Department of Molecular Pharmacology, Alcon Laboratories Inc., 6201 South Freeway, Fort Worth, TX, 76134-2099, USA.

D. J. S. SIRINATHSINGHJI
Merck Sharp & Dohme Research Laboratories, Neuroscience Research Centre, Terlings Park, Eastwick Road, Harlow, Essex CM20 2QR, UK.

D. KIRK WAYS
Department of Medicine, East Carolina University School of Medicine, Greenville, NC 27858-4354, USA.

R. ZOREC
Institute of Pathophysiology, University of Ljubljana, Ljubljana, Slovenia.

1

In situ hybridization histochemistry with radioactive and non-radioactive cRNA and DNA probes

MICHAEL E. LEWIS, ELAINE ROBBINS, and FRANK BALDINO, Jr.

1. Introduction

In situ hybridization histochemistry has been established as a uniquely powerful tool for the study of gene expression at the level of the single neurone. Although this technique has also frequently been used to confirm immunoreactive neurones as the actual site of biosynthesis of a specific peptide or protein, the greatest utility is apparent in studies of altered gene expression in development, pathology, or due to physiological or pharmacological manipulations. The purpose of this chapter is to provide the reader with established protocols for carrying out *in situ* hybridization histochemistry with either chemically synthesized DNA probes or biosynthesized cRNA probes, using either autoradiographic or enzyme histochemical methods for detecting hybridized neurones.

Historically, *in situ* hybridization was developed using radiolabelled, cloned cDNA probes, with methods for radiolabelling cRNA probes and synthetic oligonucleotide probes following thereafter. Probe labelling reactions have been carried out using substrates incorporating ^{3}H, ^{32}P, ^{35}S, and ^{125}I; however, because ^{35}S appears to provide the best compromise between the conflicting requirements for speed, sensitivity, and low background labelling, this chapter describes how probes can be labelled with this radioisotope.

In recent years, several non-radioactive procedures have been developed to detect specific nucleotide sequences under a variety of hybridization conditions. These non-radioactive probes are particularly useful for *in situ* hybridization studies where they overcome several limitations normally associated with the use of radiolabelled sequences as probes. For example, hybrids can be detected after only a short period of time, usually less than 24 hours with most non-radioactive systems, compared with the weeks typically required for autoradiographic detection of probes labelled with low energy beta emitters (e.g. ^{35}S and ^{3}H). Furthermore, the degree of cellular resolution, which is crucial for *in situ* hybridization, is significantly better than that normally achieved with autoradiography. Finally, the use of non-radioactive probes avoids the biohazards and additional expenses normally associated with the use of radioisotopes. Recently, novel *in situ* hybridization methods have been developed based on the enzymatic incorporation of digoxigenin-conjugated nucleotides into synthetic oligonucleotide probes (1) or cRNA probes (2). The digoxigenin-labelled probe is hybridized and then detected with an alkaline phosphatase-conjugated IgG which is highly specific for the digoxigenin molecule. This chapter provides detailed protocols for carrying out probe labelling and hybridization with this system.

Further methodological considerations are determined by whether the experimenter has access to the required cDNA clone and relevant molecular biological and microbiological techniques. If not, the choice of probe will be restricted to synthetic oligonucleotides, which are frequently adequate for mRNA detection. If these limitations do not apply, the abundance of the target mRNA may be a deciding factor. Since cRNA probes can be labelled to a higher specific activity than synthetic oligonucleotide probes, cRNA probes may be required for the detection of very rare target mRNAs. This chapter provides protocols for using either synthetic oligonucleotide or cRNA probes.

Regardless of the type of hybridization probe used, the investigator is obliged to establish the specificity of hybridization, an issue which has been addressed extensively elsewhere (3). In addition, if the goal of the experiment goes beyond establishing the localization of gene expression, i.e. in determining whether gene expression changes as a function of pathology, development, or some treatment, the investigator will need to establish and validate a method of quantifying hybridization. This issue, again, has been addressed for the use of radioactive probes (4) as well as digoxigenin-labelled probes (5).

2. General comments

Preservation of RNA and cellular morphology within thin tissue sections is of utmost importance when analysing gene expression using *in situ* hybridization histochemistry. The following are some considerations which must be taken into account when planning experiments.

2.1 Tissue preparation

Removing tissue from the animal and freezing it in powdered dry ice provides the maximum preservation of RNA. For *in situ* hybridization, the time between tissue extraction, blocking, and freezing should be minimized. Tissue should be stored at −70 °C prior to sectioning in a cryostat, and should not be thawed before cutting. It is best to section tissue as close as possible to the actual day of the hybridization. Thin sections (4–12 μm) increase the number of sections derived from a given tissue and enhance autoradiographic resolution. The increased surface area of thin sections also increases the likelihood of probe penetration to the mRNAs of interest. Thaw-mounting cryostat cut sections to slides subbed with porcine gelatin is essential for insuring adherence of thin tissue sections to glass slides throughout the lengthy *in situ* hybridization procedures. Once tissue has been collected by thaw-mounting on to slides it should be transported on dry ice to a −70°C freezer for storage.

2.2 Fixation

Tissue fixation is one of the more critical aspects of *in situ* hybridization. The fixation must preserve tissue morphology so that mRNA is retained within the cells. Fixatives such as paraformaldehyde, glutaraldehyde, and formalin, which are known to cross-link proteins, are acceptable for retention of mRNA. Post-fixation in paraformaldehyde is superior because it allows for the use of a variety of probe sizes. This solution should be freshly prepared, however, because as it ages, the formaldehyde breaks down into several substances, one of which is formic acid. When tissue has been perfused with paraformaldehyde (6), some loss of mRNA can be expected, but this method does allow for the localization of peptides or proteins on adjacent sections by immunocytochemistry (7).

2.3 RNase contamination

RNA molecules are very susceptible to degradation due to cleavage by ribonucleases (RNases) which can be accidentally introduced in the laboratory. Precautions to avoid this problem include:

- the wearing of disposable gloves during the preparation of materials and solutions and for the hybridization procedure
- treatment of solutions with 0.1% diethyl pyrocarbonate (DEPC), a strong inhibitor of RNases

After solutions have been allowed to stand at room temperature overnight, they can be autoclaved to remove residual DEPC. Sterile, disposable plasticware is essentially free of RNases and can be used for the storage of solutions without pretreatment.

3. Procedures for the use of ^{35}S-labelled probes

3.1 Use of cRNA probes

Use of radiolabelled cRNA probes synthesized from cloned cDNAs provides the maximum sensitivity available for *in situ* hybridization studies because cRNAs can be labelled to a higher specific activity than synthetic oligonucleotide probes, and may be required for the detection of very rare target mRNAs or when the target sequence is not known. They are, however, technically more demanding and require molecular biological/microbiological expertise for their preparation and use. Preparation of cDNAs has been discussed extensively elsewhere (8). cRNA probes may also require an additional step after synthesis: alkaline hydrolysis of the probe into shorter fragments for more effective tissue penetration. Despite some drawbacks associated with the use of cRNA probes, some of these difficulties have been overcome by the introduction of well established protocols which have been used for the detection of a wide variety of mRNAs with noted success.

Protocol 1. Synthesis of a radiolabelled cRNA probe using [^{35}S]UTP

1. To a 1.5 ml autoclaved microfuge tube, add the following in the order listed:

• 5 × transcription buffer[a]	4 μl
• 100 mM dithiothreitol	2 μl
• RNAsin (40 units)	1 μl
• 10 mM GTP	1 μl
• 10 mM ATP	1 μl
• 10 mM CTP	1 μl
• 100 mM UTP cold	2 μl
• 1 μg/μl DNA (linearized)	2 μl
• [^{35}S]UTP (sp. act. > 1000 Ci/mmol)	5 μl
• RNA polymerase (20 units)	1 μl

2. Mix well by tapping tube, and centrifuge briefly. Incubate for 1 h at 37°C. Optional: add an additional 1 μl RNA polymerase after 1 h and let the reaction continue for one more hour.

3. Digest the DNA vector and template by adding the following:

• RQ 1 DNase (1 unit)	1 μl
• RNasin (40 units)	1 μl

Incubate for 10 min at 37°C (should be accurate).

4. Stop the reaction by adding:

• 100 mM EDTA[b]	10 μl (mix well)
• 10 × STE[c]	10 μl (mix well)

5. Adjust volume to 100 μl with 20 mM dithiothreitol diluted from stock with 0.1% DEPC-treated H_2O.

6. Extract with phenol:chloroform.[d]

(a) Add an equal volume of phenol, vortex gently and let sit for 3 min. Centrifuge at 1000 *g* for 3 min. Remove and discard bottom organic phase with pipette set for the volume of phenol that was initially added. The RNA probe is in the top aqueous phase.
(b) Add an equal volume of phenol:chloroform. Repeat as above.
(c) Add an equal volume of chloroform. Repeat as above.

7. Separate the single-stranded RNA probe from the labelling reaction by precipitating in ethanol as follows:
(a) Add the following:
- 0.1 vol 4.5 M sodium acetate pH 6.0 — 10 μl
- 3 vol ice cold 100% ethanol — 300 μl
- 10 mg/ml yeast tRNA — 1 μl

(b) Mix by inversion and tapping the tube.
(c) Centrifuge briefly for 1–2 sec.
(d) Let chill on dry ice mixed with ethanol for 30 min or overnight.
(e) Centrifuge for 30 min at 12 000 *g* (minimum speed).
(f) Pour off supernatant, keeping the tube inverted.
(g) Allow pellet to dry and resuspend pellet in 50 μl 0.1% DEPC-treated H_2O.

8. Alkaline hydrolysis can be done at this point (see *Protocol 3*).

[a] 5 × transcription buffer is 200 mM Tris–HCl, pH 7.5, 30 mM $MgCl_2$, 10 mM spermidine, 50 mM NaCl.
[b] A stock solution of 0.5 M EDTA pH 8.0 is prepared by dissolving 186.1 g disodium ethylenediaminetetra-acetate (2 H_2O) into 800 ml of water. Adjust the pH to 8.0 with NaOH (approximately 20 g of NaOH pellets). EDTA will not go into solution until the pH of the solution is adjusted to approximately 8.0. Treat solution with 0.1% DEPC and sterilize by autoclaving.
[c] 10 × STE is prepared by combining 100 ml 1.0 M Tris–HCl pH 8.0, 100 ml of 0.1 M EDTA pH 8.0, and 58.44 g NaCl to 800 ml of water. Adjust pH to 8.0 and adjust volume to 1000 ml. This solution, as with all solutions containing Tris, should be made up with 0.1% DEPC-treated water which has been autoclaved. Solution can then be filter sterilized before use.
[d] Ultra pure phenol (Gibco/BRL) is stored in 500 μl aliquots at −20°C. Defrost at 60°C and equilibrate by adding 500 μl TE buffer pH 7.6. Mix well and centrifuge for 3 min. Chloroform: isoamyl alcohol (24:1) is stored at room temperature.

Protocol 2. Radioactive *in situ* hybridization histochemistry with $[^{35}S]$UTP-labelled cRNA probes

1. Pre-hybridization washes
(a) Remove slides from −70°C freezer and quickly warm to room temperature under a stream of cool air.
(b) Proceed with the following washes:
- 3% paraformaldehyde in 0.1 M PBS (plus 0.02% DEPC) for 5 min (made fresh)

Protocol 2. *Continued*

- 0.1 M PBS: rinse 2 × 1 min
- 2 × SSC:[a] rinse 2 × 1 min
- 0.1 M triethanolamine pH 8.0 with 125 μl acetic anhydride per 50 ml (acetic anhydride should be added immediately prior to use): 10 min
- 2 × SSC: rinse for 1 min
- 70% ethanol (DEPC-treated water): 1 min
- 80% ethanol (DEPC-treated water): 1 min
- 95% ethanol (DEPC-treated water): 1 min

(c) Air dry on rack.

2. Hybridization

Dilute radiolabelled probe (see *Protocol 1*) in hybridization buffer and apply 20–25 μl per section. Coverslip with two layers of Parafilm and hybridize slides in humid boxes[b] for 3.5–4 h at 50°C

Hybridization buffer[c] (final concentration):

- 40% nucleic acid grade, deionized formamide[d]
- 10% dextran sulfate
- 1 × Denhardt's[e]
- 4 × SSC + 10 mM dithiothreitol
- 1 mg/ml tRNA (yeast)
- 1 mg/ml denatured salmon sperm DNA[f]

3. Post-hybridization washes: (all steps under mild agitation)

(a) Let slides cool to room temperature

(b) Remove coverslips in 2 × SSC and place slides in 2 × SSC at room temperature

(c) Proceed with the following washes

- In 52°C water bath, wash in 50% deionized formamide in 2 × SSC for 5 min
- In 52°C water bath, wash slides in 50% deionized formamide in 2 × SSC for 20 min
- 2 × SSC: Rinse 2 × 1 min at room temperature
- In 37°C water bath, wash slides in 100 μg/ml RNase A in 2 × SSC for 30 min
- 2 × SSC: rinse 2 or 3 × 1 min at room temperature
- In 52°C water bath, wash slides in 50% deionized formamide in 2 × SSC for 5 min
- 2 × SSC + 0.05% Triton X-100: overnight on gentle shaker

(d) On the next day, transfer slides through the following washes
- 2 × SSC + 0.05% Triton X-100: two quick rinses
- 300 mM ammonium acetate: 1 min
- 70% ethanol in 300 mM ammonium acetate: 1 min
- 80% ethanol in 300 mM ammonium acetate: 1 min
- 95% ethanol: 1 min
- 100% ethanol: 2 × 5 min
- Xylene: 1 × 5 min
- Xylene: 1 × 30 min
- 100% ethanol: 2 × 5 min

(e) Air dry.

4. Place dried slides into an X-ray cassette with Kodak XAR-5 film. After developing the film (3 days' exposure is a good starting point), the slides can be dipped in liquid nuclear emulsion[g] to obtain cellular resolution.

[a] The composition of SSC is 0.015 M NaCl, 0.015 M trisodium citrate
[b] Humid boxes (ideally plastic with a tightly fitting lid) are used to prevent evaporation when sections are being incubated in a small amount of liquid. Moisture can be provided by sterile H_2O; slides can rest on plastic supports during the incubation.
[c] Hybridization buffer can be prepared in advance and stored in aliquots at −70°C.
[d] Deionized formamide is prepared as follows. Stir 5 g formamide (nucleic acid grade) into 100 ml Dowex AG-501 mixed bed resin (Bio-Rad) for 1–2 h at room temperature. Filter and store at −20°C.
[e] 50 × Denhardt's solution can be purchased from Sigma Chemical Co.
[f] Salmon sperm DNA can be denatured by boiling for 5–10 min and cooling quickly on ice.
[g] Nuclear emulsion NTB2 or NTB3 can be used. See Section 6 for procedure.

3.2 Alkaline hydrolysis of a cRNA probe

For cRNA probes greater than 1.5 kb in length, partial alkaline hydrolysis may be necessary to improve the hybridization signal by improving probe penetration. This will have to be determined by experiment with the length of the cRNA probe being used, but investigators have suggested that fragments of 150–300 bases have improved hybridization signal by at least 10%. The calculated time for the reaction t, can be determined by using the formula

$$t = L_o - L_f / K L_o L_f$$

where L_o and L_f are the initial and final fragment lengths in kilobases (kb) and K is the rate constant for hydrolysis (0.11 kb/min).

Protocol 3. Partial alkaline hydrolysis of a cRNA probe to make short segments

1. Following synthesis of the cRNA probe and resuspension in 50 µl of 20 mM dithiothreitol, add the following:

Protocol 3. *Continued*

30 μl of 0.2 M Na_2CO_3
20 μl of 0.2 M $NaHCO_3$
(Both DEPC-treated and autoclaved.)

2. Mix well by tapping tube, centrifuge briefly and incubate for the calculated time at 60°C. Temperature and time MUST BE PRECISE.
3. To terminate the reaction, add immediately on ice:

   2.5 μl 4.5 M sodium acetate pH 6.0
   5 μl 10% glacial acetic acid
4. Ethanol precipitate the probe by adding:
   - 7.5 μl 4.5 M sodium acetate pH 6.0
   - 300 μl ice cold 100% ethanol
   - 1 μl 20 mg/ml glycogen[a]

   Let chill in dry ice mixed with ethanol for 30 min or place at −70°C overnight. Centrifuge for 30 min at 12 000 *g* (minimum speed). Pour off supernatant, keeping the tube inverted. Allow pellet to dry and resuspend pellet in 50 μl DEPC-treated H_2O.

[a] Other carriers, such as yeast tRNA, can be used.

3.3 Use of synthetic oligonucleotide probes

Frequently, a viable alternative to the use of cRNA probes for *in situ* hybridization studies is the use of synthetic oligonucleotide probes. Many investigators have used these very successfully for the detection of a wide variety of target mRNAs. When the target nucleic acid sequence is known, probe design is straightforward (3). The published sequence is written 5′ to 3′ (left to right, e.g. 5′-GCAT-3′), so the probe sequence will be complementary from 3′ to 5′ (e.g. 3′-CGTA-5′) although written in the reverse order (e.g. 5′-ATGC-3′). Probes of 30–50 bases in length are frequently used and form thermally stable hybrids when the G + C content is between 50 and 60%. Once the probe sequence has been decided, many academic and commercial institutions have suitable facilities and trained personnel to carry out a custom synthesis and purification for a reasonable fee. Several labelling options are available: 5′ end-labelling, primer extension, and 3′ end labelling. The third procedure, 3′ end-labelling, uses the enzyme terminal deoxynucleotidyl transferase to catalyse the sequential addition of radioactive deoxynucleoside monophosphates to the free 3′ hydroxyl end of a synthetic oligonucleotide. Since this enzyme will continue adding deoxynucleoside monophosphates to the 3′ end, the specific activity of the probe can be controlled by reaction conditions such as time and substrate concentration. This procedure is technically easy to perform, and probes labelled with [^{35}S]dATP provide the best compromise between conflicting requirements for speed, sensitivity, and low background labelling.

Protocol 4. Radioactive *in situ* hybridization histochemistry with [^{35}S]dATP-labelled oligonucleotide probes

1. Labelling reaction: to a sterile 1.5 ml microfuge tube, add:
   - tailing buffer[a] — 5 μl
   - $CoCl_2$ solution[a] — 1.5 μl
   - oligonucleotide (35 pmol) — 1 μl
   - [^{35}S]dATP (sp. act. > 1000 Ci/mmol) — 9 μl
   - terminal transferase (50 units) — 1 μl
   - adjust volume to 25 μl with sterile H_2O

   Mix gently by tapping the tube, do not vortex. Briefly microfuge tube and incubate at 37°C for 5 min.
2. The reaction can be terminated by incubation for 10 min at 65°C. Adjust the volume to 500 μl with 0.1 M Tris–HCl pH 8.0 and keep on ice.
3. Unincorporated nucleotides can be separated using a NENSORB 20 column[b] as follows:
   (a) Settle each column (one per reaction) by tapping the side of tube until all column matrix is at the bottom.
   (b) Remove cap and attach column to ring stand over disposable beaker.
   (c) Fill with 3 ml absolute methanol (HPLC grade if possible), and force through column with a 10 ml syringe attached to a column adaptor.
   (d) Apply gentle pressure to the syringe (1 drop/2 sec) until solution is through.
   (e) Wash column with 1.5 ml Tris–HCl pH 8.0, forced through with a syringe.
   (f) Add reaction mixture to the column and apply gentle pressure to the syringe (1 drop/2 sec) until solution is through.
   (g) Elute tailed oligonucleotide by adding 0.5 ml of 20% ethanol by applying gentle pressure to column with a syringe and slowly collect first 12 drops in a sterile microfuge tube.
   (h) Count an aliquot by liquid scintillation counting; should get ~500 000 c.p.m./μl.
   (i) Add 0.1 vol freshly prepared 100 mM dithiothreitol (final concentration 10 mM) and store probe on ice until hybridization step.
4. Pre-hybridization washes: remove slides from the −70°C freezer and quickly warm them to room temperature under a stream of cool air before proceeding through the following washes:
   - 3% paraformaldehyde in 0.1 M PBS (plus 0.02% DEPC) for 5 min

Protocol 4. *Continued*

- 0.1 M PBS: rinse 3 × 5 min
- 2 × SSC:[c] rinse 1 × 10 min

Optional: Place slides in a humid box[d] and pipette ~300 μl of hybridization buffer (see below) on to each slide. Incubate slides at room temperature for 1 h.

5. Hybridization: Dilute the radiolabelled probe in hybridization buffer, allowing 500 μl of buffer containing probe per slide with 1.5×10^6 c.p.m. of activity. A higher concentration of probe may be helpful in some cases. If desired, apply Parafilm coverslips to reduce the volume per slide as explained in *Protocol 2*, Step 2. Prepare fresh hybridization buffer (10 ml) by adding:
   - 5 ml nucleic acid grade, deionized formamide[e]
   - 2 ml 20 × SSC
   - 0.2 ml Denhardt's (50× solution)[f]
   - 0.5 ml 10 mg/ml denatured salmon sperm DNA[g]
   - 0.25 ml 10 mg/ml yeast tRNA
   - 2 ml dextran sulfate
   - 0.050 ml probe + H_2O

   Remove excess pre-hybridization buffer from the slides by a quick dip in 2 × SSC and carefully dry the glass surrounding the tissue with a Kimwipe. Apply the probe, and then incubate at 37°C overnight (this temperature can be increased to 42°C for oligonucleotides greater than 36-mer).
6. Post-hybridization washes: wash the slides with gentle shaking as follows:
   - 2 × SSC for 1–2 h at room temperature
   - 1 × SSC for 1–2 h at room temperature
   - 0.5 × SSC for 0.5–1 h at 37°C
   - 0.5 × SSC for 0.5–1 h at room temperature

   NOTE: For ^{35}S-labelled probes add 2-mercaptoethanol (14 mM final concentration) and sodium thiosulfate (1% final concentration) to all washes.
7. Dry slides rapidly by blowing a gentle stream of cool air over them.
8. Place dried slides into an X-ray cassette with Kodak XAR-5 film. After developing the film (3 days' exposure is a good starting point), the slides can be dipped in liquid nuclear emulsion[h] to obtain cellular resolution.

[a] Reaction buffer contains 200 mM potassium cacodylate; 50 mM Tris–HCl, pH 6.6; 2 mM dithioerythritol (DTE), 500 μg/ml BSA; 1.5 mM $CoCl_2$ (Boehringer Mannheim Biochemicals).
[b] Column supplied by DuPont NEN.
[c] See *Protocol* 2, footnote *a* for composition.
[d] See *Protocol* 2, footnote *b* for description.
[e] See *Protocol* 2, footnote *d* for preparation.
[f] See *Protocol* 2, footnote *e* for composition.
[g] See *Protocol* 2, footnote *f* for preparation.
[h] Nuclear emulsion NTB2 or NTB3 (Kodak) can be used. See Section 6 for procedure.

4. Controls for specificity

Control procedures to validate the specificity of *in situ* hybridization histochemistry vary somewhat with the type of probe that has been selected; however, the following information can be used as a basic guide when instituting control procedures with the noted limitations. Most investigators often find it beneficial to implement a combination of control procedures to demonstrate probe specificity reliably. Further discussion of these and other control procedures is provided elsewhere (3, 9).

4.1 Colocalization of the hybridization signal with immunohistochemical staining for the encoded protein

This method can be used regardless of whether the probe selected is a cRNA probe or a synthetic oligonucleotide. Some difficulties with this approach include:

- animals usually require transcardial perfusion which results in some loss of mRNA
- conditions compatible for detection of the encoded protein may be incompatible for the detection of mRNA in the same tissue section
- peptide precursors are synthesized in cell bodies and transported to distant sites in the nerve terminals, often resulting in very low or undetectable levels of perikaryal immunostaining in untreated animals.

4.2 Colocalization of the hybridization signal with another previously characterized mRNA

This method has been used successfully when one probe has been radiolabelled while the other incorporates a non-radioactive label (12, 13). It also works very well when the non-radioactive probe is labelled with digoxigenin and both probes are of the same type (e.g. when both probes are cRNAs or synthetic oligonucleotides). The basic method for this type of colocalization procedure is discussed in Section 7. As a control procedure, it may be employed to colocalize two mRNAs that should be present in the same neurone, e.g. those encoding tyrosine hydroxylase and dopamine-β-hydroxylase in noradrenergic neurones. Alternatively, the two probes can be complementary to different regions of the same mRNAs, and should again colocalize in the same neurones.

4.3 Thermal dissociation of the hybrid at a temperature (Tm_{obs}) consistent with perfect complementarity (Tm_{calc})

This method is more successful when synthetic oligonucleotide probes are selected since the exact probe length can be determined. This is often difficult with cRNA probes unless care is taken to separate full length probe from incomplete or partly degraded probe following synthesis.

4.4 Pretreatment of the tissue with RNase to destroy target mRNA

RNase pretreatment is more effective when oligonucleotide probes are being used. Digestion of the RNase with proteinase K prior to hybridization is necessary when a cRNA probe is being used because residual enzyme could digest the probe itself.

5. Procedures for the use of digoxigenin-labelled probes

The development of non-radioactive markers for the detection of specific gene sequences in recent years has allowed investigators to overcome several limitations normally associated with the use of radiolabelled sequences as probes. As noted in the introduction, signal detection is faster and of higher resolution with non-radioactively labelled probes. Recently, novel *in situ* hybridization methods have been developed based on the enzymatic incorporation of digoxigenin-UTP or digoxigenin-dUTP (Boehringer Mannheim Biochemicals) into either cRNA probes or synthetic oligonucleotide probes, respectively. The digoxigenin-labelled probes are hybridized under conditions very similar to those used for ^{35}S-labelled probes and then detected with an alkaline phosphatase-conjugated IgG which is highly specific for the digoxigenin molecule. Moreover, the sensitivity of these probes reportedly exceeds that which can be obtained with other non-radiolabelled probe hybridization methods.

Since the following protocols are very similar to the preceding ones using radiolabelled probes, information concerning composition of buffers and preparation of reagents can be obtained from the footnotes at the end of each procedure, unless otherwise indicated.

Protocol 5. Synthesis of a non-radiolabelled cRNA probe using digoxigenin-UTP

1. To a 1.5 ml autoclaved microfuge tube, add the following in the order listed:

• 5 × transcription buffer	4 μl
• 100 mM dithiothreitol	2 μl
• RNAsin (40 units)	1 μl
• 10 mM GTP	1 μl
• 10 mM ATP	1 μl

- 10 mM CTP 1 μl
- 1 mM UTP 2 μl
- 1 μg/μl DNA (linearized) 2 μl
- 2 mM digoxigenin-UTP[a] 4 μl
- Sterile distilled H_2O 1 μl
- RNA polymerase (20 units) 1 μl

2. Mix well by tapping tube, and centrifuge briefly. Incubate for 1 h at 37°C. Optional: add an additional 1 μl RNA polymerase after 1 h and let the reaction continue for one more hour.

3. Digest the DNA vector and template by adding the following:
- RQ 1 DNase (1 unit) 1 μl
- RNAsin (40 units) 1 μl

Incubate for 10 min at 37°C (should be accurate).

4. Stop the reaction by adding:
- 100 mM EDTA 10 μl (mix well)
- 10 × STE 10 μl (mix well)

5. Adjust volume to 50 μl with 20 mM dithiothreitol diluted from stock with 0.1% DEPC-treated H_2O.

6. Separate the single-stranded RNA probe from the labelling reaction by precipitating in ethanol (see *Protocol 1*, step 7).

7. Alkaline hydrolysis can be done at this point (see *Protocol 3*).

[a] Digoxigenin-UTP can be purchased from Boehringer Mannheim Biochemicals.

Protocol 6. Non-radioactive *in situ* hybridization histochemistry with digoxigenin-labelled cRNA probes

1. Pre-hybridization washes

(a) Remove slides from −70°C freezer and quickly warm them to room temperature under a stream of cool air before proceeding through the following washes:
- 3% paraformaldehyde in 0.1 M PBS (plus 0.02% DEPC) for 5 min (made fresh)
- 0.1 M PBS: rinse 2 × 1 min
- 2 × SSC: rinse 2 × 1 min
- 0.1 M triethanolamine pH 8.0 with 125 μl acetic anhydride per 50 ml (acetic anhydride should be added immediately prior to use): 10 min
- 2 × SSC: rinse 1 × 1 min

Protocol 6. *Continued*

- 0.1 M PBS – rinse 1 × 1 min
- 0.1 M Tris/glycine pH 7.0:[a] 30 min
- 2 × SSC—rinse 2 × 1 min
- 70% ethanol (DEPC-treated water): 1 min
- 80% ethanol (DEPC-treated water): 1 min
- 95% ethanol (DEPC-treated water): 1 min

(b) Air dry on rack.

2. Hybridization
Dilute digoxigenin-labelled probe (see *Protocol 5*) in hybridization buffer and apply 20–25 μl per section. Cover with two layers of Parafilm and hybridize slides in humid boxes for 3.5–4 h at 50°C.
Hybridization buffer (final concentration):

- 40% nucleic acid grade, deionized formamide
- 10% dextran sulfate
- 1 × Denhardt's solution
- 4 × SSC + 10 mM dithiothreitol
- 1 mg/ml tRNA (yeast)
- 1 mg/ml denatured salmon sperm DNA

3. Post-hybridization washes (all steps under mild agitation):
(a) Let slides cool to room temperature
(b) Remove coverslips in 2 × SSC and place slides in 2 × SSC at room temperature
(c) Proceed through the folowing washes:

- In 52°C water bath: 50% deionized formamide in 2 × SSC (5 min)
- In 52°C water bath: 50% deionized formamide in 2 × SSC (20 min)
- 2 × SSC: rinse 2 × 1 min at room temperature
- In 37°C water bath: 100 μg/ml RNase A in 2 × SSC (30 min)
- 2 × SSC: rinse 2 or 3 × 1 min at room temperature
- In 52°C water bath: 50% deionized formamide in 2 × SSC (5 min)
- 2 × SSC + 0.05% Triton X-100 containing 2% normal sheep serum (overnight on gentle shaker)

4. Immunological detection:
(a) Wash slides 2 × 5 min in Buffer 1 (100 μM Tris–HCl pH 7.5, 150 mM NaCl).
(b) Incubate in an alkaline phosphatase conjugated anti-digoxigenin antibody[b] (diluted 1:1000 in Buffer 1 containing 1% normal sheep serum and 0.3% Triton X-100 for 5 h at room temperature (or 24–48 h at 4°C).

(c) Wash slides for 30 min in Buffer 1.
(d) Wash slides for 10 min in Buffer 2 (100 mM Tris–HCl, pH 9.5, 100 mM NaCl, 50 mM $MgCl_2$).
(e) Incubate slides in chromagen (prepared by adding 45 μl NBT solution, 35 μl of X-phosphate solution,[c] 2.4 mg levamisole to 10 ml Buffer 2). Place slides in a light-tight box on buffer-wetted filter backing. Slides must be kept in the dark during incubation (2–24 h), but can be checked periodically for colour development.
(f) Reaction can be stopped in Buffer 3 (10 mM Tris–HCl, pH 8, 1 mM EDTA).

Tissue can then be quickly dehydrated, cleared in xylene and covered with a coverslip. (Colour precipitate can be better protected from fading by using an aqueous mountant.)

[a] Tris buffers should be made up with 0.1% DEPC-autoclaved H_2O. Do not add DEPC to buffers containing Tris before autoclaving due to the instability of this reagent in the presence of Tris.
[b] Alkaline phosphatase conjugated anti-digoxigenin antibody can be purchased from Boehringer Mannheim Biochemicals.
[c] NBT/X-phosphate is provided as a substrate for alkaline phosphatase. The amounts of 45 μl and 35 μl respectively are based on substrate obtained from Boehringer Mannheim Biochemicals.

Protocol 7. Non-radioactive *in situ* hybridization histochemistry with digoxigenin-labelled oligonucleotide probes

1. Labelling reaction
(a) To a sterile 1.5 ml microfuge tube, add:

• tailing buffer	4 μl
• $CoCl_2$ solution	6 μl
• digoxigenin-11-dUTP[a]	2.5 μl
• oligonucleotide	1 μl (35 pmol)
• dATP (10 mM)	2 μl of 1:50 dilution made in sterile H_2O
• sterile H_2O	1 μl
• terminal transferase (50 units)	1 μl

(b) Incubate the reaction at 37°C for 5 min.
(c) Purify the tailed oligonucleotide from the labelling reaction by ethanol precipitation:
- adjust volume to 100 μl with sterile H_2O
- add 1:10 (10 μl) 4.5 M sodium acetate pH 6.0
- add 3 vol ice cold 100% ethanol (300 μl)
- add 1 μl of 20 mg/ml glycogen

Protocol 7. *Continued*

- mix by inversion and by tapping the tube
- centrifuge briefly (1–2 sec)
- let chill in dry ice mixed with ethanol for 30 min
- centrifuge for 30 min at 12 000 *g* (minimum speed)
- pour off supernatant, keeping the tube inverted
- allow pellet to dry and resuspend pellet in 20 μl sterile H_2O

2. Pre-hybridization washes: remove slides from −70 °C and quickly warm to room temperature under a stream of cool air before proceeding through the following washes:

- 3% paraformaldehyde in 0.1 M PBS (+ 0.02% DEPC) (5 min)
- 0.1 M PBS (3 × 5 min)
- 2 × SSC (1 × 10 min)

Optional: place slides in a humid box and pipette ~ 300 μl of hybridization buffer (see below) on to each slide. Incubate slides at room temperature for 1 h.

3. Hybridization: Dilute the digoxigenin-labelled probe in hybridization buffer to a volume of 1 ml. The required concentration of probe will depend upon mRNA abundance and must be determined experimentally. Prepare fresh hybridization buffer by adding:

- 5 ml nucleic acid grade, deionized formamide
- 2 ml 20 × SSC
- 0.2 ml Denhardt's (50× solution)
- 0.5 ml 10 mg/ml denatured salmon sperm DNA
- 0.25 ml 10 mg/ml yeast tRNA
- 2 ml dextran sulfate
- 0.050 ml probe + H_2O

Remove excess pre-hybridization buffer from the slides by a quick dip in 2 × SSC and carefully dry the glass surrounding the tissue with a Kimwipe. Apply the probe allowing 25 μl per section and then apply a Parafilm coverslip. Incubate the slides at 37 °C overnight (this temperature can be increased to 42 °C for oligonucleotides greater than 36-mer)

4. Post-hybridization washes: wash the slides with gentle shaking as follows:

- 2 × SSC for 1 h at room temperature
- 1 × SSC for 1 h at room temperature
- 0.5 × SSC for 0.5 h at 37°C
- 0.5 × SSC for 0.5 h at room temperature

5. Immunological detection:
   (a) Wash slides for 1 min in Buffer 1 (100 mM Tris–HCl; 150 mM NaCl; pH 7.5)
   (b) Incubate sections with 2% normal sheep serum plus 0.3% Triton X-100 in Buffer 1 for 30 min to 1 h at room temperature
   (c) Dilute alkaline phosphatase conjugated anti-digoxigenin antibody[b] (1:500) with Buffer 1 containing 1% normal sheep serum and 0.3% Triton X-100. Pipette diluted anti-digoxigenin antibody conjugate on to sections and incubate at room temperature for 3–5 h in a humid chamber. Do not allow sections to dry out. NOTE: for lower abundance mRNA, sections can be incubated at 4°C for 24–48 h as needed.
   (d) Wash slides for 30 min in Buffer 1 with shaking.
   (e) Wash slides for 10 min in Buffer 2 (100 mM Tris–HCl; 100 mM NaCl; 50 mM $MgCl_2$, pH 9.5).
   (f) Incubate slides in chromagen (prepared by adding 45 μl NBT solution, 35 μl X-phosphate solution,[c] and 2.4 mg levamisole to 10 ml Buffer 2) in a light-tight box. Slides must be kept in the dark during incubation (2–24 h), but can be checked periodically for colour development.
   (g) Reaction can be stopped in Buffer 3 (10 mM Tris–HCl, pH 8.0, 1 mM EDTA). Tissue can then be quickly dehydrated, cleared in xylene, and covered with a coverslip. (Colour precipitate can be better preserved by using an aqueous mountant.)

[a] Digoxigenin-dUTP can be purchased from Boehringer Mannheim Biochemicals.
[b] Alkaline phosphatase conjugated anti-digoxigenin antibody can be purchased from Boehringer Mannheim Biochemicals
[c] NBT/X-phosphate is provided as a substrate for alkaline-phosphatase. The amounts of 45 μl and 35 μl respectively are based on substrate obtained from Boehringer Mannheim Biochemicals.

6. Signal detection

With ^{35}S-labelled probes, hybrids can be visualized by film autoradiography or liquid emulsion autoradiography. Film autoradiography, although lacking cellular resolution, provides rapid information concerning the regional distribution of mRNAs and a guideline for the exposure time needed when using liquid emulsion, usually three to five times the amount of time required for film exposure. Kodak XAR-5 film can be used and developed manually according to manufacturer's instructions or in a Kodak film processor. For single cell resolution, tissue sections are subsequently dipped in nuclear emulsion (Kodak NTB2 or NTB3, or the Ilford series of emulsions) and kept in the dark at 4°C for 7–60 days depending on the probe label and message abundance. Tissue sections shown in photomicrographs in *Figures 1* and *3* were dipped in Kodak NTB3 and NTB2 emulsion, respectively. In both cases the emulsion was diluted 1:1 with 300 mM ammonium acetate at 42°C,

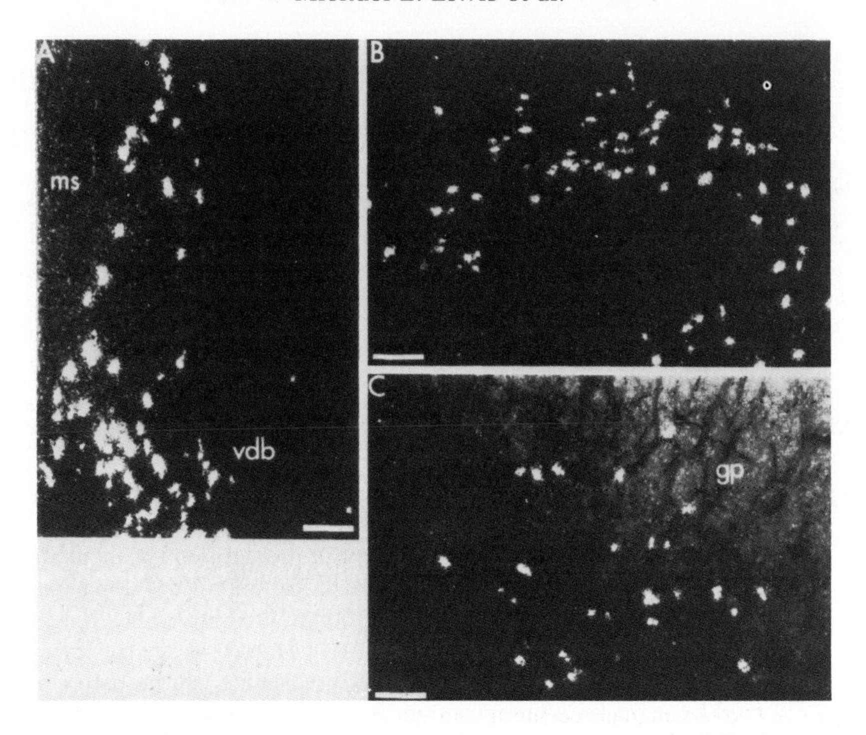

Figure 1. Dark-field photomicrographs demonstrating *in situ* hybridization for nerve growth factor receptor (NGFR) mRNA in coronal sections of the rat basal forebrain. Emulsion-dipped sections revealed silver grain clusters of NGFR mRNA in (**A**) the medial septum (ms) and vertical limb of the diagonal band of Broca (vdb), (**B**) the horizontal limb of the diagonal band of Broca, and (**C**) ventral regions of the globus palidus (gp), or nucleus basalis. Calibration bars equal 100 μm. (Reproduced from Springer *et al.* (11), with permission.)

developed in undiluted Kodak D19 for 3 min at 14°C, rinsed briefly, and fixed in Kodak Rapid Fix for 5 min. Following a 30 min rinse in running water, tissue sections were stained with haematoxylin and eosin (10) for the detection of cell bodies.

Visualization of hybrids using digoxigenin-labelled probes is depicted in the photomicrographs shown in *Figures 2* and *4*. Hybridization signal appears as a blue precipitate over the cytoplasm of the cell. Cellular resolution is frequently obtained within 24 h following hybridization.

7. Double-labelling methods

As noted above, double-labelling procedures may be used as a control, but more generally will find utility in answering questions about the synthesis of different gene products in the same neurone. At a semi-quantitative level, at

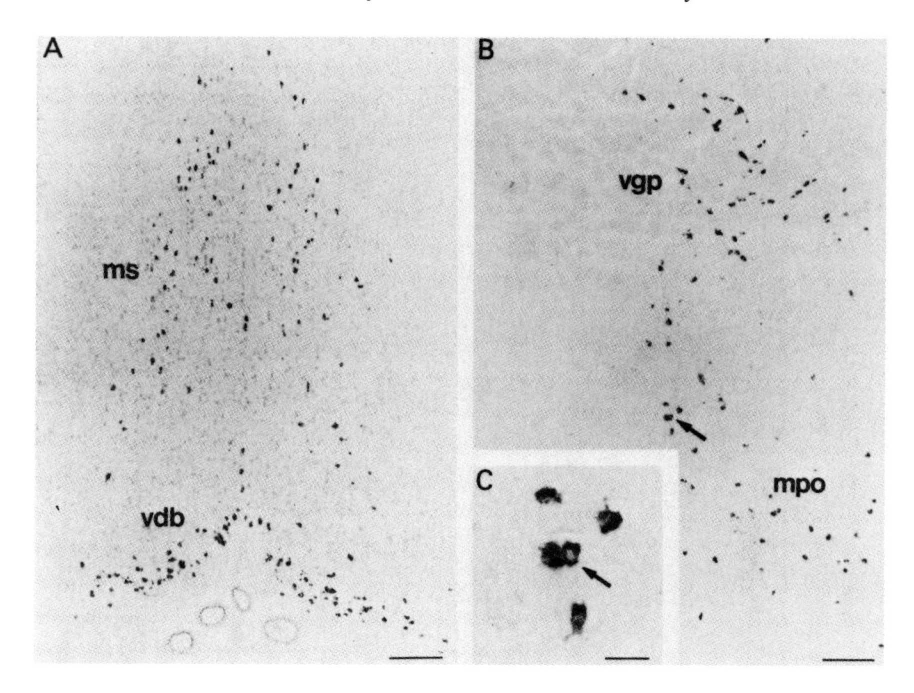

Figure 2. Photomicrograph demonstrating the localization and resolution of cells expressing NGFR mRNA using the non-radioactive detection system. Cells are distributed throughout the basal forebrain region and include: (**A**) the medial septum (ms) vertical limb of the diagonal band (VDB) of Broca, (**B**) the magnocellular preoptic area (mpo) and ventral globus pallidus-nucleus basalis region. Note the minimal background obtained using this technique and the dense cell staining confined to cytoplasmic regions of the cells (arrow in C). Bars: A, B = 200 μm; C = 50 μm. (Reproduced from Springer *et al.* (2), with permission.)

least, it may be possible to ask whether different gene products are co-regulated under particular experimental conditions.

The protocols provided in this chapter can be combined to colocalize two different mRNAs within a single cell. One requirement is that both probes are of the same type, e.g. both cRNA probes or both synthetic oligonucleotide probes. This method has been used successfully when one probe was labelled with ^{35}S and the other with digoxigenin. Each probe can be synthesized with no change in the labelling reaction. As a rule, it is best to choose a radioactive label when probing for the sequence which is suspected to be the least abundant mRNA.

For double labelling experiments, sections can be processed for pre-hybridization beginning with the 5 min post-fixation as previously described for either a cRNA probe or oligonucleotide probe. The hybridization medium, however, should contain both the radiolabelled probe and the digoxigenin-labelled probe, followed by hybridization at the appropriate temperatures.

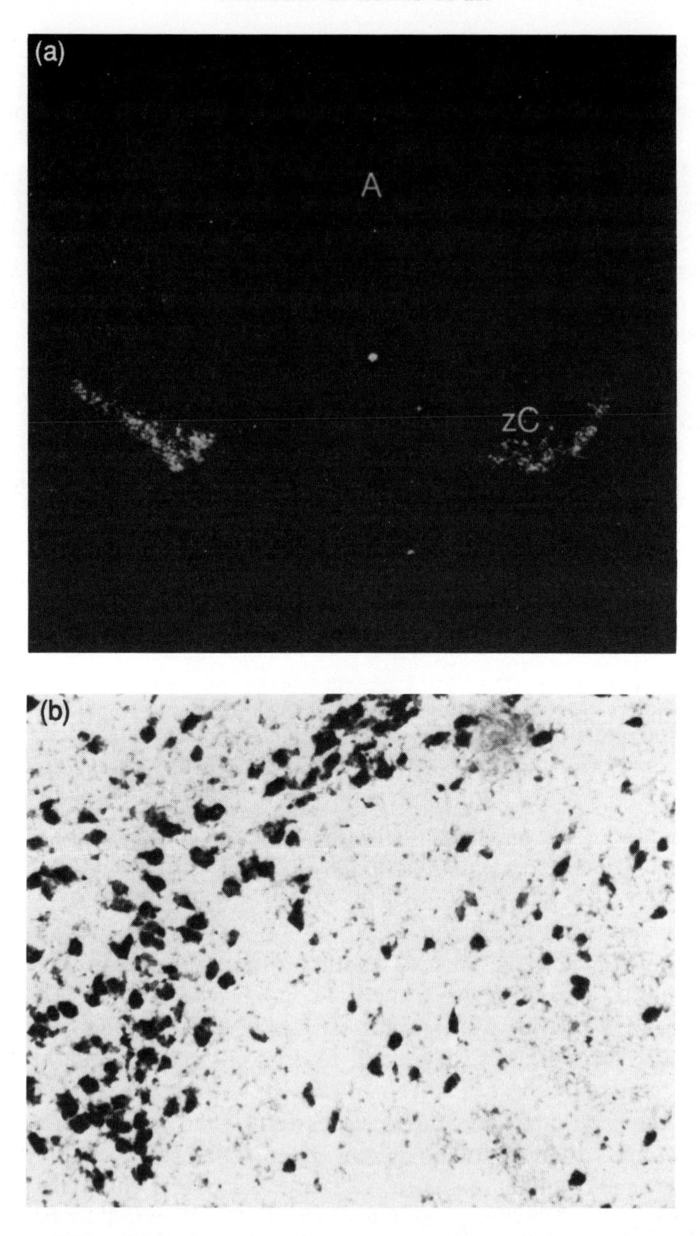

Figure 3a and 3b. Photomicrographs illustrating labelling of tyrosine hydroxylase (TH) synthesizing neurones in the substantia nigra pars compacta SNc. A 48-base synthetic oligonucleotide probe labelled with [^{35}S-]dATP (a) and non-radioactively labelled with digoxigenin-dUTP (b). Only a few scattered TH-positive neurones are observed in the zona reticulata. Intense cytoplasmic labelling can be seen over both small and large diameter neurones (b). (Bottom photomicrograph reproduced from Baldino *et al.* (12) with permission.)

If cRNA probes are being used, after the overnight wash in 2 × SSC and Triton X-100, sections are processed for detection of the digoxigenin-labelled hybrids with an anti-digoxigenin antibody coupled to alkaline phosphatase as described, except that the enzymatic reaction is stopped in 100 mM Tris–HCl, pH 9.5, 100 mM NaCl and 50 mM $MgCl_2$, followed by dehydration, clearing in xylene, and coating with Ilford nuclear research emulsion K5D (Polysciences, Warrington, PA). After exposure, the sections are developed as described in Section 6 and mounted with an aqueous medium (Polysciences).

If synthetic oligonucleotide probes are being used, post-hybridization washes should include:

- 2 × SSC for 1 h at room temperature
- 1 × SSC for 1 h at room temperature
- 0.5 × SSC for 0.5 h at 37°C
- 0.5 × SSC for 0.5 h at room temperature

Sections are then processed for detection of the digoxigenin-labelled hybrids prior to coating with Ilford K5D nuclear research emulsion as described above.

8. Conclusions

In situ hybridization histochemistry has yielded valuable insights into the distribution and regulation of mRNAs within neurones in a variety of experimental and pathological studies. In this chapter, we have provided the reader with established, well tested protocols for carrying out this method with either chemically synthesized oligonucleotides or biosynthetically derived cRNA probes, using either autoradiography or enzyme histochemistry to detect the hybrids in tissue sections. Thus, we hope that this chapter will be of use to a wide range of investigators studying the brain, so that we may yet understand how genes are regulated in the normal and aberrant functioning of this extraordinary tissue.

References

1. Lewis, M. E., Robbins, E., Grega, D., and Baldino, F., Jr. (1990). *Ann. N.Y. Acad. Sci.*, **579**, 246.
2. Springer, J. E., Robbins, E., Gwag, B. J., Lewis, M. E., and Baldino, F., Jr. (1991). Non-radioactive detection of nerve growth factor receptor (NGFR) mRNA in rat brain using *in situ* hybridization histochemistry. *J. Histochem. Cytochem.*, **39**, 231.
3. Lewis, M. E., Sherman, T. G., and Watson, S. J. (1985). *Peptides*, **6** (Suppl. 2), 75.
4. Lewis, M. E., Rogers, W. T., Krause, R. G. II, and Schwaber, J. S. (1989). *Methods Enzymol.*, **168**, 808.

5. Robbins, E., Baldino, F., Jr., Roberts-Lewis, J. M., Meyer, S. L., Grega, D., and Lewis, M. E. (1991). *Anat. Record,* **231,** 559.
6. Preece, A. (ed.) (1992). *A Manual for Histologic Techniques,* Little, Brown & Co., Boston, MA.
7. Kierman, J. A. (ed.) (1981). *Histological and Histochemical Methods, Theory and Practice.* Pergamon Press, New York.
8. Gubler, U. and Hoffman, B. J. (1983). *Gene,* **25,** 263.
9. Baldino, F., Jr., Chesselet, M.-F., and Lewis, M. E. (1989). *Methods Enzymol.,* **168,** 761.
10. Clark, G. (ed.) (1981). *Staining Procedures.* Williams & Wilkins, Baltimore, MD.
11. Springer, J. E., Robbins, E., Meyer, S., Baldino, F. Jr., and Lewis, M. E. (1990). *Cell. Mol. Neurobiol.,* **10,** 33.
12. Baldino, F. Jr., Robbins, E., and Lewis, M. E. (1992). In *Non-radioactive Labelling and Detection of Biomolecules* (ed. C. Kessler), pp. 367–72. Springer-Verlag.
13. Young, W. S. III (1989). *Neuropeptides*, **13,** 271.

2

Comparison of the localization of protein kinase C subspecies by *in situ* hybridization and immunocytochemical methods

JACQUELINE F. McGINTY, MARTA E. COUCE, and D. KIRK WAYS

1. Introduction

A variety of immunocytochemical protocols have been developed over the last 15 years which provide us with the ability to map the heterogeneous distributions of proteins and/or peptides of significance to neurotransmitter signalling. This major advance in histochemical technology has been joined recently by the advent of standardized *in situ* hybridization histochemical protocols that allow us to map the mRNAs that give rise to protein (e.g. receptors, enzymes) or peptide (transmitters, e.g. enkephalins) products. The cellular resolution and complementary RNA/protein information provided by both methods yield a unique combination for addressing complex neurochemical questions at a new level of detail. Furthermore, quantification by computerized image analysis is beginning to be applied to these histochemical techniques enabling them to become powerful quantitative, and not just qualitative, methodologies (see Chapter 4, this volume). Using these combined techniques, we are now able not only to map the distribution of peptide neurotransmitters, transmitter synthesizing and metabolizing enzymes, receptors, and, most recently, signal transduction molecules, but also to investigate the regulation of these molecules in response to physiological, pharmacological, and pathological challenges.

The accurate and sensitive application of immunocytochemistry and *in situ* hybridization histochemistry to the distribution and regulation of the protein kinase C (PKC) superfamily in the brain has become possible with the synthesis of oligopeptides and oligonucleotide probes with sequences based on the variable regions in each PKC isoform. PKC comprises a gene family of highly homologous enzymes, serine/threonine protein kinases (1, 2). This kinase family is part of a ubiquitous signal transduction pathway that is utilized by a variety of neurotransmitters, hormones, cytokines, growth factors, and oncogenes. Alterations in intracellular diacylglycerol (DAG) content induced by phosphatidyl inositol or phosphatidyl choline hydrolysis, or *de novo* synthesis, modulate PKC activity. By binding to a specific region on the PKC molecule, DAG or phorbol esters, which mimic DAG, induce a conformational change that results in kinase activation. Upon binding of DAG, PKC associates with phospholipids and, in the case of certain isoforms, with Ca^{2+}. PKC activation is often associated with translocation of the kinase from the cytosol to the inner leaflet of the cell membrane. Activated PKC directly phosphorylates its substrates which alters their activity and leads to cellular responses by stimulating other kinase families and initiating a cascade of kinases (MAP-2 kinase, c-raf, etc.).

On the basis of functional and primary sequence differences, the PKC gene family has been divided into two subfamilies, Groups A and B. Group A includes α, β1, β2, and γ isoforms while Group B is comprised of the δ, ε, ε′, ζ, and η/L isoforms. Each isoform is divided into regions of high homology (conserved regions C1–C4) which are interspersed with areas of lower homology (variable regions V1–V5). The conserved regions have various regulatory and catalytic functions. The C1 domain contains two cysteine-rich repeats, termed zinc fingers, except for that of ζ, which has only one zinc finger. This motif is necessary for binding and activation by DAG and phorbol esters. The recognition sequence for phospholipid binding is also located in the C1 domain, as is the pseudosubstrate domain. In the basal state, this site interacts with the substrate binding site of PKC and inhibits kinase activity. Upon binding of DAG, this interaction is inhibited and the kinase is activated. Investigators have taken advantage of this interaction by designing peptides identical to the pseudosubstrate domain which inhibit kinase activity, and peptides that contain a serine residue which are excellent substrates for PKC, and by preparing antisera against the conserved C1 domain which, by interfering with the pseudosubstrate site and substrate domain interaction, activate the kinase. The C2 domain contains the information necessary to impart Ca^{2+} dependence. Group A, but not Group B isoforms, contain the C2 domain. This has an important functional consequence: Group A isoforms require Ca^{2+} for activation whereas Group B isoforms are Ca^{2+} independent. Finally, the C3 domain contains an ATP consensus binding site and the C4 region contains the substrate binding site. PKC can also be divided into an N-terminal regulatory domain (C1 and C2 regions) and a C-terminal catalytic domain (C3 and C4) which are separated by the V3 region. Trypsin or calpain hydrolyse PKC in the V3 region, yielding a lipophilic regulatory domain and a constituitively activated, lipid-independent catalytic domain, termed m kinase. The physiologic role of PKC activated in this manner has yet to be elucidated.

There are marked differences in substrate specificity between the Group A and B families. Group A isoforms phosphorylate a variety of exogenous and endogenous substrates, whereas the substrate specificity of the Group B kinases appears to be more restricted. For instance, histone fractions are readily phosphorylated by the Group A isoforms but are poor substrates for the Group B isoforms. Differences in substrate specificity could provide the biochemical mechanism whereby individual isoforms selectively mediate specific cellular events by phosphorylating different substrates in response to phorbol esters or DAG (3).

Discrepancies exist between early reports from different laboratories describing the distribution of PKC immunoreactivity (IR) within the CNS by immunocytochemistry using monoclonal or polyclonal antisera raised against purified PKCs (4–7). More recent immunocytochemical studies have employed polyclonal antisera raised against synthetic oligopeptides with sequences from the variable regions of Group A or B PKC isoforms to map the distribution of isoform-specific PKC immunoreactivity in the brain (8–12). The distribution of Group A and B PKC mRNAs in the brain also has been examined by *in situ* hybridization. Isoform-specific oligonucleotides are generated by synthesizing DNA sequences, as determined by genetic cloning (13, 14), that code for the variable regions in each PKC isoform. The results of the hybridization studies (12, 15) largely substantiate the latest immunocytochemical reports of the distribution of PKC isoform IR cited above. These studies and others have shown that isoforms differ in their cellular expression. Within a specific cell type, PKC isoform content is not fixed and can be modulated. Thus, the array of PKC isoforms expressed by a given cell type and modulated by external stimuli could be an important variable in determining the cellular response to PKC activation.

In summary, differences in activation requirements, substrate specificity, and tissue distribution provide the molecular and biochemical mechanisms by which individual PKC isoforms can selectively mediate specific cellular responses elicited by phorbol esters and DAG. This diversity necessitates that studies designed to elucidate the role of PKC in signal transduction must directly examine each individual isoform. Using the immunocytochemical and *in situ* hybridization techniques outlined in this paper, such studies have been initiated.

2. Generation and characterization of antisera

2.1 Generation of polyclonal antisera

The primary antisera were raised and characterized by one of us (16) in rabbits against synthetic oligopeptides with sequences contained within the variable regions of each PKC subspecies. For illustrative purposes, this chapter will focus on the use of polyclonal antisera raised against the V3 regions of PKC α and β of the Group A family or the V5 region of PKC ε of the Group B family. The antigens were coupled to keyhole limpet haemocyanin (KLH) (Calbiochem, CA) either via a cysteine residue with *m*-maleimidobenzoyl-*N*-hydroxysuccinimide ester (MBS) (α and β) (17), or by glutaraldehyde in the

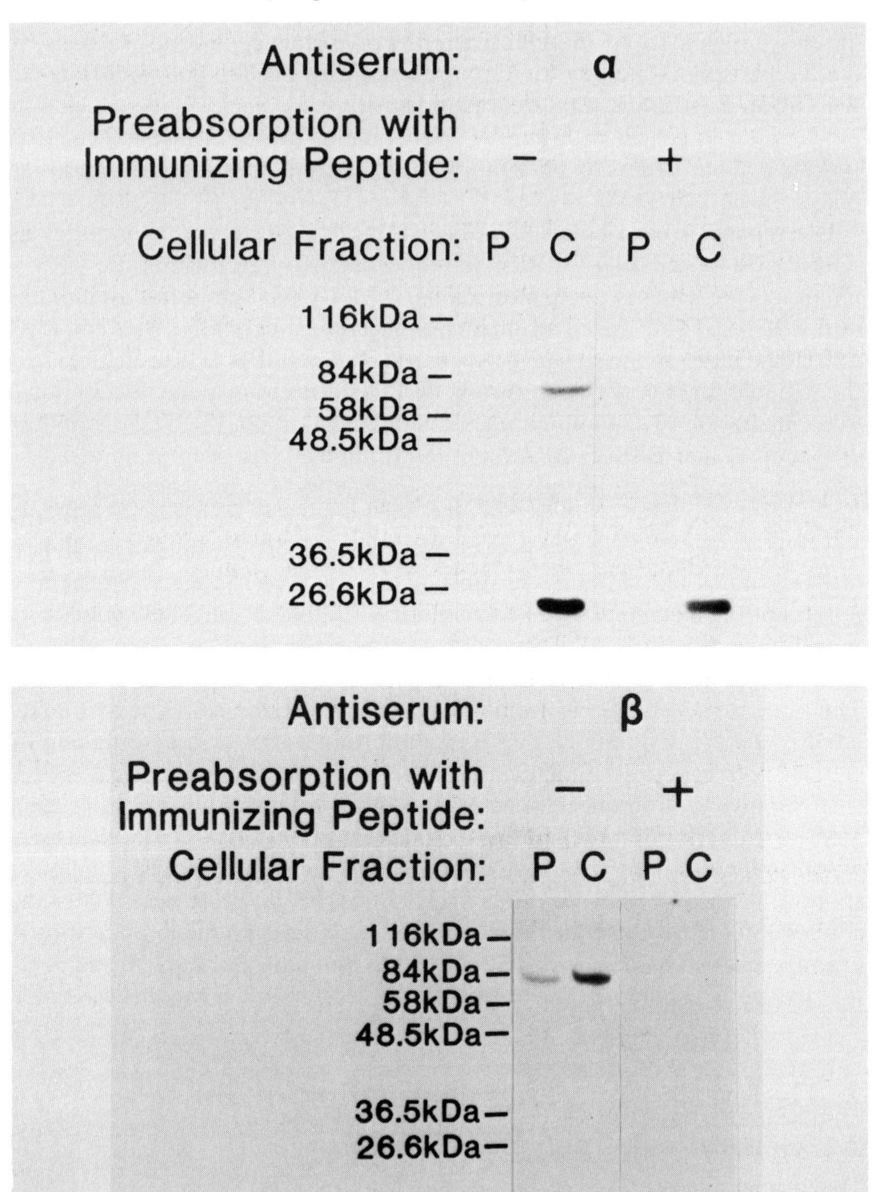

Figure 1. Western blot analysis using PKC isoform specific antisera. Equal concentrations of particulate (P) or cytosolic (C) extracts prepared from U937 cells were subjected to SDS–PAGE, transferred to nitrocellulose filters, incubated with α or β isoform-specific antisera (top and bottom panels, respectively) in the absence (−) or presence (+) of the peptide used for immunization, hybridized with ^{125}I labelled protein A, and subjected to autoradiography. Molecular weight markers are shown on the left side. Each antiserum recognized a protein with a molecular weight of ~80 kd. The ability of the immunizing peptide to abolish detection of this protein confirms its specificity.

case of ε, following the method of Schaap and colleagues (3). Preadsorption of the α and β antisera with the respective peptide sequences against which each antiserum was raised, abolished detection by western blot analysis of proteins of approximate molecular weights 84–87 kd (*Figure 1*). After fractionation of the isoforms by hydroxyapatite chromatography (18), western blot analysis using the α and β antisera demonstrated proteins with a molecular weight of 84 kd eluting at peaks III and II of PKC activity, respectively (*Figure 1*). In western blot analysis (not shown), the PKC ε antiserum recognized a protein with a molecular weight (87 kd) slightly higher than that of the α and β isoforms; this protein coeluted with PKC β on hydroxyapatite chromatography as has been demonstrated by others (18).

Protocol 1. Generation of antisera

1. Dissolve 1.7 mg of conjugate (peptide-KLH) in 0.2 M phosphate-buffered saline (PBS) mixed 1:1 (v/v) with complete Freund's adjuvant (Sigma, MO) and administer i.m. to New Zealand white rabbits.
2. Boost with i.m. injections of conjugate (1 mg) in incomplete Freund's adjuvant at 6 week intervals.
3. Perform phlebotomy 10–14 days after each boost, spin down blood to obtain serum and screen each antiserum by western blot analysis (see *Protocol 2*).

2.2 Characterization of antisera

Protocol 2. Western blotting

1. Add equal protein concentrations of sample, generally 100 μg in 50 μl, to an equal volume of Laemmli buffer. The buffer is prepared by mixing the following ingredients and heating to 110°C for 5 min:
   - 2% SDS
   - 100 mM dithiothreitol (DTT)
   - 60 mM Tris, pH 6.8
   - 0.01% bromophenol blue
   - 0.075% glycerol
2. Load 20 μl of the denatured sample or prestained molecular weight markers (α_2-macroglobulin, β-galactosidase, fructose 6-phosphate kinase, pyruvate kinase, and fumarase (Sigma, MO)) to SDS–polyacrylamide gel using a 10% running gel and transfer to nitrocellulose filters at 100 V for 60 min.

Protocol 2. *Continued*

3. Block non-specific sites by incubating the filter in 0.2 M PBS, pH 7.2, containing 0.05% Tween 20 and 3% bovine serum albumin (BSA) (w/v) for 120 min at room temperature (RT).
4. Incubate filters with antisera at the appropriate dilution overnight at 4°C in the same buffer.
5. Wash filters four times in PBS with 0.05% Tween 20.
6. Incubate filters with [^{125}I]protein A, 0.03 μCi/ml, in PBS with 0.05% Tween 20 and 3% BSA (w/v) for 90 min at RT.
7. Rinse filters four times in PBS with 0.05% Tween 20, dry, and expose to Kodak XAR film in a light-tight cassette with intensifying screens for 1–14 days at −70°C.
8. Determine antiserum specificity by incubating antiserum with 4 μg of the immunizing peptide for 60 min at RT prior to incubation on the filter.[a]

[a] Each antiserum is considered to be specific for its respective PKC isoform if immunodetection is abolished by adsorption with its own antigen and not with any others.

3. Immunocytochemistry

3.1 Small animal perfusion

For immunocytochemistry, the rats were deeply anaesthetized with 4% chloral hydrate and perfusion-fixed (see below) with chilled 4% buffered paraformaldehyde as described (19). The brains were removed, post-fixed for 1.5 h, and stored in 15% sucrose at 4°C until cut on a sliding microtome.

Protocol 3. Animal perfusion

1. Equip a Masterflex peristaltic pump (catalogue no. 7520-00) with Masterflex Tygon tubing (catalogue no. 6409-14), inside diameter 0.064 inches, and attach 1 ml tuberculin syringes with glue at both ends of the tubing.
2. Attach a 14G stainless steel blunt-ended needle on to the end to be inserted into the heart, and run 0.9% NaCl (4°C) through the tubing to ensure that no air bubbles are present.
3. Anaesthetize rat with 1 ml/100 g body weight of a 4% chloral hydrate solution (400 mg/kg) and place rat on a Rubbermaid box.[a]
4. Open the thoracic cavity by cutting horizontally through abdominal skin just below the xiphoid process and expose the heart.
5. Cut the diaphragm and extend the excision laterally up through the ribs. Clamp the descending aorta to force perfusate through carotid arteries.

6. Turn on the perfusion pump at low speed, having one end of tube inserted in ice cold saline, allowing saline to run through tube. Immediately insert a 14G needle into left ventricle, enter the ascending aorta, and clamp the needle in place.
7. Increase pump speed to 2.5 on dial (150–180 r.p.m.) once the right atrium is cut, and perfuse with ~50–75 ml of 0.9% saline.
8. Clamp tube with haemostats at perfusate end, with pump at same speed, and transfer into cold 4% paraformaldehyde in 2 × PO_4 solution.[b] Release haemostats and perfuse animal with 250 ml 4% paraformaldehyde.
9. Remove brain and post-fix it in 4% paraformaldehyde for 1.5 h at 4°C, then transfer it into 15% sucrose at 4°C for cryoprotection.

[a] The top of the box is ventilated to allow drainage of perfusate.
[b] Paraformaldehyde solution is made by adding 40 g paraformaldehyde to 2 × PO_4 solution (0.244 M: 7.7 g NaOH and 33.66 g $NaH_2PO_4 \cdot H_2O$ q.s. to 1 litre distilled H_2O) heated to 60–65°C.

3.2 Immunoperoxidase methodology

The immunocytochemical reaction was performed using avidin–biotin–peroxidase reagents (Elite kit, Vector Labs, Burlingame, CA) and 0.5% 3,3′-diaminobenzidine tetrahydrochloride and 0.003% H_2O_2 (19).

Protocol 4. Immunoperoxidase methodology

1. Cut 50 μm sections on a sledge microtome equipped with a freezing stage (American Optical Co., NY) throughout the brain area of interest (dorsal hippocampus in this case). Collect sections free-floating in 12 mM PBS.
2. Pretreat sections in 0.5% Triton X-100 for 5 min.
3. Rinse sections three times with PBS (10 min each) with continuous agitation.
4. Incubate sections in the appropriate primary antiserum (e.g. PKC α, β, or ε) diluted in 0.1% BSA in PBS at the appropriate dilution (1:1000 and 1:4000) for 24 h at 4°C with continuous agitation (Environ shaker, LabLine Instruments, IL).
5. Wash the sections three times in PBS for 10 min each.
6. Incubate sections with biotinylated goat anti-rabbit IgG (Elite Vectastain kit, Vector Labs, Burlingame, CA) at a dilution of 0.44% in 0.3% Triton X-100/PBS for 60–90 min at RT with continuous agitation.
7. Wash sections again three times in PBS for 10 min each.
8. Incubate sections with avidin–biotin–horseradish peroxidase at a concentration of 0.88% in 0.3% Triton X-100/PBS at RT.

Protocol 4. *Continued*

9. Develop the immunohistochemical reaction, after three PBS washes, using 0.05% 3,3′-diaminobenzidine tetrahydrochloride (Sigma, MO) and 0.003% H_2O_2 for a length of time that will depend upon the primary antibody used (2 min in the case of the particular antibodies used in this study).

10. Mount sections from 0.3% gelatin/95% ethanol/50% PBS on gelatin-subbed slides,[a] air dry and cover slip out of xylene with GBX mounting medium (Fluka).

[a] Note that slides are soaked in 70% ethanol/1 M acetic acid then submerged once in gelatin solution (5 g gelatin 275 bloom (Fisher), 0.5 g chromium potassium sulfate (Fisher), 1 litre distilled H_2O) and dried.

For examples of the distribution of PKC α, β, and ε IR in the dorsal hippocampus of the rat, see *Figures 2–4*. Figures 2–4 are reproduced with permission from McGinty *et al.* (12). © 1991, Churchill Livingstone, New York.

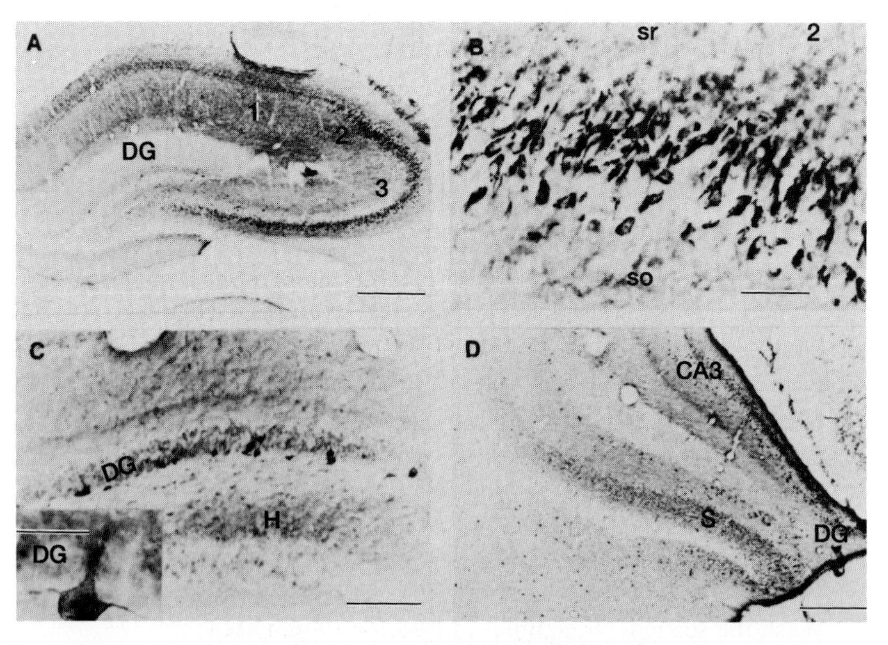

Figure 2. PKC α IR in coronal sections of rat hippocampus. **A.** PKC α IR is prominent in hilar interneurones and CA3–2 (2, 3) pyramidal cells with lower intensity in CA1 (1) pyramidal cells. Bar = 500 μm. **B.** Higher magnification of PKC α IR in CA2 (2) pyramidal cells. sr = stratum radiatum; so = stratum oriens. Bar = 50 μm. **C.** Higher magnification of PKC α IR in hilar (H) neurones immediately subjacent to dentate granule (DG) cell layer. Bar = 150 μm. **Inset:** Homogeneous staining of putative pyramidal basket cell in subgranular zone. Bar = 25 μm. **D.** PKC α IR in interneurones and CA3 pyramidal cells of the ventral hippocampus. S = subiculum. Bar = 500 μm.

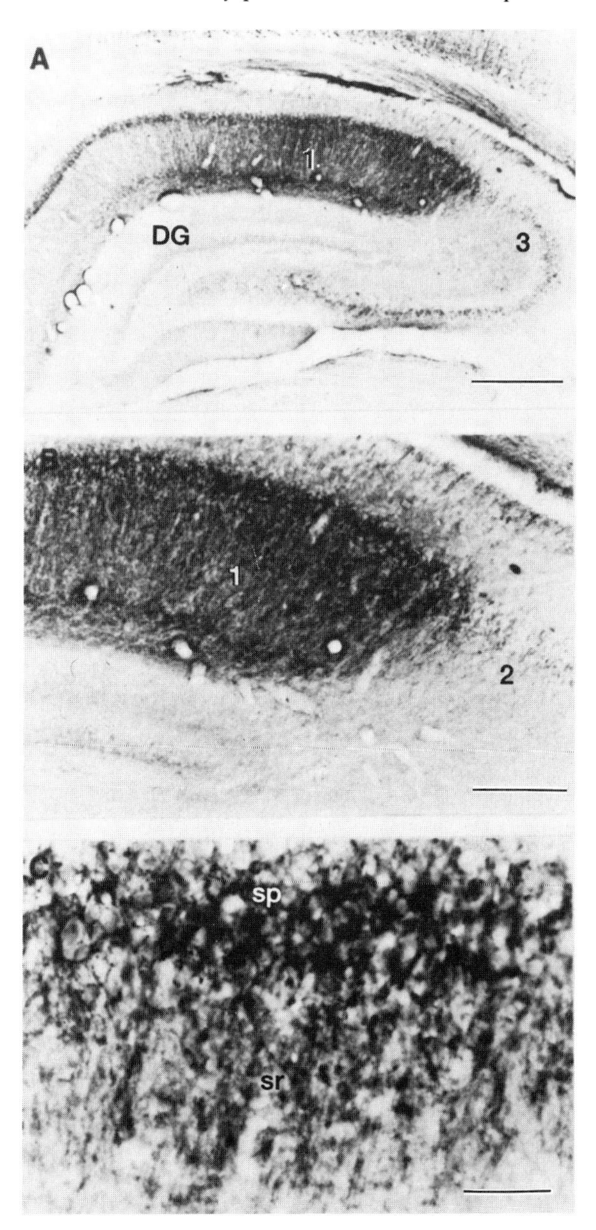

Figure 3. PKC β IR in coronal sections of rat hippocampus. **A** PKC β IR is most prominent in CA1 (1) pyramidal cells and less intense in CA2–3 (2, 3) pyramidal cells. Bar = 500 μm. **B.** Higher magnification of PKC β IR in pyramidal cells at the CA2–1 junction. Bar = 225 μm. **C.** Higher magnification of PKC β IR in CA1 pyramidal cell bodies and dendrites in stratum radiatum (sr). sp = stratum pyramidale. Bar = 50 μm.

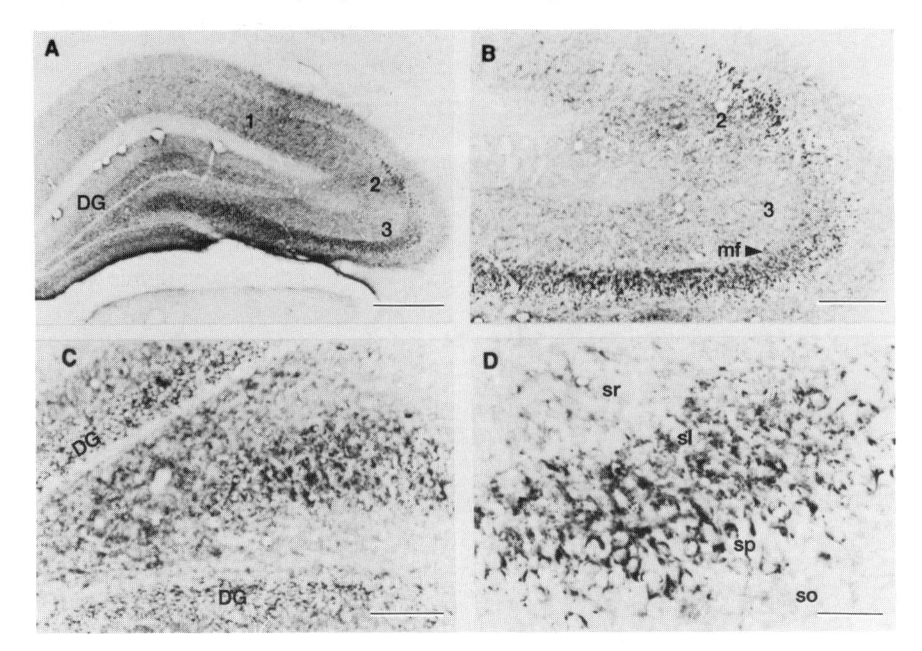

Figure 4. PKC ε IR in coronal sections of rat hippocampus. **A.** PKC ε IR is prominent in dentate granule (DG) cells, mossy fibres, and CA3 pyramidal cells. 1, 2, 3 = CA fields. Bar = 500 μm. **B.** Higher magnification of PKC ε IR in putative mossy fibres (mf) and CA3–2 pyramidal cells. Bar = 175 μm. **C.** Higher magnification of PKC ε IR in dentate granule (DG) cells and hilar neurones. Bar = 100 μm. **D.** Higher magnification of PKC ε IR in putative mossy fibres in stratum lucidum (sl) and CA3 pyramidal cells (sp). so = stratum oriens; sr = stratum radiatum. Bar = 40 μm.

3.3 Antibody/antigen adsorption controls

Adjacent sections are incubated with each primary antiserum preadsorbed with 0.1, 1.0, or 10 μM of its own antigen or with each of the other peptide subspecies. This protocol is employed to determine the specificity of each antibody to the peptide antigen against which it was raised. The antibody is adsorbed to the peptide for 4 h at RT or overnight at 4°C prior to incubating tissue sections with the mixture. If the antiserum is specific for the peptide, no tissue staining will occur.

Protocol 5. Antibody/antigen adsorption controls

1. Add 500 ml of 200 μM peptide solution to 500 μl of anti•erum diluted to twice the final concentration[a] to yield 1 ml solution of 100 μM peptide/ antiserum at working dilution.
2. Dilute remaining volume from the 200 μM peptide solution to 20 μM. Add 500 μl of this solution to 500 μl of antiserum, diluted as in step 1, to yield 10 μM peptide/antiserum at working dilution.

3. Dilute 20 μM peptide solution to 2 μM; add 500 μl to 500 μl of the diluted antiserum to yield 1 μM peptide/antiserum working dilution.
4. Add 500 μl of PBS to 500 μl of the diluted antiserum to yield 0 μM peptide/antiserum at working dilution.
5. Incubate antibody/antigen mixture in a test tube for 4 h at RT or for 24 h at 4°C.
6. Incubate tissue sections at 4°C overnight with peptide/antibody solutions, along with sections incubated in freshly prepared antibody alone at an appropriate concentration.

[a] i.e. 1:8000 for ε or 1:2000 for α and β antisera to yield 1:4000 and 1:1000 final dilutions, respectively.

4. *In situ* hybridization histochemistry

4.1 *In situ* hybridization methodology

This technique provides the localization of a specific mRNA in its original anatomical context and offers single cell resolution. Different types of probe, that specifically hybridize to the mRNA of interest may be used. We have chosen small oligodeoxynucleotide probes (48-mer) since they are easy to obtain and readily penetrate tissue without enzymatic pretreatment. Once a particular DNA sequence has been published, the choice of an appropriate oligonucleotide sequence is straightforward and can be designed to differentiate families of closely related transcripts (as in the case of the PKC family) (also see Chapter 1, this volume).

Frozen sections are cut on a cryostat at different levels throughout the hippocampal formation and thaw-mounted on to gelatin-chrom alum coated slides (see *Protocol 7*). Sections are post-fixed with 4% buffered paraformaldehyde, and then acetylated, defatted in alcohol and chloroform, and dried. We currently post-fix the sections since unperfused tissue is easier to cut in the cryostat and works optimally with this technique. However, for a combination with immunocytochemistry on adjacent sections, perfusion fixation is recommended. Probes are labelled at the 3′ end using [α-^{35}S]dATP and terminal deoxynucleotidyl transferase (tdt) (see *Protocol 6*). Sections are incubated overnight with the probe in a humid atmosphere (20) and subsequently washed under highly stringent conditions.

The use of gloves during the different steps as well as diethylpyrocarbonate (DEPC)-treated, autoclaved water for the different solutions is important in order to avoid contamination with RNases. The use of molecular biology grade reagents is highly recommended.

4.1.1 Oligonucleotide probes

The specific 48-mer oligodeoxynucleotide exonic probes for α (bases 1195–1242

(14)), β (bases 2167–2214 (13)), and ε (bases 1373–1420 (21)) PKCs were synthesized on a DNA synthesizer and characterized by Northern blot analysis (15). Lyophilized probe was dissolved in Tris EDTA (TE) buffer, pH 7.6. The purity and concentration of deoxyoligonucleotide probes were checked spectrophotometrically by calculating the concentration of the probes in μg/ml from optical density readings (Ultrospec II, LKB, Biochrom, Sweden) at a wavelength of 260 nm (OD_{260}) assuming that 1 OD_{260} unit corresponds to a DNA concentration of 37 μg/ml and that 1 M DNA is equivalent to a concentration of 300 g/l of each nucleotide. The purity of the probe was also checked spectrophotometrically by analysing the ratio obtained between the readings at wavelengths of 260 and 280 nm. Final concentrations were adjusted to a 5 μM working solution and 40 μM stock.

The probes were labelled in a tailing reaction using [α-^{35}S]dATP (1075 Ci/mmol) and tdt, with cobalt chloride as a cofactor, to a specific activity of 4–6 × 10^5 d.p.m./μl.

Protocol 6. Oligonucleotide labelling

1. Add the following components in order, to a final volume of 50 μl in 1.5 ml tubes:

• sterile H_2O	enough for final volume of 50 μl
• oligonucleotide probe (0.1 μM)	1.0 μl (from a 5 μM stock kept refrigerated)
• 5 × tdt reaction buffer	10 μl (Boehringer Mannheim, IN) (frozen; thaw prior to use)
• cobalt chloride	3 μl from a 25 mM solution (Boehringer Mannheim)
• [α-^{35}S]dATP	0.05 nmol (1 μM), with a specific activity of ~1.075 Ci/mmol (New England Nuclear, MA)
• tdt enzyme	1.0 μl (25 U) (added last to the solution on ice)

2. Vortex tubes at low speed and centrifuge them briefly (13 600 *g* for 1 min). Then, allow the enzymatic reaction to proceed by placing the tubes at 37°C for 5 min.
3. Add 400 μl of TE pH 7.6 (10 mM Tris, 1 mM EDTA; stored at 4°C) to the tubes at RT to stop the reaction. Immediately add 2 μl of tRNA (25 μg/ml) (frozen; thaw prior to use) and vortex briefly.
4. Extraction of the labelled probe is performed in two steps. First, add an equal volume (450 μl) of phenol[a]:chloroform:isoamyl alcohol (50:49:1 stock, refrigerated) to the mixture, vortex vigorously, and microfuge for 10 min.
5. Transfer aqueous phase into new tubes. Subsequently add an equal volume

(450 μl) of chloroform:isoamyl alcohol (49:1, stored at room temperature), vortex the mixture vigorously, and centrifuge the tubes for 10 min.

6. Transfer aqueous phase into new tubes.
7. To recover the probe by precipitation, add 0.05 vol (23 μl) of 4 M NaCl (made with sterile water and filtered) and 2.5 vol (1 ml) of cold 100% ethanol.
8. Put tubes in a dry ice–100% ethanol bath at −70 °C for 30 min to precipitate.
9. Discard supernatant after centrifugation for 30 min at 4°C and save the pellet.
10. Keeping the tube inverted, wash the pellet with 1 ml cold 100% ethanol, and dry it for 2–5 min.
11. Reconstitute the pellet in 100 μl TE plus 1 μl 1 M dithiothreitol (DTT) (frozen; thaw prior to use) in order to prevent disulfide bridge formation.
12. Let the tubes sit for ~15 min and then count 1 μl in a β counter in 10 ml of scintillation fluid.

[a] Use redistilled quality phenol only and equilibrate it with Tris buffer 0.1 and 0.5 M, pH 7.6–8.

4.1.2 Tissue preparation

For *in situ* hybridization, the rats are deeply anaesthetized with 4% chloral hydrate (400 mg/kg) and decapitated. The brains are removed immediately and dipped in cold isopentane (−40°C) for 15 sec before storing at −70°C. Thin (12 μm) frozen sections are cut on a cryostat microtome (Hacker Instruments, NJ) and thaw mounted on to gelatin-chrom alum coated slides. Slides are stored with desiccant at −70°C. Prior to hybridization, tissue sections are pretreated as described in *Protocol 7*.

Protocol 7. Tissue section pretreatment

1. Thaw slides[a] for 15–30 min.
2. Fix sections in 4% paraformaldehyde in PBS for 10 min in a Coplin jar or glass histology dish, depending on the number of slides.
3. Transfer the slides to a new dish or discard the previous solution, then add 0.25% acetic anhydride (125 μl/50 ml) in sterile 0.1 M triethanolamine/ 0.9% NaCl (pH 8.0) and leave for 10 min.[b]
4. Discharge previous solution or transfer the slides to a series of increasing concentrations of ethanol in H_2O: 70% (1 min), 80% (1 min), 95% (1 min), and 100% (1 min).
5. Defat the tissue by dipping the slides in chloroform for 5 min. Decant and repeat for another 5 min.
6. Discard previous solution or transfer the slides to a series of decreasing concentrations of ethanol: 100% (2 × 1 min), 95% (1 min), 80% (1 min), 70% (1 min), and then into sterile water (1 min).

Protocol 7. *Continued*

7. Dry slides under cool blown air.
8. Proceed to the hybridization step or store the slides at −70°C.

[a] Slides should be gelatin-subbed twice, prior to collecting the sections, as follows: first soak the slides in hot soapy water, then rinse in running H_2O, soak in 80% ethanol, and dip twice in filtered gelatin solution (4 g porcine pig gelatin 300 bloom (Sigma), 0.4 g chromium potassium sulfate and 800 ml distilled H_2O).

[b] The acetylation step is performed to avoid non-specific binding of the oligonucleotide probe with charged groups in tissue.

4.1.3 Hybridization

Protocol 8. Hybridization

1. Dissolve probe in hybridization buffer (kept at −20°C):
   - 50% formamide
   - 4 × SSC (3 M sodium chloride, 0.3 M sodium citrate) pH 7.2
   - 500 μg/ml sheared salmon sperm DNA
   - 250 μg/ml yeast tRNA
   - 1 × Denhardt's solution[a]
   - 10% dextran sulfate
   - DEPC-treated water

   in order to obtain 1 × 10^6 c.p.m./25 μl of hybridization buffer/section.
2. Add 1 μl freshly made 5 M DTT stock per 50 μl of hybridization buffer (giving 100 mM DTT).
3. Add 25 μl of the mixture (probe + hybridization buffer + DTT) per section and immediately cover with a small piece of Paraffilm. Transfer the slides to a humid incubator at 37°C for 20 h (overnight).
4. Preventing the slides from drying, remove pieces of Paraffilm by submerging each slide in a beaker with 1 × SSC. Wash the slides three more times with fresh 1 × SSC for ~30 sec each and once with 50% formamide/ 2 × SSC at RT.
5. Transfer the slides to 2 × SSC/50% formamide at 40°C with continuous agitation and wash four times for 15 min each.
6. Finally, transfer the slides to fresh 1 × SSC twice for 30 min each at RT.
7. Rinse them briefly in water, then in 70% ethanol, and dry.

[a] A 50× stock solution is prepared as follows: 5 g Ficoll, 5 g polyvinylpyrrolidone, 5 g BSA, and H_2O to make up to 500 ml.

4.1.4 Autoradiography

The dried slides are apposed to Kodak X-OMAT AR film in a light-tight cassette. The film will be used to provide information on the intensity of the hybridization signal in the region of interest (dorsal hippocampus in this case); it will also provide information about the length of exposure needed for nuclear emulsion as well as the level of background. After 2–4 days the films are developed and the slides are dipped in NTB-3 photographic emulsion, which detects β radiation very efficiently and gives a fine grain resolution when developed. After 7 or 14 days, slides are developed in Dektol, and fixed in Kodak fixer (see *Protocol 9*). Care must be taken when managing films or photographic emulsion in the dark room, since each of them needs different conditions. When apposing films to slides in cassettes or developing them manually, use a red light with GBX-2 filter. For working with photographic emulsion, the darkroom should have an ambient temperature of 20–22°C and not less than 60% humidity (22). It also should be equipped with the following:

- safelight with a #2 Wrattan red filter (Kodak)
- water bath set at 42°C
- black autoradiography exposure boxes
- mini-glass Coplin jar
- slide drying racks
- refrigerator for storage of NTB-3 emulsion and slides
- 0.1% Dreft detergent in water (filtered)
- slides (experimental, control, and blank)

Protocol 9. Autoradiography

1. Tape slides to poster board inside a light-tight cassette and, under red safelight (GBX-2 filter, Kodak), place XAR-2 film on top of the slides. Tuck in overlying sheet to minimize light exposure. Expose in the dark for 2–4 days for PKCs.
2. Develop films in darkroom using GBX developer (Kodak) diluted 1:1 with water at 20°C for 5 min with an initial agitation.
3. Transfer film to the stopbath (Kodak) for 30 sec with continuous agitation.
4. Fix the film using Kodak Rapid Fix at full strength for 3 min, agitating initially and within the last minute.
5. Wash the film for 5 min under running water.
6. Allow the film to dry in a dust-free environment with a temperature not exceeding 40°C.

Protocol 9. *Continued*

7. Subsequently, dip slides in photographic emulsion as follows:
   (a) Melt Kodak NTB-3 emulsion (stored at 4°C) in 42°C water bath for ~30 min. Then, dilute emulsion 1:1 with 0.1% Dreft detergent/ sterile water (filtered and warmed for 30 min) and let the mixture stand for 20–30 min in the water bath.
   (b) Submerge a test slide into the emulsion to dispel any air bubbles before dipping experimental and control slides.
   (c) Dip slides individually into the Coplin jar, slowly remove them from emulsion, and wipe the back with a Kim-wipe.[a]

8. Let slides dry for ~2 h and then transfer them to exposure boxes with two or three Humi-capsules (United Desiccants, NJ). Tape lids on with electrical tape and store boxes in the dark at 4°C until ready to develop.

9. When required, take slides out of the refrigerator and let them reach RT under safelight (#2 Wrattan red filter).

10. Fill dishes with Dektol (filtered, diluted 1:1 with sterile H_2O), sterile H_2O, and Kodak fixer (full strength, filtered). Place dishes on top of ice water to cool to 17°C.

11. Develop slides for 2 min when temperature is at 17°C.[b]

12. Wash slides, vigorously, for 1 min in water and immediately transfer them to the dish containing Kodak Fix for 3 min.

13. Rinse the slides under running water 30 min prior to dehydrating and coverslipping.

14. Wait until the slides are completely dried (1–2 days). Wash them then with 1% Sparkleen detergent, heated to 40°C, to get rid of emulsion residues that may remain on the back of the slides.

[a] Always maintain the slides in a vertical position.
[b] Longer times and higher ambient temperatures increase background markedly.

For examples of the distribution of PKC mRNA hybridization signal in the dorsal hippocampus of the rat, see *Figures 5–7*. Figures 5–7 are reproduced with permission from McGinty *et al.* © 1991, Churchill Livingstone, New York.

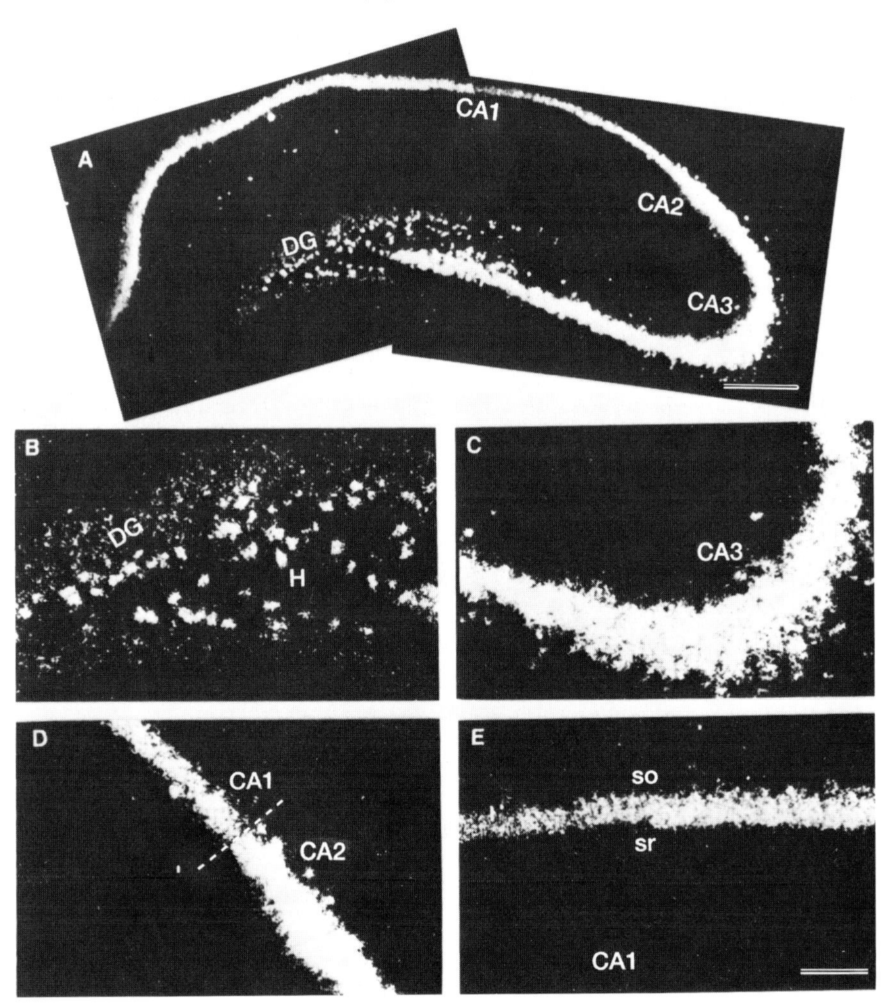

Figure 5. PKC α mRNA in coronal sections of rat hippocampus. **A.** PKC α mRNA is prominent in hilar interneurones and CA3 pyramidal cells with moderately high intensity in CA1 pyramidal cells and low intensity in dentate granule (DG) cells. Bar = 1 mm. **B.** Higher magnification of PKC α mRNA in hilar (H) neurones and dentate granule (DG) cells. **C.** Higher magnification of PKC α mRNA in CA3 pyramidal cells. **D.** Higher magnification of PKC α mRNA in pyramidal cells at the junction of CA1–2. **E.** Higher magnification of PKC α mRNA in CA1 pyramidal cells. so = stratum oriens; sr = stratum radiatum. Scale bar in B–E = 500 μm.

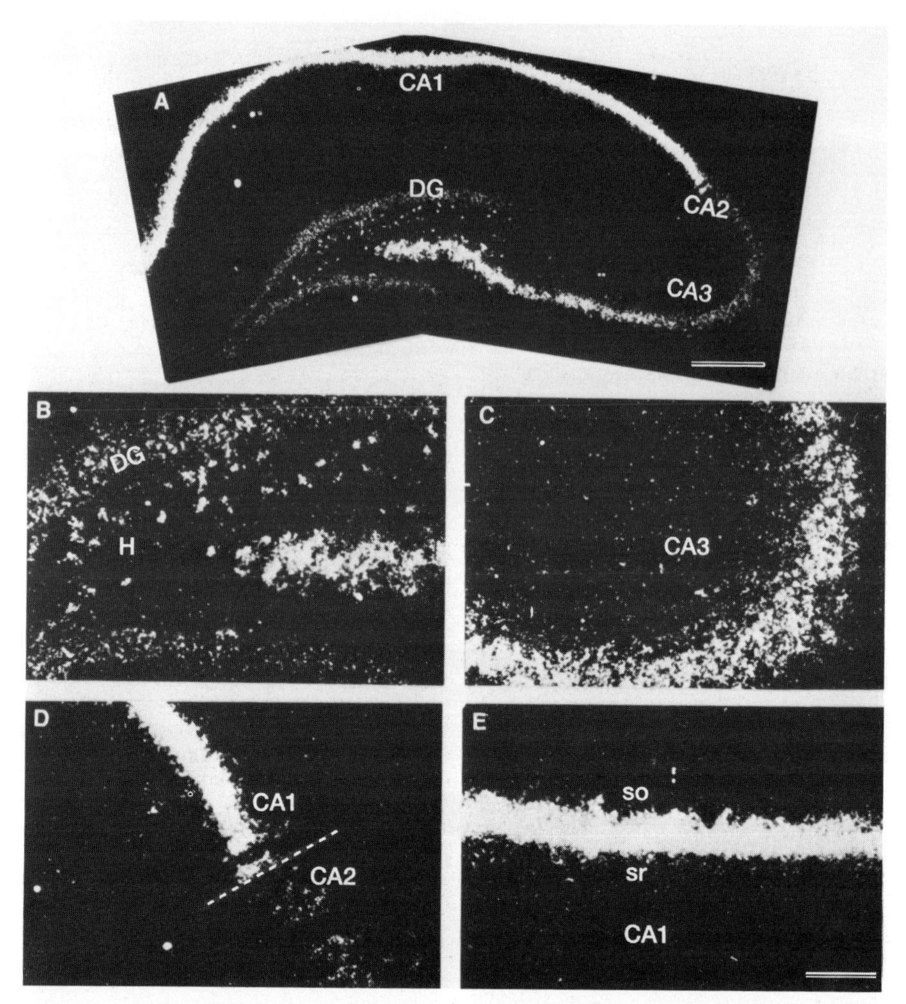

Figure 6. PKC β mRNA in coronal sections of rat hippocampus. **A.** PKC β mRNA is most prominent in CA1 pyramidal cells and is less intense in dentate granule (DG) cells, hilar interneurones and CA2 pyramidal cells. Bar = 1 mm. **B.** Higher magnification of PKC β mRNA in hilar (H) neurones and dentate granule (DG) cells. **C.** Higher magnification of PKC β mRNA in CA3 pyramidal cells. **D.** Higher magnification of PKC β mRNA in pyramidal cells at the junction of CA1–2. **E.** Higher magnification of PKC β mRNA in CA1 pyramidal cells. so = stratum oriens; sr = stratum radiatum. Scale bar in B–E = 500 μm.

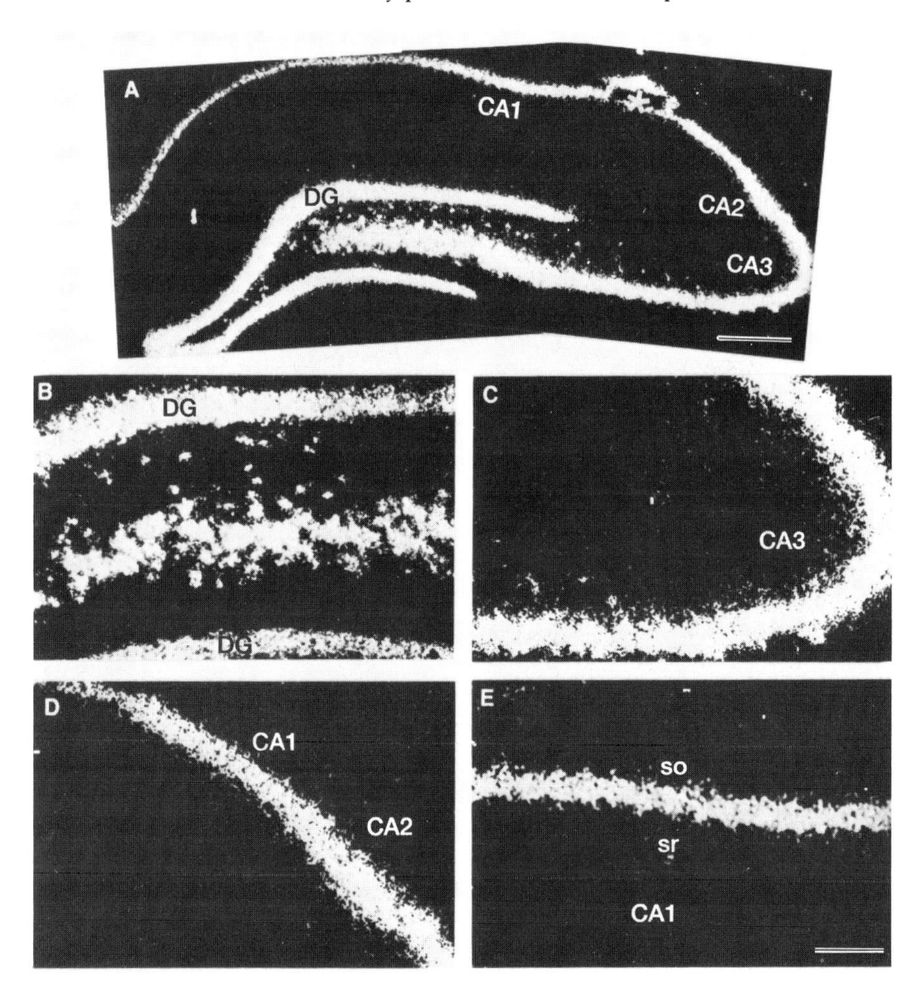

Figure 7. PKC ε mRNA in coronal sections of rat hippocampus. **A.** PKC ε mRNA is most prominent in dentate granule (DG) cells, CA3–1 pyramidal cells, and hilar interneurones. *, Tissue artefact. Bar = 1 mm. **B.** Higher magnification of PKC ε mRNA in hilar neurones and dentate granule cells. **C.** Higher magnification of PKC ε mRNA in CA3 pyramidal cells. **D.** Higher magnification of PKC ε mRNA in pyramidal cells at the junction of CA1–2. **E.** Higher magnification of PKC ε mRNA in CA1 pyramidal cells. Scale bar in B–E = 500 μm.

References

1. Nishizuka, Y. (1988). *Nature*, **334,** 661.
2. Parker, P. J., Kour, G., Marais, R. M., Mitchell, F., Pears, C. Schaap, D., Stabel, S., and Webster, C. (1989). *Mol. Cell. Endocrinol.*, **65,** 1.
3. Schaap, D., Parker, P. J., Bristol, A., Kriz, R., and Knopf, J. (1989). *FEBS Lett.*, **243,** 351.
4. Wood, J. G., Girard, P. R., Mazzei, G. J., and Kuo, J. F. (1986). *J. Neurosci.*, **6,** 199.
5. Mochly-Rosen, D., Basbaum, A. I., and Koshland Jr., D. E. (1987). *Proc Natl Acad. Sci. USA*, **84,** 4660.
6. Huang, F. L., Yoshida, Y., Nakabayashi, H., Young III., W. S., and Huang, K. P. (1988). *J. Neurosci.*, **8,** 4734.
7. Stichel, C. C. and Singer, W. (1988). *Exp. Brain Res.*, **72,** 443.
8. Saito, N., Kikkawa, U., Nishizuka, Y., and Tanaka, C. (1988). *J. Neurosci.*, **8,** 369.
9. Hosoda, K., Saito, N., Kose, A., Ito, A., Mori, M., Hirata, M., Ogita, K., Ono, Y., Igarashi, K., Kikkawa, U., Nishizuka, Y., and Tanaka, C. (1989). *Proc. Natl Acad. Sci. USA*, **86,** 1393.
10. Ito, A., Saito, N., Hirata, M., Kose, A., Tsujino, T., Yoshihara, C., Ogita, K., Kishimoto, A., Nishizuka, Y., and Tanaka, C. (1990). *Proc. Natl Acad. Sci. USA*, **87,** 3195.
11. Saito, N., Kose, A., Ito, A., Hosoda, K., Mori, M., Hirata, M., Ono, Y., Igarashi, K., Ogita, K., Kikkawa, U., Nishizuka, Y., and Tanaka, C. (1989). *Proc. Natl Acad. Sci. USA*, **86,** 3409.
12. McGinty, J. F., Couce, M. E., Bohler, W. T., and Ways, D. K. (1991). Protein kinase c subspecies distinguish major cell types in rat hippocampus: an immunohistochemical and *in situ* hybridization histochemical study. *Hippocampus*, **1,** 293.
13. Knopf, J. L., Lee, M.-H., Sultzman, L. A., Kriz, R. W., Loomis, C. R., Hewick, R. M., and Bell, R. M. (1986). *Cell*, **46,** 491.
14. Ono, Y., Fuji, T., Igarashi, K., Kikkawa, U., Ogita, K., and Nishizuka, Y. (1988). *Nucleic Acids Res.*, **16,** 5199.
15. Young III, W. S. (1988). *J. Chem. Neuroanat.*, **1,** 177.
16. Fletcher, D. J. and Ways, D. K. (1991). *Diabetes*, **40,** 1496.
17. Lerner, R. A., Green, N., Alexander, H., Liu, F.-T., Sutcliffe, J. G., and Shinnick, T. M. (1981). *Proc. Natl Acad. Sci. USA*, **78,** 3404.
18. Konno, Y., Ohno, S., Akita, Y., Kawasaki, H., and Suzuki, K. (1989). *J. Biochem.*, **106,** 673.
19. McGinty, J. F., Van der Kooy, D., and Bloom, F. E. (1984). *J. Neurosci.*, **4,** 1104.
20. Young III, W. S., Bonner, T. I., and Brann, M. R. (1986). *Proc. Natl Acad. Sci. USA*, **83,** 9827.
21. Ono, Y., Fujii, T., Ogita, K., Kikkawa, U., Igarashi, K., and Nishizuka, Y. (1988). *J. Biol. Chem.*, **263,** 6927.
22. Rogers, A. W. (1979). *Techniques of Autoradiography*. Elsevier/North-Holland Biomedical Press, Amsterdam.

3

Imaging gene expression in neural grafts

D. J. S. SIRINATHSINGHJI and S. B. DUNNETT

1. Introduction

In recent years considerable attention has been devoted to the development of techniques for the implantation of embryonic neural tissue into the adult mammalian brain as a tool for replacing damaged host tissue and repairing damaged neuronal circuits. Indeed, several studies have shown that embryonic neurones grafted into the mammalian brain can survive, grow, differentiate, and reverse many of the structural, biochemical, physiological, and behavioural deficits associated with experimental or genetic neuronal loss (1). This method thus has immense relevance to the possible treatment of human neurodegenerative disease. In addition, it is a powerful technique for the neurobiologist for the exploration of the basic cellular, receptor, and molecular mechanisms regulating the development and functional organization of specific neuronal systems within the brain.

One graft model which we have used to great advantage to study the above processes is the rat model of Huntington's disease (HD) which is characterized histopathologically by a profound degeneration of intrinsic striatal neurones and their axonal projections to striatal efferent sites, the globus pallidus and pars reticulata of the substantia nigra. This histological picture is reproduced in the rat by the injection of excitotoxic amino acids (e.g. kainate, ibotenate, and quinolinate) into the neostriatum (2–4); the toxins destroy neuronal cell bodies but spare axons en passage (5–7) e.g. striatal dopamine (DA) fibres originating from DA neurones in the pars compacta of the substantia nigra. In this chapter we have chosen to describe the grafting procedure in the rat model of HD because we believe that this graft model is yielding more information on the cellular and molecular organization of neural tissue grafts than any other system yet studied. Grafts of embryonic striatal tissue have been shown to reliably survive transplantation to the excitotoxic amino acid lesioned rat neostriatum, develop into a new striatum like structure, express many of the neuronal markers characteristic of the normal striatum, and reconstruct many features of intact striatal neuronal circuitry (8–21). However, these grafts contain some abnormal features which must be considered in terms of the organization of neuronal systems within the grafts. At the structural level the embryonic striatal eminence contains non-striatal (cortical, amygdaloid, and pallidal) as well as striatal precursors (22, 23) and it is not possible to dissect striatal cells selectively for grafting. As a consequence, immunocytochemical analyses (24–26) indicate that the striatal grafts manifest a heterogeneous organization containing discrete 'patches' of cells expressing distinctive striatal features interspersed in a non-striatal matrix. The striatal and non-striatal compartments within the striatal grafts can be easily distinguished by staining brain sections for acetylcholinesterase (AChE) (27). Thus, striatal neuronal markers within striatal grafts are expressed predominantly within AChE-rich zones whereas non-striatal markers are predominantly expressed in AChE-poor zones (19, 24–26). Importantly, the afferent and efferent connections of the grafts are seen to become established within the striatal-like compartments (11, 19) (see *Figure 1*).

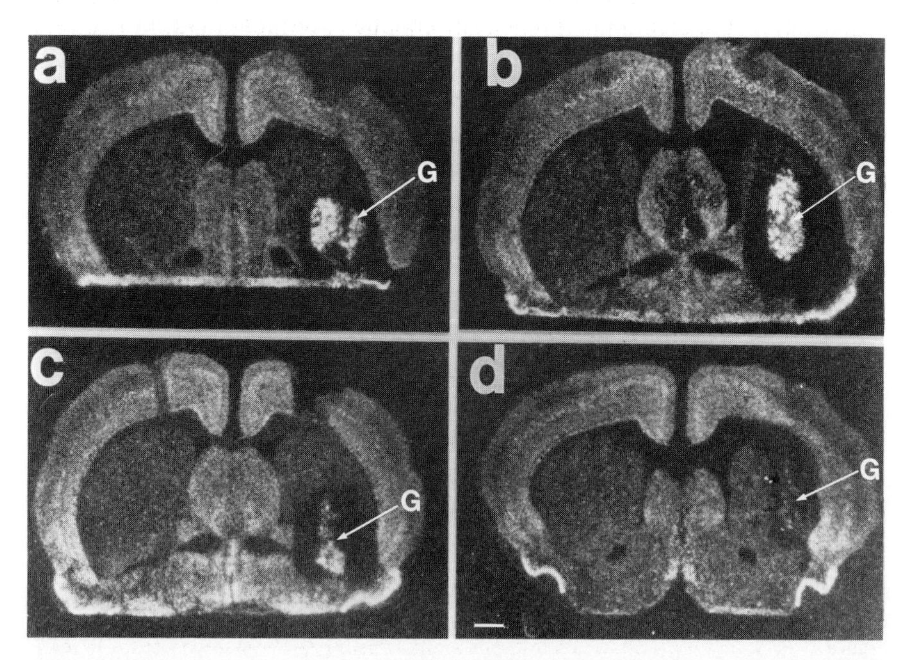

Figure 1. Negative prints of X-ray film autoradiograms showing the expression of the mRNA for the growth associated phosphoprotein (GAP–43) in developing primordial striatal tissue grafts (G). (**a**, 7 days; **b**, 15 days; **c**, 30 days; **d**, 3 months) following *in situ* hybridization of brain sections with a ^{35}S-labelled antisense oligonucleotide probe (complementary to the rat GAP-43 gene) using the procedures outlined in this chapter. Note the decrease in GAP-43 mRNA at 30 days indicating that most of the synaptic contacts are possibly established at about this time when the grafts have reached maturity. In 3 month old (mature grafts) GAP-43 mRNA expression in the grafts is undetectable and similar to the control striatum. Exposure time = 7 days; scale bar = 1 mm.

Recent advances in molecular biology as exemplified by molecular cloning and DNA sequencing techniques, nucleic acid manipulation, the availability of automated DNA synthesizers, and the development of histochemical techniques based on the new methods of molecular biology e.g. *in situ* hybridization have opened new avenues for research in the neurosciences for the anatomical localization and control of specific neuronal genes within the brain. Since mRNA availability is frequently the rate-determining step in peptide synthesis, the cellular content of the neurotransmitter-specific mRNA provides an index of gene expression and consequently of the rate of neurotransmitter synthesis and thus neurotransmitter turnover. We have developed an *in situ* hybridization histochemical technique which has proved to be an easy, reliable, and extremely specific and sensitive approach for studying at the molecular level the expression and functional organization of specific neuronal markers within striatal grafts and for assessing the dynamic changes occurring in specific cells within the grafts during development and in response to specific trophic, neurochemical, or behavioural stimuli (see *Figure 2*). However, the technique is obviously equally applicable to studies on gene expression in the normal brain as well as in other types of embryonic neural tissue grafts e.g. the localization of cells expressing tyrosine hydroxylase (TH) mRNA in embryonic nigral DA-rich tissue grafts implanted into the DA-

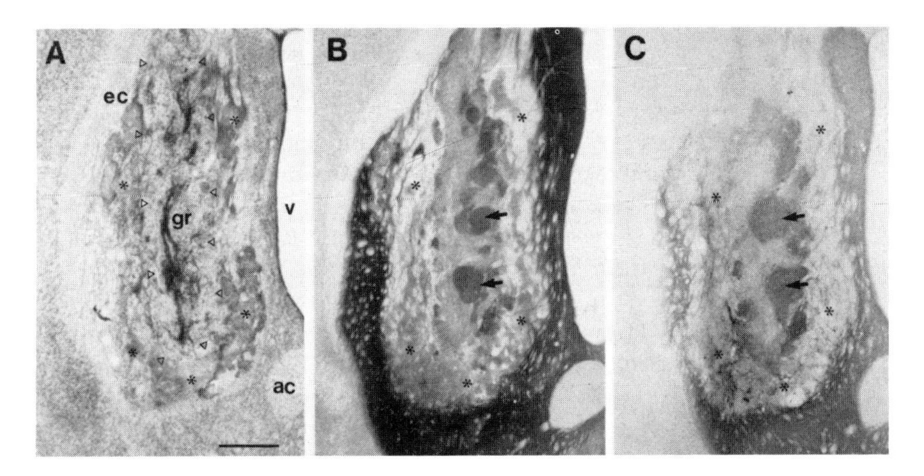

Figure 2. Photomicrographs of a primordial striatal tissue graft implanted into the ibotenic acid lesioned striatum. Adjacent sections are stained with cresyl violet (**A**), acetylcholinesterase (AChE) histochemistry (**B**), or tyrosine hydroxylase (TH) immunohistochemistry (**C**). The extent of the lesion is marked by gliosis and loss of neurones, and a decrease in AChE and TH staining of the neuropil (indicated by asterisks in A, B, and C respectively). The graft–host border is most apparent at higher magnification in the nissl stained sections and is marked by open arrowheads in A. The grafts show characteristic patch zones that correspond in both the TH and AChE sections (arrows in B and C). Abbreviations: ac, anterior commissure; ec, external capsule; gr, graft; v, lateral ventricle. Scale bar in A = 500 μm.

denervated rat neostriatum (28). It also becomes an even more powerful approach for studying the above mechanisms when used in combination with other techniques such as receptor autoradiography and/or immunohistochemistry or histochemical staining which can be easily performed on sections adjacent to those taken for *in situ* hybridization (29–31).

In this chapter we describe the two principal procedures: neural grafting in the rat model of Huntington's disease, i.e. the implantation of primordial striatal neurones in the ibotenic acid lesioned rat neostriatum, and the *in situ* hybridization technique for the visualization of neuronal mRNAs within such grafts.

2. Neural transplantation

2.1 Factors to be considered

Successful transplantation of neural tissues involves identification of an appropriate source of graft tissue, dissection and preparation of that tissue for transplantation, and surgery to implant the graft tissue into the host brain. Several parameters need to be considered at the outset, depending on the experimental questions to be investigated:

i. Type of tissue

Tissues from a wide variety of brain areas, peripheral nerve, and neuroendocrine glands have been successfully transplanted into the mammalian brain, as well as a variety of non-neural tissues such as skin, heart, or muscle.

ii. Source of donors

Whereas cells and tissues derived from peripheral nervous system (PNS) and other peripheral organs can well survive transplantation into the brain, central nervous system (CNS) tissues only survive when derived from the developing organism. In a few cases (e.g. neocortex), CNS donor tissues can be derived successfully from the neonatal brain, but more typically fetal/embryonic donors are required. Optimal graft survival is obtained when the graft tissue is dissected from donor embryos taken with 1–2 days of mitosis of the neurones of experimental interest.

iii. Host animals

The age of the host is less critical. Embryonic grafts can survive transplantation well in hosts from birth to old age. However, the formation of reciprocal graft–host connections is generally better when implants are made into the brains of neonatal hosts, and in particular the stimulation of host axon sprouting to innervate the grafts appears to be less pronounced in adult than in neonatal brain.

iv. Implantation site

The selection of implant site can be important. First, the grafts must be placed in the host brain where the vasculature is sufficiently rich to nourish the graft and incorporate it into the host blood supply. Second, outgrowing graft axons generally do not penetrate over long distances in the adult brain, so for some experimental purposes they need to be placed close to their normal targets. Third, the grafts may need to be placed in a position that allows easy subsequent access (e.g. in the cortex, a ventricle, or the anterior eye chamber, for visual monitoring or implantation of cannulae or electrodes into the grafts).

v. Implantation procedure

A variety of experimental implantation procedures may be selected once the primary requirements of embryonic donor tissue and suitable vascular supply have been met. One major class of techniques involves implantation of pieces of embryonic tissue, either as small plugs into the host parenchyma, by placement into natural spaces of the brain (e.g. the ventricles), or by placement into artificial cavities (e.g. in the neocortex). The other major class of procedures involves injection of neural tissues either in small minced fragments or as dissociated cell suspensions, directly into the host parenchyma. This latter technique has the greatest flexibility because virtually any target site in the brain is accessible, multiple deposits can be placed into different sites, the procedure causes little additional surgical damage, and stereotaxic injection provides accurate placement even into deep brain structures.

vi. Immunological issues

Although immunological factors severely limit transplantation of peripheral organs, the brain manifests a substantial degree of immunological privilege. Although the prevention of rejection needs to be considered when transplanting between animals of different species or widely divergent strains, for most experimental purposes using outbred strains of rats or primates, immunological reactions do not cause any problems.

A comprehensive account of a range of neural transplantation techniques has been published as a separate volume in this series (31). Detailed protocols for breeding rats and monkeys, accurate determination of the stage of pregnancy, dissection of a wide variety of embryonic CNS nuclei, and the procedures for implantation into the brain, spinal cord, or anterior eye chambers of neonatal and adult rats may be found there, along with additional information on how to label the grafts and engineer them in culture to express particular gene products. This account is restricted to the particular procedure that has been most widely investigated in experiments on gene expression in neural grafts, viz. transplantation of embryonic striatum into the denervated neostriatum, in an animal model of Huntington's chorea (9–15). Specifically, we describe the placement of a single graft deposit midway between two excitotoxic lesion sites in the right neostriatum of adult rats.

2.2 Neostriatal lesions

Striatal grafts survive and grow best when the host neostriatum is lesioned prior to implantation. Although a variety of excitotoxins may be used, including kainic acid and quinolinic acid (2–4), we have obtained the best results using ibotenic acid. You require:

- Stereotaxic frame (David Kopf Inc.) with 45° ear bars and an electrode carrier
- Injection cannula of 10 cm, 30 gauge stainless steel tubing, connected via 60 cm narrow polyethylene tubing to a 10 μl Hamilton glass microsyringe mounted in a microdrive pump (Harvard Instruments Ltd).
- Instruments: dental drill with a no. 3 drill bit, scalpel, cotton buds, gauze swabs, suture needle, needle holder, and scissors.
- Toxins: make up a 0.06 M solution of ibotenic acid in phosphate buffer and adjust to pH 7.4. Divide into 50 μl aliquots in 0.4 ml plastic centrifuge tubes, and store at −20°C.

Protocol 1. Ibotenic acid lesion of the neostriatum

1. Thaw one aliquot of the toxin. Flush the connected cannula, tubing, and syringe with distilled water. Load toxin by depressing the syringe to the 0.5 μl mark, draw up ~ 0.2 μl air,[a] followed by 8–9 μl toxin. Once correctly loaded, check flow. Mount the cannula in the stereotaxic electrodc carrier and the syringe in the microdrive pump.
2. Anaesthetize the rat. Place in the stereotaxic frame with the nose bar set 2.3 mm below the interaural line. Open the skull by a single midline scalpel incision, and clean the bone surface. Measure the anterior and lateral location of the skull burr holes, and make small drill holes through the skull.
3. Injection co-ordinates. Two injections are to be made, one placed at $A = 0.2$ mm, $L = 3.4$ mm, and $V = 5.0$ mm, and the other at $A = 1.6$ mm, $L = 2.6$ mm, and $V = 5.0$ mm, with A measured anterior to bregma, L lateral to the midline, and V vertical below dura. Measure the anterior and lateral locations for the skull burr holes, and drill through the skull.
4. Check the pump and cannula for smooth flow of the toxin. Set the microdrive pump to a flow rate of 0.25 μl/min, and lower the injection cannula to the correct co-ordinates for the first injection. Turn on pump and infuse 0.5 μl over 2 min. Turn off pump and allow a further 3 min for diffusion of the toxin. Raise the cannula to withdraw from the brain. Repeat for the injection at the second set of co-ordinates.

5. Clean and suture the wound. Remove the rat from the stereotaxic frame. This is a convenient time while the animal is anaesthetized to add an indelible identification mark, e.g. using an ear punch code. Inject 10 ml glucose saline s.c. to prevent dehydration, and place the rat on a warming blanket in a recovery cage. No special post-operative care is required.

[a] The small air bubble between the distilled water and the toxin stops the dilution of the toxin by diffusion. Moreover, blockages in the system are easily detected by a failure of the bubble to move during injection.

Behavioural testing to characterize functional asymmetries induced by the lesions may be conducted over the following days. Graft surgery is generally conducted 1–2 weeks following the initial lesions. This interval may be extended for much longer periods, but caution is required for shorter intervals as the toxic process may compromise the viability of the grafts if still ongoing.

2.3 Collection of donor tissue

Graft surgery can be considered to involve three separate components, requiring adjacent surgical stations. The embryos have to be removed and the relevant areas of the brain dissected (this section). The embryonic tissue is prepared as a dissociated cell suspension (Section 2.4) and then the suspension is implanted into the ibotenic acid lesioned host striatum (Section 2.5). The following protocols are based on the method of Schmidt *et al.* (32) and more detailed accounts of the protocol can be found elsewhere (33–35). To collect the donor tissue you need:

- Instruments for the Caesarean section: an ether jar, barbiturate anaesthetic, sterile forceps, and pointed scissors.
- Instruments for the embryonic dissection: Dumont #5 forceps, scalpel handle with #11 blade, Vannas ultrafine spring iridectomy scissors, disposable 10 cm Petri dishes
- Solutions: basic medium, 100 ml 0.6 % glucose in sterile 0.9% saline.
- Dissecting microscope, preferably with a zoom objective in the range ×6 to ×20.
- Donors: pregnant rats at embryonic (E) 15–16 days of gestation. This may be achieved by ordering accurately staged pregnant rats from the animal supplier, or by breeding in house, confirming pregnancy by observing a vaginal plug. In either case, the stage of the pregnancy should be confirmed on the day of surgery by palpation of the pregnant rat under ether anaesthesia (34).

Protocol 2. Dissection of embryonic striatum

1. Sterilize all instruments in an autoclave and use sterile solutions throughout.
2. Anaesthetize the pregnant rat with an overdose of barbiturate. Lay her on

Protocol 2. *Continued*

her back, clean the abdomen with alcohol, open the abdomen, remove the uterine horns (which typically contain 10–15 embryos) and transfer to 10 ml basic medium in a Petri dish.

3. Remove the embryos from the uterus with small pointed scissors, separate from the amniotic sac and placenta, and transfer to a second Petri dish. Measure the crown–rump length of the embryos which should be ~ 13–16 mm at E15–16 days of age.
4. Remove the brains from the embryos with fine scalpel, fine scissors, and Dumont forceps, and transfer them to a third Petri dish.
5. Dissect the striatal primordium bilaterally from each brain in the basic medium, using the Dumont forceps and iridectomy scissors (see *Figure 3*). Open the hemisphere by a longitudinal incision along the dorsal surface and fold the cortical mantle outwards. Identify the double ridge of the striatal primordium in the floor of the lateral ventricle. Remove with a single tangential scissor cut (see *Figure 3*).
6. Collect the individual striatal pieces from each embryo in a small Petri dish containing basic medium at room temperature.

2.4 Preparation of dissociated cell suspensions

To prepare the dissociated cell suspension you need:

- Instruments: Dumont #5 forceps, two 400 μl Durham tubes (i.e. small glass test tubes), six glass Pasteur pipettes and a rubber pipette bulb, two 1 ml syringes with 25 mm, 23 gauge needles.
- Solutions: trypsin solution (5 ml 0.1% trypsin grade II (Sigma) in basic medium) and dissociation medium (0.04% DNase I (Sigma) in basic medium)
- Incubator: an incubator or water bath able to take 1.5 ml Eppendorf tubes, set at 37°C.

Protocol 3. Preparation of the dissociated cell suspension

1. Sterilize glassware and instruments. Fire-polish the pipettes in a Bunsen flame to smooth the tips in a range of different diameters (~ 0.25–1 mm).
2. Transfer the striatal tissue pieces into 300 μl trypsin solution in a Durham tube. Place the Durham tube in a 1.5 ml Eppendorf tube containing 100 μl water (to aid heat conduction) and incubate at 37°C for 20 min.
3. Remove the Durham tube from the incubator. Wash the cells by four replacements with the dissociation medium. Tap the cell clumps down to the bottom of the tube. Use one syringe to draw off excess fluid (taking

care not to take up any of the tissue pieces) and discard. Use the other syringe to add fresh dissociation medium. Repeat four times.

4. On the final wash add dissociation medium to bring the volume to a level to yield a concentration of one striatal primordium per 3 μl medium (i.e. 6 μl per embryo). This can best be done by adding medium to the washed tissue pieces against a measured volume in a second Durham tube.
5. Dissociate the tissue pieces mechanically by drawing into a fire-polished Pasteur pipette followed by expulsion several times. It is easiest to use two or three pipettes of different diameters, starting with the coarsest. The goal is to achieve a relatively even suspension of dissociated cells without causing excessive damage to the integrity of individual cells. This is achieved using no more than 15–20 gentle pipette strokes in total to dissociate the tissue.
6. The Durham tube containing the cell suspension can conveniently be held in a small handling block made from Perspex and drilled with a 7 mm diameter, 2 cm deep hole. Draw off aliquots for transplantation as required. The suspension is good for up to 3–6 h, and may be kept at room temperature (36).

2.5 Stereotaxic implantation of cell suspensions

For implantation of the dissociated cell suspensions, you need:

- Stereotaxic frame (David Kopf Inc.) with 45° ear bars and an electrode carrier
- Injection syringe: a 10 μl glass microsyringe with removable, thin-walled needle (o.d. 0.5 mm, i/d 0.25 mm) cut square or with a maximum 45° bevel

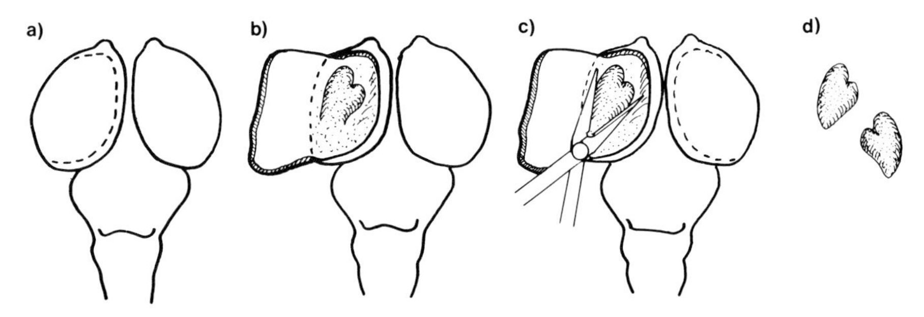

Figure 3. Dissection of striatal eminence from within the lateral ventricular cavity of a 15 mm (E16) embryo. From Dunnett and Björklund (31), with permission. The brain is laid on its ventral surface with the dorsal cortex upwards. A longitudinal cut is made through the medial cortex (a) and folded aside to expose the striatal primordium in the floor of the lateral ventricle (b). The striatal eminence is snipped off by a superficial horizontal cut using the iridectomy scissors (c). The striatal pieces are collected from both hemispheres (d).

(Hamilton or Scientific Glass Engineering). Mount the syringe directly in the electrode carrier of the stereotaxic frame. Use a cuff cut from polyethylene tubing to protect the syringe

- Instruments: dental drill with #3 drill bit, scalpel, cotton buds, gauze swabs, suture needle, needle holder, and scissors.

Protocol 4. Stereotaxic implantation of the dissociated cell suspension

1. Flush the injection syringe. Load by drawing up 10 μl cell suspension (sufficient for three grafts). Take care that the needle does not block and that the suspension includes no bubbles.
2. Anaesthetize the rat. Place in the stereotaxic frame with the nose bar set 2.3 mm below the interaural line. Reopen the skull by a single midline scalpel incision, and clean the bone surface. Make a small drill hole through the skull, midway between the previous two burr holes.
3. Lower the syringe needle to the injection co-ordinates at $A = 0.9$ mm, $L = 3.0$ mm, $V = 5.0$ mm.
4. Inject the graft suspension by gently tapping the plunger. Inject 0.2–0.4 μl aliquots at the rate of 1 μl/min so as to inject a total graft volume of 3 μl (equivalent to one striatal primordium) over 3 min. Allow a further 3 min to aid diffusion of the cells before withdrawing the injection needle.
5. Clean and suture the wound. Remove the rat from the stereotaxic frame and place on a warming blanket in a recovery cage. No special post-operative care is required.

The above protocols provide a simple, efficient and reliable procedure for implanting striatal grafts in adult rats. The procedure generally yields large grafts with a survival rate that is close to 100%. The fact that striatal tissues may expand tenfold or more in volume makes this model particularly suitable for the study of internal organization histologically or autoradiologically, which is a major reason why this model has been investigated so widely (10, 14, 15, 18, 22–26).

3. *In situ* hybridization

3.1 The laboratory environment

Two principal issues must be taken into serious consideration when a laboratory is being set up for mRNA studies. Firstly, it is necessary to have a clean dust-free environment which is also ribonuclease (RNase)-free to prevent degradation of mRNA. Secondly, since the *in situ* hybridization procedure described in this chapter involves the use of radioactively labelled (^{35}S) oligonucleotide probes, great care must be exercised when handling and disposing of radioactive material to limit radiation exposure to the investigator and his colleagues and contamination of the environment. As regards the first issue, all glassware must be bought new and used solely for the *in situ* hybridization procedure. They should be washed, treated with 0.1% diethylpyrocarbonate (DEPC) (Section 3.2), autoclaved, and then wrapped in aluminium foil and baked in an oven at 180°C for a minimum of 4 h (overnight preferably) to destroy remaining RNases. Sterile disposable RNase-free plasticware e.g. syringes, Falcon tubes, and Petri dishes, should be used whenever possible. Eppendorf tubes, micropipette tips, etc. should be autoclaved before use. Wear clean disposable gloves at all times when handling glassware, reagent-containing bottles, oligonucleotides, and freezers/fridges where reagents are stored.

As regards the second issue the safety regulations imposed by the investigator's institution concerning the storage, handling, and disposal of radioisotopes and radioactive waste must be followed strictly. Store radiolabels according to the manufacturer's instructions (−70°C for ^{35}S). Allow to thaw in a fume hood behind proper shielding screens immediately before use. ^{35}S is a weak β-emitter and emissions are easily blocked by most container walls and thus it does not penetrate the bare skin readily. However, inhalation or ingestion of volatile ^{35}S by-products could be a problem. Thus it is advisable to handle this isotope in a fume hood. Wear disposable gloves and a laboratory coat at all times to prevent contamination of hands, skin, personal clothing, laboratory stock solutions, and equipment. Carry a personal radiation monitor badge to assess exposure to radiation. All radioactive material must be disposed of according to the institution's regulations. It is wise to monitor with a Geiger monitor the laboratory bench immediately after using the radiolabel to check for contamination; decontaminate immediately if necessary. In summary, good common sense and good laboratory practice are essential. The reader should refer to Blumberg (37) and Zoon (38) for ideas regarding laboratory facilities and equipment and maintaining an RNase-free laboratory and safety procedures for radiolabelled compounds.

3.2 Solutions

All solutions must be free from RNases and stored in sterile (RNase-free) containers. Gloves must be worn at all times when handling containers. Thus, all solutions must be treated with DEPC (0.1%) and then autoclaved; stock solutions are diluted with DEPC-treated water. DEPC is an alkylating agent and inactivates any protein present in the solutions.

3.2.1 DEPC-treated water (0.1%)

Add 2 ml DEPC (Sigma D5758) to 2.0 litres of nanopure (Millipore) water in a 2.5 litre bottle. Shake vigorously for 10 min and allow to stand overnight. This procedure should be performed in a fume hood since DEPC is a potential carcinogen. The solution is then autoclaved. Upon autoclaving, DEPC is broken down to CO_2 and ethanol.

3.2.2 Phosphate-buffered saline (PBS)

Phosphate-buffered saline (PBS) is made up as a stock solution (10 × PBS) which is 1.3 M NaCl, 70 mM Na_2HPO_4 and 30 mM NaH_2PO_4. Filter, treat with DEPC (1.0 ml DEPC per 1.0 litre PBS) and autoclave. 1 × PBS is prepared by diluting the stock with DEPC-treated water.

3.2.3 (Poly)L-lysine

Dissolve 25 mg (poly)L-lysine hydrobromide (mol. wt > 350 000) (Sigma P1524) in 5.0 ml of DEPC-treated water. Store 1.0 ml (5 mg/ml) aliquots in sterile Eppendorf tubes at −20°C.

3.2.4 Ethanol

Absolute ethanol is diluted with DEPC-treated water to give the required concentration (for example, 50, 70, 95%).

3.2.5 4% Paraformaldehyde in PBS

Paraformaldehyde powder is toxic and harmful to mucous membranes; thus always wear gloves and a face mask when handling it and perform the following procedure in a fume hood.

(a) Weigh 40 g paraformaldehyde (BDH), add slowly to 500 ml DEPC-treated water in a sterile 1.0 litre volumetric flask.
(b) Heat to 60°C with continuous stirring; do not heat above 65°C.
(c) Gradually add drops of 1 M NaOH to the milky suspension until it clears.
(d) Add 500 ml of 2 × PBS and mix well.
(e) Filter with 0.2 μm Millipore filter into a sterile 1 litre volumetric flask and allow to cool.
(f) Check that the pH is about 7.0; store in refrigerator or cold room at +4°C.
(g) Use paraformaldehyde solution within 1 month since it may polymerize.

3.2.6 20 × SSC

20 × SSC is a 20-fold stock solution of 3 M sodium chloride and 0.3 M sodium citrate. Two litres of 20 × SSC are prepared by adding 438.25 g sodium chloride and 220.5 g trisodium citrate to a sterile 2.5 litre bottle and adding nanopure water to a 2.0 litre volume. Filter into a sterile 2.5 litre bottle, add 2.0 ml DEPC, shake vigorously for 10 min, leave overnight, and autoclave the next day. Dilutions of 20 × SSC are made with DEPC-treated water.

3.2.7 Dithiothreitol

Dithiothreitol (DTT) is a reducing agent and when using ^{35}S-labelled probes, it will reduce or prevent cross-linking of sulfur residues. Make up 1 M DTT by dissolving 3.09 g DTT (Sigma D9779) in 29 ml DEPC-treated water. Store in 1.0 ml aliquots in sterile Eppendorf tubes at −20°C.

3.2.8 TENS buffer

TENS buffer is 0.14 M sodium chloride, 20 mM Tris–HCl (pH 7.5), 5 mM EDTA and 0.1% sodium dodecylsulfate (SDS). TENS buffer is prepared by adding the following to a sterile 500 ml bottle:

- 20 ml Tris–HCl (pH 7.5)
- 5 ml 0.5 M EDTA (pH 7.0)
- 14 ml 5 M NaCl
- 5 ml 10% SDS
- 466 ml DEPC-treated water

Filter with a Millipore 0.2 μm filter into a sterile baked bottle.

3.2.9 Sephadex G50 solution

(a) Add 5 g Sephadex G50 medium (Pharmacia cat no. 17-00 45-01) to a baked glass bottle (100 ml).
(b) Add 120 ml filtered TENS buffer and leave for 1–2 h.
(c) Wash several times with TENS buffer to remove soluble dextran.
(d) Autoclave (10 lb/sq. in) for 20 min.
(e) Pour off TENS supernatant and store at +4°C.

3.2.10 Ammonium acetate (300 mM)

Dissolve 20.81 g of ammonium acetate in 900 ml nanopure water. Adjust to pH 7.0 with acetic acid.

3.2.11 Tailing buffer

5 × Tailing buffer is obtained from Pharmacia. It is supplied free of charge with purchase of terminal deoxynucleotidyl transferase enzyme. 5 × Tailing buffer is 0.5 M potassium cacodylate (pH 7.2), 10 mM cobalt chloride, and 1.0 mM DTT). Pipette 1.25 μl aliquots into sterile 1.0 ml Eppendorf tubes. Store at −20°C.

3.2.12 Salmon sperm DNA

Salmon sperm DNA (Sigma type III, sodium salt, cat. no. D1626) is prepared according to the procedure described by Sambrook *et al.* (39) with some modifications:

(a) After the alcohol precipitation step, the DNA is recovered by centrifugation and redissolved at a concentration of 4 mg/ml.
(b) The OD_{260} is determined and the exact concentration of DNA estimated and adjusted to 4 mg/ml.

(c) Filter with Millipore filter to remove any particles.
(d) Boil for 10 min and store in aliquots (4–5 ml) at −20°C.
(e) Before use heat the solution for 5 min in boiling water, chill rapidly on ice, and then use for making up the hybridization buffer (this section).

3.2.13 Heparin

Dissolve the heparin (sodium) (BDH, cat no. 28470) at a concentration of 120 mg/ml of DEPC-treated water. Store in 50 μl aliquots in sterile Eppendorf tubes at −20°C.

3.2.14 Polyadenylic acid

Dissolve 100 mg polyadenylic acid (potassium salt) (Sigma P9403) in 10 ml of DEPC-treated water. Store in 1.0 ml aliquots (5.0 mg/ml) in sterile Eppendorf tubes at −20°C.

3.2.15 50 × Denhardt's

(a) To 500 ml of DEPC-treated water add the following:
   - 5 g Ficoll 400 (Sigma F-2637)
   - 5 g polyvinylpyrrolidone (Sigma P-5288)
   - 5 g bovine serum albumin (Sigma A-8022) fraction V.
(b) Stir until substances are dissolved completely and store in 25 ml aliquots in sterile Falcon tubes at −20°C.

3.2.16 0.5 M sodium phosphate

Mix 0.5 M Na_2HPO_4 and 0.5 M NaH_2PO_4 until pH reaches 7.0. Filter, treat with DEPC (1.0 ml DEPC/litre) and autoclave.

3.2.17 Deionized formamide

(a) To a 5 g of BDH or Sigma Amberlite ion-exchange resin MB-1A add 50 ml formamide (Gibco BRL cat no. 540-5515 UA (100 g), 540-5515 UB (550 g)).
(b) Stir the mixture at room temperature for 30 min until the pH is approximately 7.0.
(c) Filter through Whatman paper and store as 25 ml aliquots in sterile Falcon tubes (50 ml) at −20°C.

Formamide is a potential carcinogen: thus take care when using it. Wear gloves and do the procedure in a fume hood.

3.2.18 Oligonucleotide hybridization buffer

(a) The *in situ* hybridization buffer is:
   - 10% dextran sulfate
   - 50% deionized formamide

- 4 × SSC
- 5 × Denhardt's solution
- 200 μg/ml sheared salmon sperm DNA
- 100 μg/ml long chain polyadenylic acid
- 120 μg/ml heparin
- 25 mM sodium phosphate, pH 7.0
- 1 mM sodium pyrophosphate

(b) To make 50 ml of the buffer, put the following in a sterile 50 ml Falcon propylene tube:
- 5 g dextran sulfate
- 25 ml 100% deionized formamide
- 10 ml 20 × SSC
- 5 ml Denhardt's solution
- 2.5 ml 4 mg/ml sheared salmon sperm DNA
- 1.0 ml 5 mg/ml polyadenylic acid
- 50 μl 120 mg/ml heparin
- 2.5 ml 0.5 M sodium phosphate pH 7.0
- 0.5 ml 0.1 M sodium pyrophosphate

Adjust to 50 ml with DEPC-treated water. Mix by vortexing; the dextran takes a long time to dissolve. However, it goes easily into solution when left overnight. So, wrap the propylene tube in aluminium foil and store overnight at +4°C. On the following morning gently shake the tube and vortex for a few minutes to mix the dissolved dextran. Wait for 0.5–1 h to allow the air bubbles to disperse. The hybridization buffer is then ready for use.

3.3 Preparation of glass slides

Proper pre-treatment of glass sides is necessary to avoid (1) sticking of labelled probe to glass and thus causing undesired background, (2) degradation of mRNA, and (3) the sections falling off the slides during the hybridization and post-hybridization procedures. Thus, it is most important to use clean sterile slides.

Protocol 5. Preparation of glass slides

1. Soak the required number of slides (BDH slides are recommended) in absolute alcohol in a clean container (e.g. a clean plastic food box with a tightly fitting lid) for at least 24 h.
2. Remove slides individually, wipe dry with a clean towel, and place on aluminium foil. Work in a dust-free area.

Protocol 5. *Continued*

3. Wrap slides in aluminium foil and bake in an oven at 180°C for a minimum of 4 h to destroy RN.
4. Remove slides from oven and allow to cool to room temperature.
5. Wear clean disposable gloves.
6. Thaw out 1.0 ml (poly)L-lysine hydrobromide (5 mg/ml) and dilute in 50 ml DEPC-treated water (to obtain 0.01% (poly)L-lysine) in a sterile 50 ml Falcon propylene tube; mix well and pour into a small sterile Sterilin Petri dish).
7. Completely immerse each glass slide (one at a time) into the (poly)L-lysine solution for ~ 5 sec; remove with sterile (autoclaved) forceps, drain off the excess (poly)L-lysine solution by blotting the bottom of the slide with a paper towel, place in a sterile slide rack, and allow to air-dry in a dust-free area.
8. When the slides are dry, place in a slide box with silica gel and store at +4°C until use. They can be used up to 1 month after coating with (poly)L-lysine but it is advisable to coat slides immediately prior to use and not to store them for prolonged periods.

3.4 Cryostat sectioning of the brain

Procedures must be adopted to minimize (1) the effects of stress on the animal both in terms of animal welfare and the possible changes in gene expression in the brain and peripheral tissues e.g. pituitary and adrenals, and (2) the possible degradation of mRNA. Animals must be killed quickly and the brains removed and frozen as rapidly as possible without damage.

Protocol 6. Sectioning of brain tissue

1. Wear clean disposable gloves.
2. Quickly decapitate the animal, remove brain carefully and rapidly, and either freeze it on powdered dry ice or snap-freeze in isopentane (5 sec) cooled in dry-ice or freeze it directly on to a clean cryostat chuck on dry-ice using RNase-free mounting medium (Cryo-M Bed, Bright Instruments).
3. Wrap each frozen brain thoroughly in two or three strips of Parafilm to prevent desiccation and keep on dry-ice. If cryostat sections cannot be cut the same day, transfer the brains to −80°C.
4. Prior to cutting sections, transfer the brain to the cryostat and allow to equilibrate to −15 to −20°C.
5. Sterilize cryostat microtome knife with a piece of tissue soaked in absolute alcohol.

6. Wear clean gloves.
7. Cut sections (10–12 μm) at −15 to −20°C and mount (three or four per slide) on to the (poly)L-lysine-coated slides. During the cutting session store the slides with the mounted sections in a slide box containing silica gel. Sections must be thoroughly dried (minimum 1–2 h) before the fixation step (*Protocol 7*).
8. If sections cannot be fixed the same day store the slides in an air-tight clean slide box with silica gel at −80°C. Seal the seams of the slide box with tape.
9. The next day remove the slide box from the −80°C freezer, allow to come to room temperature, and then fix the sections in 4% paraformaldehyde (*Protocol 7*).

3.5 Fixation of brain tissue sections

Efficient fixation procedures are important for maintaining good tissue morphology, for stabilization and retention of tissue mRNAs, and for destroying residual RNases. Over-fixation may inhibit accessibility of probe to mRNA. Several pre- and post-fixation procedures have been described (40). However, a procedure adopted in our laboratories is described below (*Protocol 7*) which consistently gives reliable results.

Protocol 7. Fixation of brain tissue sections

1. Wear clean disposable gloves.
2. Place the slides (with the mounted sections) in sterilized, RNase-free stainless steel slide racks and treat the sections as follows:
   (a) 5 min in ice-cold 4% paraformaldehyde in PBS
   (b) 1 min in 1 × PBS
   (c) 1 min in 1 × PBS
   (d) 5 min in 50% ethanol/DEPC-treated water
   (e) 5 min in 70% ethanol/DEPC-treated water
   (f) 5 min in 95% ethanol/DEPC-treated water

3.6 Storage of fixed sections

After fixation in 4% paraformaldehyde/PBS (*Protocol 7*) the sections may be stored in clean slide boxes at −70°C or in 95% alcohol/DEPC-treated water at +4°C in a spark-proof cold room or refrigerator. The latter method is the preferred one in our laboratories. Storage of the sections at −70°C can lead to tissue desiccation and thus poor histology and mRNA signal and also moisture condensation may occur on the sections on removal from −70°C if sections are not allowed to come to room temperature properly. Moisture may cause leakage of cellular RNase and thus mRNA degradation. These problems are not encountered when sections are stored in alcohol. Storage in alcohol in addition to the easy access to and visualization of the appropriate sections needed for *in situ* hybridizations also helps to defat the sections. This latter point is important since some probes bind to white matter and may give increased background/non-specific labelling. Under our laboratory conditions, mRNA is stable for years in 95% ethanol/DEPC-treated water at +4°C.

Protocol 8. Storage of fixed brain sections

1. Wear clean disposable gloves.
2. Make up 2 litres of 95% ethanol/DEPC-treated water in a sterile RNase-free graduated cylinder (baked at 180°C).
3. Pour into a clean plastic food box (250 × 220 × 70 cm) (which has been DEPC-treated, rinsed with DEPC-treated water and then with absolute alcohol). A food box of this size will hold four stainless steel slide racks and 2 litres of alcohol will completely immerse the slides.
4. Transfer the fixed sections to the plastic food container, cover lightly, seal the seams with plastic tape, and label the container appropriately.
5. Ensure that the slides are always completely immersed in the alcohol; top up with 95% alcohol if necessary.

3.7 Design, synthesis, and purification of oligodeoxyribonucleotide probes

Various types of nucleic acid probes have been used to detect mRNAs in tissues by *in situ* hybridization, for example, single stranded complementary DNA, double stranded complementary DNA, complementary RNA, and synthetic oligodeoxyribonucleotide probes. All of these probes work quite well but it is not within the scope of this chapter to assess their relative advantages and disadvantages. In our laboratories we routinely use synthetic oligonucleotide probes which we have found to be extremely easy to use, very sensitive, and extremely specific. Oligonucleotide probes can be easily synthesized on DNA synthesizers now available in most molecular biology institutes; however, several private biotechnology firms will make synthetic oligonucleotides upon request, the cost depending upon the length (number of bases) and purification procedures. Several factors must be considered when designing oligonucleotide probes (39) but the probe length and GC/AT ratio are important aspects to consider since they may affect the stability of the hybrid. In general, probes with greater GC content will form more stable hybrids since GC base pairs are stabilized by three hydrogen bonds in contrast to the two hydrogen bonds which stabilize AT/AU base pairs. However, if the GC content of the oligonucleotide is too high (>65%), non-specific labelling may occur since the thermal stability of the probe will be greater. The probe length and GC content, in addition to the formamide and salt concentrations in the hybridization buffer, will also determine the appropriate conditions for hybridization and post-hybridization treatments. In our laboratories we design probes of a standard length of 45 bases and with a GC content of ~50–60%. By using such probes the hybridization, post-hybridization, and histochemical procedures and conditions are standardized in the laboratory (Sections 3.8 and 3.9).

3.8 Radioactive labelling of oligonucleotide probes and purification of radiolabelled probes

The commonly used radioisotopes for labelling oligonucleotides are ^{32}P and ^{35}S. These produce labelled probes of high specific activity allowing for fast regional localization of mRNA in tissue sections. In our laboratories we routinely use ^{35}S which provides probes of high specific activities, produces good resolution on X-ray film and excellent cellular resolution after liquid emulsion autoradiography, and has a longer half-life (87.4 days) and is less hazardous to use than ^{32}P (37).

There are two methods of labelling synthetic oligonucleotides: (1) kinasing reaction with T4 polynucleotide kinase which catalyses the transfer of the γ-phosphate of ATP to the 5′-OH terminus of the molecule. This enzyme transfers only one ^{32}P residue to the oligonucleotide and therefore probes so labelled are not usually of high enough specific activity to detect mRNAs by *in situ* hybridization. (2) Tailing with terminal deoxynucleotidyl transferase enzyme which catalyses the repetitive transfer of mononucleotide units from a deoxynucleoside triphosphate to the 3′-OH terminus of the synthetic oligonucleotide with the release of inorganic pyrophosphate. This method of labelling results in probes of high specific activity (*Protocols 9* and *10*). The length of the poly(dATP) tail may be of the order of 15–25 residues and can be checked by polyacrylamide gel electrophoresis. The number of residues transferred by the terminal transferase is influenced by the concentration of the enzyme, the oligonucleotide/isotope molar ratio, and the duration of the incubation. Changes in the oligonucleotide/isotope ratio result in changes in tail length and specific activity.

Protocol 9. Radioactive ^{35}S-labelling of oligonucleotide probes

1. Wear clean disposable gloves.
2. Remove from −20 °C freezer an Eppendorf tube containing 1.25 μl of 5 × tailing buffer and ensure that it is all at the bottom of the tube by gently tapping the tube on the bench top. Keep on ice.

Protocol 9. *Continued*

3. To the tailing buffer add the following with a 20 μl Gilson micropipette using a fresh sterile micropipette tip at each step:
   (a) 1.0 μl oligonucleotide (5 ng) (we routinely make up stock solutions of oligonucleotides as 1 μg/μl DEPC-treated water; 1.0 μl of a 1/200 dilution (5 ng) is used for labelling).
   (b) 8.25 μl DEPC-treated water.
   (c) 1.0 μl [^{35}S]deoxyadenosine 5′ (α-thio)triphosphate (1000–1500 Ci/mmol, 10 mCi/ml, New England Nuclear-DuPont, cat. no. NEG 034-H). Always keep the label at −70 °C and thaw in a fume hood immediately before use.
   (d) 1.0 μl terminal deoxynucleotidyl transferase enzyme (concentration 15–20 000 U/ml; Pharmacia cat no. 27-0730-01, 27-0730-02). The enzyme is supplied as a solution in 100 mM potassium sulfate (pH 6.9), 10 mM β-mercaptoethanol and 50% glycerol. The enzyme should always be kept at −20°C, preferably in a Stratagene cooler which can be removed from the freezer and brought to the bench when the enzyme is being used. Return to −20°C freezer immediately after use.
4. Check that the total volume of the contents of the reaction tube is 12.5 μl; mix gently by pipetting up and down with the Gilson micropipette. Avoid introducing air bubbles.
5. Incubate immediately in a water-bath at 30–32°C for 1.5 h.
6. Stop the reaction by adding 40 μl DEPC-treated water.
7. Purify the labelled probe by the Sephadex G50 spin column method. This separates unincorporated nucleotides from labelled probe (*Protocol 10*).

Protocol 10. Purification of ^{35}S-labelled probe

The labelled oligonucleotide is purified using a Sephadex G50 spin column which is prepared using the procedures described (37) but with some modifications.

1. Wear clean disposable gloves.
2. Place a disposable sterile 1.0 ml syringe into a sterile 15 ml Falcon tube (Falcon 2059).
3. Remove the syringe plunger and use it to plug the bottom of the syringe with autoclaved siliconized glass wool (39).
4. Using a 1.0 ml disposable pipette with a 2.0 ml pipette holder gently pour pre-swollen G50 Sephadex in TENS buffer into the 1.0 ml syringe. The TENS buffer will elute into the Falcon tube. Keep on filling the syringe with Sephadex G50 slurry until the whole syringe is filled with Sephadex. Remove air bubbles by gently tapping the syringe. Allow the excess TENS buffer to drain into the Falcon tube and then discard the eluted buffer.

5. Cover the whole assembly (Falcon tube and syringe) with the Falcon tube cap and spin at 2000 r.p.m. for 2 min in a low speed centrifuge. Discard the eluted buffer. The packed G50 column should reach about the 0.9–1.0 cm mark on the syringe.
6. Remove the Sephadex-filled 1.0 ml syringe, insert a sterile decapped Eppendorf tube (appropriately labelled according to the probe used) into the Falcon tube and replace the Sephadex G50 syringe column.
7. Pipette the 52.5 μl of probe solution (12.5 μl reaction volume + 40 μl DEPC-treated water) on to the top of the Sephadex column and spin exactly as before (2000 r.p.m. for 2 min). The purified labelled probe in about 50 μl volume is collected in the decapped Eppendorf tube.
8. Add 2 μl 1M DTT to the purified probe solution, mix by vortexing, and keep on ice.
9. Transfer 2 μl of probe solution to a scintillation vial, add 4 ml scintillant, mix properly by vortexing, and count in a scintillation counter for 1 min. Counts are usually in the range of $150–200 \times 10^3$ d.p.m./μl i.e. $0.75–1.0 \times 10^{10}$ d.p.m./μg.
10. It is wise to check that the quality of the labelling is satisfactory before removing the sections from storage in alcohol and proceeding with the hybridization step (*Protocol 11*).

Protocol 11. Incubation of brain sections with labelled probe

1. Remove slides with the apposed sections from the alcohol, place in a slide rack and allow to air-dry thoroughly (about 1 h).
2. Dilute the labelled probe in hybridization buffer to give 2500–5000 c.p.m./μl.
3. Add 40 μl 1M DTT per 1.0 ml of hybridization buffer and mix by vortexing; leave for 10–15 min until air bubbles have dispersed.
4. Using a Gilson micropipette apply 100 μl aliquots of hybridization buffer to each slide making sure that this volume is uniformly distributed over all the brain sections.
5. Cover with a strip of Parafilm (the inner surface of the Parafilm is apposed to the sections); avoid air bubbles.
6. Place slides in a sterile Petri dish (Falcon 1058; 150 × 15 mm); this will hold four or five slides.
7. Humidify with a piece of Kleenex tissue soaked in 50% formamide/4 × SSC.
8. Wrap Petri dishes (e.g. five together) with cling film.
9. Incubate at 42°C overnight (for ~18–20 h).

3.9 Post-hybridization treatments

After incubation the labelled sections are washed to remove unhybridized probe and under stringency conditions to reduce background without loss of signal to verify the specificity of the probe for the mRNA being investigated.

Protocol 12. Post-hybridization treatments

1. 1 × SSC at room temperature; remove Parafilm coverslips by rinsing individual slides; wash each slide by agitation to remove excess hybridization buffer and unhybridized probe; put slides into slide racks and then transfer to:
2. 1 × SSC at 55°C (30 min).
3. 1 × SSC at 55°C (30 min two complete changes).
4. 1 × SSC at room temperature for 1 h. Add 1% sodium thiosulfate to the 1 × SSC solutions and then repeat steps 3 and 4 in a shaking water bath.
5. 1 × SSC at room temperature (2 sec).
6. 0.1 × SSC at room temperature (2 sec).
7. Nanopure water at room temperature (2 sec).
8. 50% ethanol (2 sec).
9. 70% ethanol (2 sec).
10. 95% ethanol (2 sec). Ammonium acetate (300 mM) is added to the alcohols to prevent denaturation of the hybrids.
11. The sections are dried thoroughly in a stream of air for ~1–2 h at room temperature before exposure to dry X-ray film (*Protocol 12*).

3.10 Autoradiography

All autoradiographic procedures must be carried out in a light-tight dark room with suitable safelight illumination with appropriate filters (e.g. CEA 4B, Kodak Wratten 6B or GB × 2, Agfa-Gaevert R1) and a 15 W bulb.

Protocol 12. Exposure to X-ray film

1. Place the dried slides (taped side by side with double-sided Sellotape on a sheet of paper) in an X-ray cassette.
2. In the dark room lay a sheet of X-ray film (Kodak XAR 5 or Amersham Hyperfilm β-max) compatible with the size of the cassette, close securely, and tape the seams of the cassette with black plastic tape.
3. Place cassette in a cool dark place (cupboard or drawer) away from any other source of radiation. The length of exposure of the labelled sections

to film depends upon several factors including the specific activity of the labelled probe, the mRNA and its abundance in the brain area being investigated. For several neuropeptide/neurotransmitter and receptor mRNAs in the rat brain exposure times between 4 and 7 days are sufficient to observe good hybridization signals under the hybridization conditions described. The length of exposure on film may indeed depend upon the question being asked and the needs of the investigator.

Protocol 13. Development of X-ray film

In the dark room under safe light illumination remove film from cassette and develop as follows:

1. Kodak D19 or Ilford D19 developer at 21–23°C for 5 min.
2. Running cool water for 30 sec.
3. Rapid fix (Kodak) for 5 min.
4. Wash in running cool water for 30 min.
5. Air dry for 30–60 min.

3.11 Emulsion autoradiography

For the cellular localization of mRNA transcripts, dip sections in liquid photographic emulsion (Kodak NTB-2, or Ilford K5) in a dark room under safe-light conditions.

Protocol 14. Dipping sections in liquid photographic emulsion

1. Use a clean glass rod to decant Ilford K5 nuclear emulsion shreds into a 50 ml propylene tube (Falcon) or a clean 50 ml glass cylinder up to the 20 ml mark.
2. Melt the decanted emulsion in a water bath at 43°C; stir gently (to avoid formation of air bubbles) with the glass rod.
3. Add 10 ml 600 mM ammonium acetate containing 0.5% glycerol (this avoids cracking of dried emulsion) to another 50 ml propylene tube or glass cylinder and warm to 43°C.
4. Add an equal volume (10 ml) of melted emulsion (from step 1) to the prewarmed ammonium acetate (in step 3); stir gently with the glass rod.
5. Pour the emulsion (from step 4) into the dipping chamber (Electron Microscopy Supplies, Washington) and allow to equilibrate in a beaker of distilled water prewarmed to 43°C.
6. Dip two or three blank slides to remove air bubbles and to check the quality of the emulsion.

Protocol 14. *Continued*

7. Hold slides (with the radioactively labelled sections) by the frosted end and slowly dip twice (~2 sec each). Dip slides one at a time. Do this with a smooth action to ensure an even coating. Blot the bottom end of the slide on a paper towel, wipe the back of the slide with a paper towel to remove excess emulsion, and then place slides on a cool horizontal aluminium surface for ~10 min to prevent emulsion from running (make sure that the coated sections are uppermost).
8. Place slides upright in a rack to dry for 1–1.5 h at room temperature. When emulsion is dry, place slides in a black plastic slide box containing a drying agent (e.g. silica gel). Seal the seams of the box with black tape, wrap in foil, and leave at room temperature for 2–3 h. Then store at +4°C until development.

3.12 Development of emulsion-coated sections

The length of exposure of radioactively labelled sections must be determined by the investigator taking into account the specific activity of the labelled probe and the level of the hybridization signal in the nuclei being investigated. It is important to remember to dip sections in liquid emulsion as soon as possible after dry X-ray film autoradiography. In general, excellent cellular resolution is obtained if the exposure time to emulsion is about five to seven times the length of exposure of the sections to dry X-ray film necessary for a clear hybridization signal.

Protocol 15. Development of emulsion-coated sections

1. Remove the black boxes (containing emulsion-coated slides) from the refrigerator and allow 30–60 min for slides to come to room temperature. Condensation on the sections may thus be avoided.
2. Place slides in slide racks and develop the sections as follows:
   (a) Ilford (PHENISOL) developer at 17°C for 5 min. PHENISOL is diluted 1:5 with nanopure water.
   (b) Stop bath (2% chromium sulfate in 2% sodium metabisulfite in nanopure water) at 17°C for 5 min.
   (c) Fixative (30% sodium thiosulfate in stop-bath solution) at 17°C for 10 min.
   (d) Nanopure water for 30 min (2 × 15 min).
   (e) Air-dry for 1–2 h.
   (f) Counterstain with thionin, cresyl violet, or methylene blue, dehydrate through an ethanol series (70, 80, 95, and 100%), clear in xylene (3 × 5 min), and coverslip with Permount; get rid of air bubbles by gently pressing coverlip and wipe excess Permount from the edges. Place slides on a slide warmer at 40°C for Permount to harden.

3.13 Controls for *in situ* hybridization experiments

To check for hybridization signal and oligonucleotide probe specificity it is essential to perform a few control experiments especially when a probe is being tested for the first time.

Protocol 16. Controls for *in situ* hybridization

1. Hybridize appropriate brain tissue sections with ^{35}S-labelled 'sense' oligonucleotide probe (complementary to the antisense oligonucleotide) under the same hybridization conditions used for the 'antisense' probe. No signal should be obtained with the 'sense' probe.
2. Treat the brain sections with RNase (100 μg/ml) prior to hybridization with the labelled 'antisense' probe to determine whether this pretreatment will abolish the hybridization signal.
3. Incubate appropriate brain tissue sections with excess of unlabelled 'antisense' oligonucleotide probe in the presence of ^{35}S-labelled 'antisense' probe to determine if 'cold' probe will block the signal obtained with labelled probe (competition experiment).
4. Perform a Northern blot analysis with ^{32}P-labelled antisense probe on total RNA or poly(A)$^+$ RNA extracted from brain tissue to determine if the antisense probe hybridizes to the correct mRNA transcript. Northern analysis could also be performed with ^{32}P-labelled 'sense' probe. No hybridization signal should be obtained.

3.14 Quantification of mRNA in brain tissue sections

Relative levels of mRNA in brain nuclei may be obtained by making optical density (OD) measurements of the hybridization signal on X-ray film autoradiograms using a computerized video image analysis system. However, more accurate quantification may be obtained if calibrations are done with external radioactive brain paste standards similar to the technique used for quantitative receptor autoradiography (see Chapter 4, this volume). Thus, for the quantification of mRNA signals on X-ray film autoradiograms, brain paste standards containing known amounts of [^{35}S]dATP must be prepared and exposed to X-ray film at the same time as the hybridized brain sections. The brain paste standards are used to generate a standard curve and OD units can be converted into radioactivity units. The results can be expressed in terms of molar quantities of label bound per unit area (e.g. attomoles/mm^2). This value reflects the number of mRNA copies in the region examined. The hybridization signal in any brain nucleus reflects the number of cells expressing the mRNA and the hybridization signal per cell as indicated by the grain density per cell. Although it is possible though tedious to count manually both the numbers of mRNA-expressing cells per unit area as well as the grain density per cell under the light microscope, it is feasible to obtain both measurements with a computer assisted image analysis system. For details on the quantitative measurements of mRNA in tissue sections, the reader should refer to several well described procedures (41, 42).

4. Concluding comments

The recent advances in molecular biology have provided us with powerful tools for the study of brain structure, organization, and function. We have described how these procedures can be applied to neural transplantation studies for the anatomical localization and functional organization of specific neuronal mRNAs within the grafts and for delineating the mechanisms regulating the survival, growth, and differentiation of implanted embryonic neurones. These mechanisms need to be elucidated since optimizing the level of neural graft function is fundamental to the full realization of the experimental and therapeutic potential of such grafts. To achieve reliable results, both the neural grafting and *in situ* hybridization techniques must be carried out under rigorous and well controlled conditions in the laboratory and with the necessary practice to achieve the required level of dexterity. Although the radioactive *in situ* hybridization procedure is a sensitive and immensely powerful research tool to answer the above questions the investigator may find it beneficial to use it in combination with non-radioactive *in situ* hybridization histochemical methods (see Chapter 1, this volume) or immunohistochemistry (Chapter 6, this volume) or receptor autoradiography (Chapter 4, this volume) not only to determine if two or more gene products are colocalized within the same cell but to assess if a specific neuronal mRNA is indeed translated and processed into a functional protein. Such a repertoire of techniques is now providing neuroscientists with the necessary expertise to understand the organization and function of specific cells in the brain of the developing and adult animal.

Acknowledgements

Our research studies have been funded by grants from the British Medical Research Council, the Parkinson's Disease Society, the Huntington's Disease Association and by financial support from our own institutions.

References

1. Dunnett, S. B. and Richards, S.-J. (ed.) (1990). *Neural Transplantation: From Molecular Basis to Clinical Application. Prog. Brain Res.,* **82**.
2. Coyle, J. T. and Schwarcz, R. (1976). *Nature,* **263,** 244.

3. Beal, M. F., Kowall, N. W., Ellison, D. W., Mazurek, M. F., Swartz, K. J., and Martin, J. B. (1986). *Nature,* **321,** 168.
4. McGeer, P. L. and McGeer, E. G. (1976). *Nature,* **263,** 517.
5. Schwarcz, R. and Coyle, J. T. (1977). *Brain Res.,* **218,** 347.
6. Schwarcz, R., Hökfelt, T., Fuxe, K., Jonsson, G., Goldstein, M., and Terenius, L. (1979). *Exp. Brain Res.,* **37,** 199.
7. Isacson, O., Brundin, P., Gage, F. H., and Björklund, A. (1985). *Neuroscience,* **16,** 799.
8. Björklund, A., Lindvall, O., Isacson, O., Brundin, P., Wictorin, K., Strecker, R. E., Clarke, D. J., and Dunnett, S. B. (1987). *Trends Neurosci.,* **10,** 509.
9. Isacson, O., Dunnett, S. B., and Björklund, A. (1986). *Proc. Natl Acad. Sci. USA,* **83,** 2728.
10. Isacson, O., Dawbarn, D., Brundin, P., Gage, F. H., Emson, P. C., and Björklund, A. (1987). *Neuroscience,* **22,** 481.
11. Clarke, D. J., Dunnett, S. B., Isacson, O., Sirinathsinghji, D. J. S., and Björklund, A. (1988). *Neuroscience,* **24,** 791.
12. Sirinathsinghji, D. J. S., Dunnett, S. B., Isacson, O., Clarke, D. J., Kendrick, K., and Björklund, A. (1988). *Neuroscience,* **24,** 803.
13. Dunnett, S. B., Isacson, O., Sirinathsinghji, D. J. S., Clarke, D. J., and Björklund, A. (1988). *Neuroscience,* **24,** 813.
14. Sirinathsinghji, D. J. S., Morris, B. J., Wisden, W., Northrop, A., Hunt, S. P., and Dunnett, S. B. (1990). *Neuroscience,* **34,** 675.
15. Sirinathsinghji, D. J. S., Wisden, W., Northrop, A., Hunt, S. P., Dunnett, S. B., and Morris, B. J. (1990). *Prog. Brain Res.,* **82,** 433.
16. Wictorin, K., Isacson, O., Fischer, W., Nothias, F., Peschanski, M., and Björklund, A. (1988). *Neuroscience,* **27,** 547.
17. Wictorin, K. and Björklund, A. (1989). *Neuroscience,* **30,** 297.
18. Wictorin, K., Clarke, D. J., Bolam, J. P., and Björklund, A. (1989). *Eur. J. Neurosci.,* **1,** 117.
19. Wictorin, K., Ouimet, C. C., and Björklund, A. (1989). *Eur. J. Neurosci.,* **1,** 690.
20. Wictorin, K., Simerly, R. B., Isacson, O., Swanson, L. W., and Björklund, A. (1989). *Neuroscience,* **30,** 313.
21. Wictorin, K., Clarke, D. J., Bolam, J. B., and Björklund, A. (1990). *Neuroscience,* **37,** 301.
22. Walker, P. D., Chovanes, G. I., and McAllister II, J. P. (1987). *J. Comp. Neurol.,* **259,** 1.
23. McAllister, J. P., Walker, P. D., Zemanick, M. C., Weber, A. B., Kaplan, L. I., and Reynolds, M. A. (1985). *Dev. Brain Res.,* **23,** 282.
24. Graybiel, A. M., Liu, F.-C., and Dunnett, S. B. (1989). *J. Neurosci.,* **9,** 3250.
25. Liu, F.-C., Graybiel, A. M., Dunnett, S. B., and Baughman, R. W. (1990). *J. Comp. Neurol.,* **295,** 1.
26. Graybiel, A. M., Liu, F.-C., and Dunnett, S. B. (1990). *Prog. Brain Res.,* **82,** 401.
27. Koelle, G. B. (1954). *J. Comp. Neurol.,* **100,** 211.
28. Sirinathsinghji, D. J. S., Dunnett, S. B., Northrop, A. J., and Morris, B. J. (1990). *Neuroscience,* **37,** 757.
29. Sirinathsinghji, D. J. S. and Dunnett, S. B. (1989). *Brain Res.,* **504,** 115.
30. Sirinathsinghji, D. J. S. and Dunnett, S. B. (1991). *Mol. Brain Res.,* **9,** 263.

31. Dunnett, S. B. and Björklund, A. (ed.) (1992). *Neural Transplantation: A Practical Approach.* IRL Press, Oxford.
32. Schmidt, R. H., Björklund, A., and Stenevi, U. (1981). *Brain Res.,* **218,** 347.
33. Björklund, A., Stenevi, U., Schmidt, R. H., Dunnett, S. B., and Gage, F. H. (1983). *Acta Physiol. Scand. Suppl.,* **522,** 1.
34. Dunnett, S. B. and Björklund, A. (1992). In *Neural Transplantation: A Practical Approach* (ed. S. B. Dunnett and A. Björklund), pp. 1–19. IRL Press, Oxford.
35. Björklund, A. and Dunnett, S. B. (1992). In *Neural Transplantation: A Practical Approach* (ed. S. B. Dunnett and A. Björklund), pp. 57–78. IRL Press, Oxford.
36. Brundin, P., Isacson, O., and Björklund, A. (1985). *Brain Res.,* **331,** 251.
37. Blumberg, D. D. (1987). *Methods Enzymol.,* **152,** 3.
38. Zoon, R. A. (1987). *Methods Enzymol.,* **152,** 25.
39. Sambrook, J., Fritsch, E. F., and Maniatis, T. (1989). *Molecular Cloning, A Laboratory Manual.* 2nd Edn. Cold Spring Harbor Laboratory Press, Cold Spring Harbor, NY.
40. Conn, P. M. (ed.) (1989). *Methods in Neurosciences,* Vol. 1, *Gene Probes.* Academic Press, New York.
41. McCabe, J. T., Desharnais, R. A., and Pfaff, D. W. (1989). In *Neuroendocrine Peptide Methodology* (ed. P. M. Conn), pp. 107–33. Academic Press, New York.
42. Uhl, G. R. (1989). In *Neuroendocrine Peptide Methodology* (ed. P. M. Conn), pp. 135–60. Academic Press, New York.

4

Quantitative autoradiography: a tool to visualize and quantify receptors, enzymes, transporters, and second messenger systems

N. A. SHARIF and R. M. EGLEN

1. Introduction

The distribution and relative concentration of endogenous neurochemicals and their receptors/acceptors has traditionally been determined by immunohistochemical, radioimmunological, and ligand binding techniques. Whilst ligand binding procedures conducted on tissue homogenates have permitted determination of receptor affinities for neurotransmitters and drugs, and gross receptor distribution analysis, they lack the high level of anatomical resolution and sensitivity that can be obtained using autoradiographic techniques.

Autoradiography can be broadly defined as a technique for localizing radioisotopes bound to a solid specimen using a layer of radiation-sensitive photographic emulsion (detector material). In more specific terms, receptor/acceptor autoradiography is a combination of radioligand binding and autoradiographic techniques. The radioactive particles (β-particles, γ-rays, and X-rays) from the radioisotope convert the silver bromide contained in the detector emulsion to metallic silver which yields a visible image after the process of development and fixation. Excellent in-depth overviews of the chemistry and physics underlying the autoradiographic processes have been published previously (1, 2).

The autoradiographic processes used in the neurosciences can be categorized into *in vitro* and *in vivo* techniques, yielding different levels of resolution either at the light- or electron-microscopic (EM) levels. In addition, autoradiography can utilize thin sections (5–30 μm) of individual organs or the whole of the organism under study. While individual organ-based autoradiography has proved most useful for biochemical and pharmacological studies, whole-body autoradiography relies on the *in vivo* administration of the radioactive compound and has mostly been used in toxicological studies to determine the disposition of the parent drug or its metabolites. The most popular of the autoradiography methods in neuroscience is the *in vitro* light-microscopic one, because it is simple, relatively quick, reproducible, relatively cheap, and quantitative, and can be very productive. The technique is now one of choice for visualizing and quantifying receptors in an anatomical framework and has been detailed by other experts in the field (3, 4). For all of the above reasons, we have decided to make it the theme of this chapter. The readers are directed to publications by Sokoloff (5), Beaudet and Szigethy (6), and Ullberg (7) for the *in vivo*, EM autoradiography and whole body autoradiography procedures, respectively.

The *in vitro* light-microscopic autoradiographic method has been employed to detect, map, and quantify many kinds of acceptors/receptors using tissues derived from the major organs of the mammalian body, in particular the central nervous systems (CNS) (8–10). The aim of this chapter, therefore, is to provide the readers with detailed practical guidelines for accomplishing the goals of visualizing and quantifying these entities without the need to refer to the papers where these methods were originally described. However, readers are encouraged to consult the original publications for more complete information whenever it is deemed necessary. I have attempted to include the most common of the assay systems currently being employed by research scientists in the academic and industrial institutions. The information is comprehensive, but not exhaustive, and unfortunately there may be some inadvertent omissions.

2. Tissue preparation

Small laboratory animals (rats, mice, and guinea pigs) are sacrificed by CO_2 asphyxiation and the tissue(s) of interest rapidly excised over ice. Large animals (rabbits, cats, dogs, and monkeys) need to be sacrificed with an overdose of an anaesthetic (e.g. Nembutal). Excess blood and adhering dura mater are removed from the organ, the latter trimmed if necessary and then frozen on to microtome chucks in dry ice with plastic embedding material (Tissue teck (Miles Labs)) or water. The frozen tissues are wrapped in Parafilm to prevent dehydration, and allowed to equilibrate at −20°C for an hour, either in the cryostat microtome or in the freezer. It is recommended that tissue sections be cut within 1–2 days in order that receptor activity is not lost, which may happen during prolonged freezing. In addition, the tissues should not be allowed to undergo freezing–thawing before sections are cut. Optimal conditions for cutting sections are shown in *Protocol 1*. Investigators should wear a laboratory coat, rubber gloves, and safety glasses during all the procedures described in this and other chapters, and are encouraged to attend a full safety training class and consult their safety officers for all institutional safety procedures including those for handling radioactivity and biohazardous materials.

Protocol 1. Optimum conditions for cutting tissue sections on a freezing microtome

1. Ensure tissue is free of or has minimum of surface dura mater.
2. Ensure tissue has been equilibrated to −12°C for at least an hour, and that the specimen and cryostat chamber temperatures are maintained at −12°C and −20°C, respectively.
3. Ensure microtome knife is sharp, undamaged, and rust-free. Set the blade angle at 12–15°.
4. Set the cutting thickness dial to 10–20 μm for good results.
5. Advance the leading edge of the anti-roll bar close to, but not touching, and just forward (1–2 mm) of the knife edge to obtain flat sections of even thickness.
6. Ensure that the tissue is mounted squarely on the chuck-holder to obtain symmetrical sections. Proceed to cut tissue sections.
7. Collect sections on gelatin-coated (see *Protocol 2*) glass microscope slides (kept at room temperature) by thaw-mounting. If sections curl over when the anti-roll bar is lifted, use a soft camel-hair brush (maintained at −20°C in cryostat) to flatten the section gently before thaw-mounting.
8. Allow the thaw-mounted sections to dry at room temperature for 15–30 min before placing them in slide boxes and transferring the latter to a −40°C or −80°C freezer.
9. When sectioning is completed, allow the tissue to thaw at room temperature and discard it safely as a biohazard item. Extra care should be taken when excising, mounting, cutting, and disposing of primate or human tissues. Check with your safety officer for proper safety guidelines and procedures for your institution.

Whilst the above procedures have generally been found to yield good tissue sections with well preserved morphology and receptor binding activity, there are many pitfalls. Some of the possible problems and their causes/solutions can be listed as follows:

- sections fold under while they are being cut. Space between anti-roll bar and blade is too large: adjust spacer screws
- sections become torn during cutting. Anti-roll bar is too close to blade: adjust spacer screws
- sections crease on one side. May be due to presence of dura: try to remove the dura with cold forceps (−12 to −20°C)
- horizontal scores or lines on sections. Tissue is too cold: adjust specimen and/or cryostat chamber temperature

- vertical scores or lines on sections. Blade is damaged, or there is existing tissue on the knife, or there is ice on the knife: rectify as necessary

---

Protocol 2. Gelatin-coating of microscope slides

1. Purchase microscope slides that have a small portion frosted at one end so that slides can be labelled with a pencil or marker pen.
2. Always hold slides by their edges and wash them in soapy water or in a dish-washer, even if they are pre-washed when purchased. The final rinse should be in distilled water and the slides allowed to dry in a dust-free environment.
3. Prepare 1% gelatin solution:
   - gelatin 5.0 g
   - chrom alum 0.25 g
   - distilled water 500 ml

   Heat water to 60°C before adding the reagents, mix thoroughly, and filter solution into slide staining dishes.
4. Load slides in metal slide-holders (50 at a time if possible) and place them in the hot gelatin solution until totally immersed.
5. Immediately after dipping, stand the slides vertically and allow to dry in a dust-free environment.

Protocol 3. Preparation of brain-paste radiation standards

1. Obtain seven rat brains (~14 g tissue).
2. Add 7 ml distilled water and homogenize brains with a Polytron (setting 5–6 for 30 sec to obtain a homogeneous paste).
3. Weigh 650 mg of the paste into each of 10 Eppendorf tubes.
4. Add 40 μl of the following dilutions of the ^{3}H- or ^{125}I-labelled ligand to the tubes:
   (a) 40 μl of undiluted stock (i.e. 2–5 μCi radioligand in 40 μl water).
   (b) 40 μl of 2/3 dilution.
   (c) 40 μl of 1/2 dilution.
   (d) 40 μl of 1/3 dilution.
   (e) 40 μl of 1/4 dilution.
   (f) 40 μl of 1/6 dilution.
   (g) 40 μl of 1/8 dilution.
   (h) 40 μl of 1/12 dilution.
   (i) 40 μl of 1/16 dilution.
   (j) 40 μl of 1/24 dilution.

5. Add one drop of silicone oil or equivalent to each tube and mix the contents thoroughly (up to 8–10 min) using a vortex mixer.
6. Centrifuge tubes at 3000 *g* for 1 min to remove air bubbles.
7. Freeze tube contents in liquid nitrogen (1 min).
8. Remove paste pellets (tap tubes on bench) and store at −20°C.
9. Mount each paste pellet (cone down) on microtome chucks, surround with embedding material or water, cut 10–20 μm sections on a freezing microtome, and thaw-mount on slides.
10. Collect one section from each of 10 standards on each slide and allow to dry at room temperature for at least 30 min before transferring them to slide boxes for storage (−20°C or room temperature).
11. Collect four sections from each standard's pellet into scintillation vials and determine the average amount of radioactivity for each radiation standard. Subsequently perform a protein assay to determine the amount of protein in each brain-paste standard section and calculate the fmol radioligand/mg protein for each standard. Also determine the amount of radioactivity per each section of the standards and express this as moles/section.

3. Receptor labelling and film autoradiogram generation

Before the receptors/acceptors on the tissue sections can be radiolabelled they must be thawed slowly at room temperature and allowed to dry completely. This procedure ensures that the sections will continue to adhere strongly to the gelatin on the slides and will not be washed away during the rinsing procedures. A typical labelling procedure is shown in *Protocol 4*.

Protocol 4. Radiolabelling of receptors or acceptors

1. After thawing and drying the tissue sections, use a pencil or a marker pen to identify each slide on the frosted surface if not already done.
2. Place the slides in the desired sequence in slide-holders and immerse them in the appropriate pre-incubation buffer (400–500 ml) to eliminate endogenous substances that may interfere with the binding assay. Use a slow agitation procedure using a rotary shaker.
3. After pre-incubation, stand the slide-holders vertically to drain off excess buffer and allow the sections to become semi-dry.
4. Place the slides in Coplin jars or horizontally over metal rods. If using Coplin jars, have the correct buffer solution, radioligand, and displacing

Protocol 4. *Continued*

agent already in the jars. If using the metal rod approach, pipette 0.3–0.8 ml of the appropriate incubation buffer solution containing the radioligand (plus any other reagents like peptidase inhibitors) at the desired concentration together with 0.1–10 μM unlabelled competing drug directly over the sections. The slides that just receive the radioligand solution will represent 'total binding', while those receiving the radioligand and the unlabelled drug will represent 'non-specific binding'.

5. Allow the incubation to proceed to equilibrium at the desired temperature (e.g. 60 min at 23 °C). Use proper lead shielding when working with ^{125}I-labelled compounds.
6. Terminate the assay by draining the incubation buffer mixture from the slides and rinsing the latter in excess (500 ml) fresh cold (4 °C) buffer using gentle agitation on a rotary shaker. The correct wash-time and overall conditions have to be determined for each new radioligand unless these are already described in the literature. Use several new rinse dishes in order to achieve maximum removal of unbound radioligand from the slides/sections and to dilute the overall free radioactivity in solution.
7. At the end of rinsing in buffer, quickly dip the sections in ice-cold distilled water to remove buffer salts, and rapidly dry the sections using a stream of cool dry air.
8. Desiccate the labelled sections overnight in order to remove all the moisture from the tissues.
9. For macro-autoradiography using sheet-film, mount the slides on cardboard (with the sections facing up) using scotch tape. For micro-autoradiography, process the slides for liquid emulsion coating or apposition of emulsion-coated coverslips following the procedures shown in *Protocol 5*.
10. For macro-autoradiography, place the slide-laden cardboard in X-ray cassettes, and secure with tape. Also attach a slide bearing a set of radiation standards (either commercially available methylpolyacrylate standards (Amersham 'Microscales') or brain-paste standards (see *Protocol 3*)) to the cardboard. In a dark-room, place a radiation-sensitive sheet-film (LKB Ultrofilm or Amersham Hyperfilm), emulsion-side face-down, on top of the labelled sections and tightly shut the X-ray cassette.
11. Allow the film–section apposition to proceed for the detailed time (a few hours to months depending on the isotope used; for ^{125}I a few hours to a few days is sufficient, while 1–6 months may be necessary for ^{3}H) before developing the autoradiograms (see *Protocol 7*).

The receptor labelling procedures described above have been successfully employed to visualize several receptors, enzymes, etc. (see *Tables 1–23* and

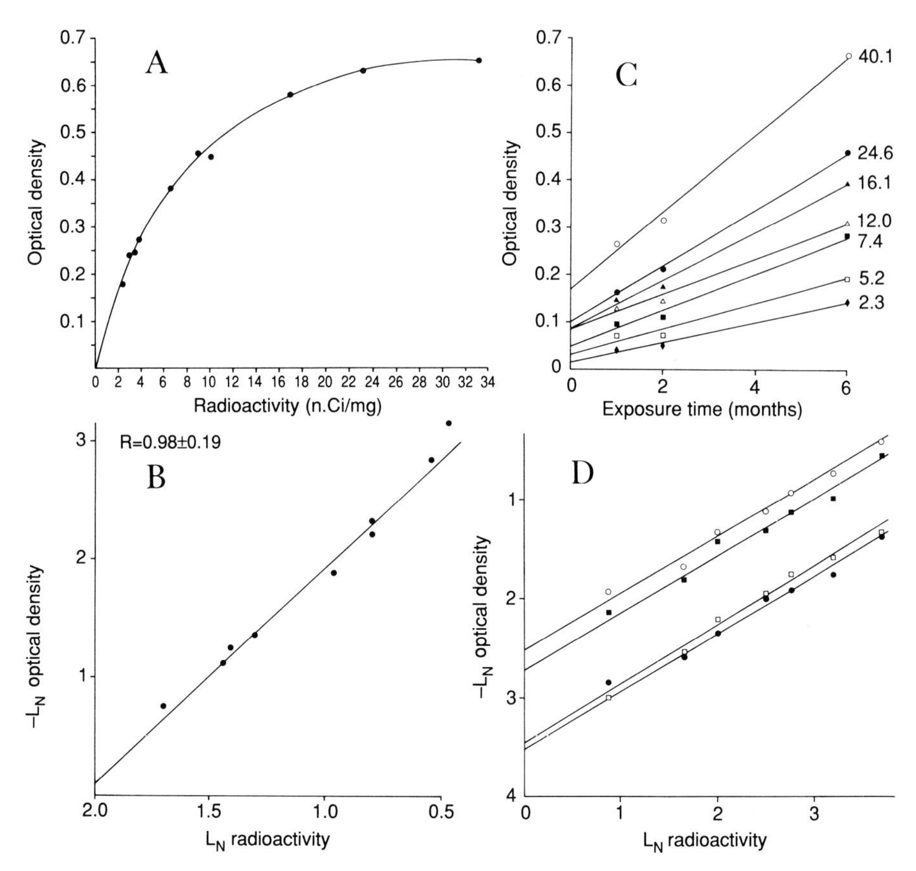

Figure 1. Typical response characteristics of a radiation-sensitive sheet film apposed to ^{3}H-containing brain paste radiation standards as determined by quantitative image analysis. (**A**) shows the non-linearity of the optical density (OD) versus radiation curve; (**B**) depicts the transformed data in (a) now represented as the Ln OD versus Ln radiation; (**C**) shows the film response parameters following exposure to different quantities of radioactivity over a 6 month period; (**D**) shows some linearized data from (C). (Reproduced with permission from Sharif and Hughes (ref. 61).)

Figures 1–12). Some of the problems that can be encountered in performing these assays, and possible causes, are as follows:

- sections fall off the slides during the pre-incubation step: the gelatin solution used for subbing slides may have been too dilute or the thaw-mounting of sections may have been incorrect or incomplete.
- the incubation buffer dries up during the incubation: either insufficient buffer was used to cover the sections, or the temperature and air movement in the room may have been too high resulting in excessive evaporation.

Table 1. Autoradiographic localization procedures for classical neurotransmitter receptors

Receptor subtype	Radioligand and concentration	Preincubation buffer	Preincubation time/temp	Incubation buffer	Incubation time/temp	Wash buffer	Stop assay time/temp	References
Adenosine-1	[^{3}H]cyclohexyl-adenosine 3 nM ± 5 μM R-phenylisopropyl-adenosine	170 mM Tris–HCl with 0.5 IU/ml of adenosine deaminase	20 min/23°C	Same as preincubation buffer	90 min/23°C	Same as preincubation buffer	2 × 3 min/4°C	13
Adenosine-2	[^{3}H]CGS-21680 5 nM ± 20 μM cold CADO	50 mM Tris–HCl + 10 mM $MgCl_2$ + 2 IU/ml adenosine deaminase	30 min/37°C	Same as preincubation buffer	120 min/23°C	50 mM Tris–HCl	5 min/4°C	14
Alpha-1 (a and b subtypes) (The proportions of α_{1a} and α_{1b} sites can be found by inactivating α_{1b} sites with chloro-ethylclonidine) ([^{3}H]WB4101 may selectively label α_{1a} sites)	[^{3}H]Prazosin 0.3–3 nM ± 10–100 μM cold phentolamine	50 mM Tris–HCl	170 min/23°C	170 mM Tris–HCl + 50 mM NaCl	60 min/23°C	50 mM Tris–HCl	2 × 12 min/4°	15; see also 16, 17 See Laporte *et al.* (18) for use of [^{3}H]5-methyl-urapidil See Young & Kuhar (19) for [^{3}H]WB-4101 binding to α_{1a} sites
Alpha-2 (a, b, and c subtypes)	[^{3}H]Rauwolscine 2nM ± 10 μM cold phentolamine	50 mM Tris–HCl	60 min/23°C	Krebs-bicarbonate	60 min/23°C	50 mM Tris–HCl	2 × 5 min/4°C	20
	[^{3}H]RS-15385-197 0.1 nM ± 1 μM phentolamine	50 mM Tris–HCl + 0.5 mM EDTA	20 min/23°C	Same as preincubation buffer	60 min/23°C	Same as incubation buffer	2 × 5 min/4°C	21
	[^{3}H]Idazoxan 1 nM ± 10 μM phentolamine	50 mM Tris–HCl	60 min/23°C	50 mM Tris–HCl	60 min/23°C	50 mM Tris–HCl	2 × 5 min/4°C	22

Beta-1 and 2 Adrenergic (Beta-1 and beta-2 sites can be selectively labelled with [125I]cyanopindolol in the presence of 0.07 μM ICI-118551 or 0.1 μM CG20712, β-2 and β-1 antagonists respectively)	[125I]cyanopin-dolol. 25–50 pM ± 200 μM cold (-)isoproterenol (Labels β-1 and β-2 sites)	25 mM Tris–HCl containing 154 mM NaCl, 0.25% polypeptide and 1.1 mM ascorbate	10 min/23°C	Same as preincubation buffer	120 min/23°C	25 mM Tris–HCl	2 × 7 min/4°C	23, 24
Dopamine-1 (D_1)	[125I]SCH23982 55 pM containing 50 nM mianserin ± 10 μM cold SC23390	50 mM Tris–HCl	30 min/23°C	Same as preincubation buffer	60 min/23°C	50 mM Tris–HCl	4–10 min/4°C	25, 26
Dopamine-2 (D_2)	[125I]sulpiride 0.2 nM ± 10 μM apomorphine	50 mM Tris–HCl	30 min/23°C	Same as preincubation buffer	60 min/23°C	50 mM Tris–HCl	4–10 min/4°C	27
GABA-A (GABA-B sites can be labelled with [3H]GABA in the presence of 10 μM isoguavacine to block GABA-A sites (Wilkin *et al.* (29))	[3H]muscimol 10–150 nM ± 100 μM cold GABA	50 mM Tris–citrate	15 min/4°C	50 mM Tris–citrate	30 min/4°C	50 mM Tris–citrate	4 × 5 sec/4°C	28
Benzodiazepine	[3H]flunitrazepam 5–50 nM ± 2 μM cold clonazepam	50 mM Tris–citrate	15 min 4°C	50 mM Tris–citrate	30 min/4°C	50 mM Tris–citrate	5 min/4°C	28
Cannabinoid	[3H]CP-55940 10 nM ± 10 μM cold CP-55244 or use [3H]WIN 55,212-2 [5,7 apthyl]	—	—	50 mM Tris–HCl + 5% BSA	150 min/37°C	50 mM Tris–HCl + 1% BSA	2 × 120 min/ 0°C	30

Table 1. *Continued*

Receptor subtype	Radioligand and concentration	Preincubation buffer	Preincubation time/temp	Incubation buffer	Incubation time/temp	Wash buffer	Stop assay time/temp	References
Glutamate sites								
1. Glutamate	[^{3}H]glutamate 100 nM + 0.5 mM cold L-glutamate	50 mM Tris–acetate	10 min/30°C	50 mM Tris–acetate	10 min/4°C	50 mM Tris–acetate	30 sec/4°C	31
2. Kainate (KA)	[^{3}H]kainate 100 nM ± 100 μM cold KA	50 mM Hepes–KOH	10 min/23°C	50 mM Hepes–KOH	30 min/4°C	50 mM Hepes–KOH	2 × 30 sec/4°C	32
3. NMDA	NMDA (100 μM)-sensitive [^{3}H]-glutamate binding	50 mM Tris–acetate	10 min/30°C	50 mM Tris–acetate	10 min/4°C	50 mM Tris–acetate	30 sec/4°C	33
4. Quisqualate	[^{3}H]AMPA 15 nM ± 1 mM glutamate	50 mM Tris–HCl + 100 mM KCl	20 min/25°C	50 mM Tris–HCl	30 min/25°C	50 mM Tris–HCl	2 × 30 sec/4°C	34
Histamine								
H_1 sites	[^{3}H]mepyramine 5 nM ± 2 μM triprolidine	—	—	300 mM Na/K phosphate buffer	40 min/4°C	300 mM Na/K-phosphate buffer	10 min/4°C.	35
H_2 sites	[^{125}I]iodoamino-potentidine 50 pM ± 3 μM cold tiotidine	50 mM Na_2/K phosphate buffer	15 min/23°C	50 mM Na_2/K phosphate	3 h/23°C	50 mM Na_2/K-phosphate	5 × 4 min/4°C	36
H_3 sites	[^{3}H]R-α-methyl-histamine 1 nM ± 1 μM cold thioperamide	—	—	50 mM phosphate	45 min/25°C	50 mM phosphate	2 × 3 min/4°C	37
Muscarinics								
M_1 sites	[^{3}H]pirenzepine 25 nM ± 5 μM cold atropine	Tris–Krebs buffer	20 min/23°C	Tris–Krebs buffer	60 min/23°C	50 nm Tris–HCl	10 min/4°C	38 Also see Cortes *et al.* (39)
M_2 sites	[^{3}H]AF-DX116 10 nM ± 1 μM atropine	Krebs-bicarbonate	15 min/4°C	Krebs-bicarbonate	60 min/4°C	Krebs-bicarbonate	3 × 4 min/4°C	40 Also see Wang *et al.* (41)
M_3 sites	[^{3}H]4-DAMP 0.3–0.6 nM ± 5 μM atropine	Tris–Krebs buffer	20 min/23°C	Tris–Krebs buffer	60 min/23°C	50 mM Tris–HCl	10 min/4°C	42, 125

Nicotinic sites	[^{3}H]acetylcholine 10 nM ± 100 μM carbachol	50 mM Tris–HCl + 1.5 mM atropine sulfate + 120 mM NaCl + 5 mM KCl + 1 mM $MgCl_2$ + 2 mM $CaCl_2$. The second preincubation also had 100 μM di-isopropyl fluorophosphate	2 × 15 min/23°C	Same as preincubation buffer	60 min/4°C	Same as preincubation buffer	2 × 2 min/4°C	43
	[^{125}I]BTX 1.4 nM ± mM L-nicotine bitartrate	50 mM Tris–HCl + 1 mg/ml BSA	30 min/23°C	50 mM Tris–HCl + 1 mg/ml BSA	120 min/23°C	50 mM Tris–HCl	6 × 30 min/4°C	43
	[^{3}H]nicotine 3.5 nM ± 10 μM L-nicotine	—	—	50 mM Tris–HCl	20 min/23°C	50 mM Tris–HCl	4 × 30 sec/4°C	43
Serotonin (5HT)								
$5HT_{1a}$	[^{3}H]8-OH-DPAT 2 nM ± 1 μM cold 5HT	170 mM Tris–HCl + 4 mM $CaCl_2$, 0.1% ascorbate	30 min/23°C	Same as preincubation buffer	60 min/23°C	Same as incubation buffer	2 × 5 min/4°C	44 Also see Verge *et al.* (45)
$5HT_{1b}$	[^{125}I]cyano-pindolol 12 pM ± 30 μM isoprenaline	170 mM Tris–HCl + 150 mM NaCl	10 min/23°C	Same as preincubation buffer	120 min/23°C	Same as preincubation buffer	2 × 15 min/4°C	46
$5HT_{1c}$	[^{3}H]mesulergine 5 nM ± 0.1 μM cold 5HT	170 mM Tris–HCl	15 min/23°C	170 mM Tris–HCl	120 min/23°C	170 mM Tris–HCl	2 × 10/4°C	47
$5HT_{1d}$	[^{3}H]5HT 2 nM + 0.1 μM 8-OH-DPAT + 0.1 μM mesulergine ± cold 1 μM 5HT + 0.1 μM 8-OH-DPAT + 0.1 μM mesulergine	170 mM Tris–HCl + 4 mM $CaCl_2$ + 0.01% ascorbate	30 min/23°C	Same as preincubation buffer	60 min/23°C	Same as preincubation buffer	5 min/4°C	48
	[^{125}I]GTI 25 pM ± 10 mM 5HT	170 mM Tris–HCl + 4 mM $CaCl_2$	30 min/23°C	170 mM Tris–HCl + 4 mM $CaCl_2$ + 10 mM pargyline + 0.05% ascorbate	60 min 23°C	Same as preincubation buffer	2 × 5 min/4°C	56

Table 1. *Continued*

Receptor subtype	Radioligand and concentration	Preincubation buffer	Preincubation time/temp	Incubation buffer	Incubation time/temp	Wash buffer	Stop assay time/temp	References
$5HT_2$	[^{3}H]ketanserin 1–2 nM ± 1 μM cold ritanserin methysergide	170 mM Tris–HCl	15 min/23°C	170 mM Tris–HCl	30 min/23°C	170 mM Tris–HCl	2 × 10 min/4°C	49
$5HT_3$	[^{3}H]quipazine 0.5 nM containing 0.1 μM paroxetine + 1 μM cold (s)zacopride	Tris–Krebs buffer	30 min/23°C	Tris–Krebs	60 min/23°C	50 mM Tris–HCl	2 × 5 min/4°C	50
	[^{3}H]GR65630 0.2–0.5 nM ± 1 μM cold (s)zacopride	Tris–Krebs buffer	30 min/23°C	Tris–Krebs	60 min/23°C	50 mM Tris–HCl	2 × 5 min/4°C	50
	[^{3}H]BRL43694 5–10 nM ± 100 μM cold GR65630	5 mM Hepes	60 min/23°C	5 mM Hepes	30 min/23°C	5 mM Hepes	2 × 3 sec/4°C	51
	[^{3}H]Zacopride/ 1 nM ± 1 μM cold granisetron	50 mM Hepes–Krebs	30 min/37°C	50 mM Hepes–Krebs	30 min/37°C	50 mM Hepes–Krebs	2 × 30 sec/4°C	52
	[^{125}I]Iodo-zacopride 0.22–0.5 nM ± 1 μM GR65630	—	—	50 mM Hepes	60 min/25°C	50 mM Hepes	2 × 5 min/4°C	53
	[^{125}I]ICS205930 1.5 nM ± 50 μM 5HT	170 mM Tris–HCl with 150 mM NaCl, 4 mM $CaCl_2$ and 0.01% ascorbate	30 min/23°C	Same as preincubation buffer + 10 mM pargyline	60 min/23°C	Same as preincubation buffer	5 min/4°C	54
	[^{3}H]RS42358-197 1–3 nM ± 1 μM (s)zacopride	50 mM Tris–Krebs	30 min/23°C	50 mM Tris–Krebs	60 min/23°C	50 mM Tris–HCl	2 × 5 min/4°C	55

Table 2. Autoradiographic localization procedures for peptide receptors

Receptor subtype	Radioligand and concentration	Preincubation buffer	Preincubation time/ temperature	Incubation buffer	Incubation time/ temperature	Wash buffer	Stop assay time/ temperature	References
Atrial natriuretic factor (ANF)	$[^{125}I]$ANF 50 pM ± 1 μM cold ANF	50 mM Hepes containing 0.005% PEI	10 min/25°C	50 mM Tris–HCl (pH 7.4) containing 100 mM NaCl, 5 mM $MgCl_2$ 0.5 μg/ml PMSF, 40 μg/ml bacitracin, 4 μg/ml leupeptin, 2 μg/ml chymostatin, 0.5% BSA,	60 min/25°C	Same as incubation buffer	4 × 2 min washes in excess buffer 4°C	57
Angiotensin II (AII) (AT-1 subtype can be labelled in the presence of 10 μM PD123177; AT-2 subtype can be labelled in the presence of 10 μM DUP-753)	$[^{125}I]Sar^1$-$Ileu^8$-AII 50–100 pM ± 1 μM cold AII	20 mM sodium phosphate buffer containing 150 mM NaCl, 10 mM $MgCl_2$, 1 mM EDTA	30 min/23°C	20 mM sodium phosphate buffer containing 150 mM NaCl, 10 mM $MgCl_2$, 1 μM captopril, 60 μg/ml bacitracin, and 1.3 μg/ml of phosphoramidon, pepstatin A, bestatin, chymostatin, antipain, and 0.13 μg/ml amstatin	90 min/23°C	50 mM Tris–HCl (pH 7.4)	2 × 5 min washes in 500 ml buffer/4°C	58, 59, 126
AII (3-8) (hexapeptide)	$[^{125}I]$AII(3–8) 0.6 nM ± 0.1 μM cold AII(3–8)	50 mM Tris–HCl + 5 mM EDTA, 150 mM NaCl, 20 μM bestatin, 50 μM Plummer's inhibitor + 100 μM PMSF + 0.1% BSA	30 min/23°C	Same as preincubation buffer	120 min/37°C	Same as preincubation buffer	2 × 5 min/4°C	60

Table 2. *Continued*

Receptor subtype	Radioligand and concentration	Preincubation buffer	Preincubation time/ temperature	Incubation buffer	Incubation time/ temperature	Wash buffer	Stop assay time/ temperature	References
Bradykinin (B_2) (BK)	[^{125}I]BK (0.1–0.2 nM); [^{3}H]BK (1–3 nM) ± 1 μM cold BK (In future, [^{3}H] Des-Arg^{10}-kallidin & [^{3}H]NPC-17731 may be useful tools for labelling B_1 and B_2 sites respectively (Burch *et al.* (63))	25 mM TES buffer	20 min/23°C	25 mM TES buffer containing 1 mM dithiothreitol, 1 mM 1,10-phenanthroline 1 μM captopril, 0.1% BSA	90 min/23°C	25 mM TES buffer	2 × 5 min washes in 500 ml buffer/4°C	61, 62
Bombesin	[^{125}I]Bombesin 100 pM ± 1 μM cold bombesin	10 mM Hepes	5 min/23°C	10 μm Hepes containing 130 mM NaCl, 4.7 mM KCl, 5 mM $MgCl_2$, 1 mM EGTA, 0.1% BSA, 100 μg/ml bacitracin	60 min/23°C	10 mM Hepes containing 0.15 BSA	4 × 2 min washes in buffer/4°C	64
Cholecystokinin (CCK) (CCK-A and CCK-B subtypes can be labelled in the presence of 0.1 μM L-365260 and L-364718, respectively) (Hill and Woodruff (66))	[^{125}I]BH-CCK-8; 0.2 nM ± 1 μM cold CCK-8	50 mM Tris–HCl	15 min/23°C	10 mM Hepes buffer containing 130 mM NaCl, 4.7 mM KCl, 5 mM $MgCl_2$, 1 mM EGTA, 0.25 mg/ml bacitracin, 5 mg/ml BSA	120 min/23°C	50 mM Tris–HCl	3 × 5 min/4°C	65

Calcitonin gene-related peptide (CGRP)	[^{125}I]CGRP 0.1 nM ± 1 μM cold CGRP	50 mM Tris–HCl	5 min/23°C	50 mM Tris–HCl containing 5 mM $MgCl_2$ and 2 mM EGTA	120 min/23°C	50 mM Tris–HCl containing 0.1% BSA	4 × 3 min/4°C	67
Corticotrophin releasing factor (CRF)	[^{125}I]CRF or analog 0.1 nM ± 1 μM cold CRF or analog	—	—	50 mM Tris–HCl containing 5 mM $MgCl_2$, 2 mM EGTA, 0.1% BSA, 100 KIU/ml aprotinin and 0.1 mM bacitracin	60 min/23°C	Dulbecco's phosphate-buffered saline containing 1% BSA	2 × 20 min/4°C	68
Endothelins (ET-1, ET-2 and ET-3; sarafotoxin, vasoactive intestinal contractor)	[^{125}I]labelled peptides 10–20 pM ± 1 μM cold peptide	10 mM Hepes containing 5 mM $MgCl_2$	15 min/23°C	10 mM Hepes containing 5 mM $MgCl_2$ and 0.1% BSA	120 min/23°C	50 mM Tris–HCl	3 × 5 min/4°C	69, 70
Galanin	[^{125}I]galanin 0.1 nM ± 1 μM cold galanin	10 mM Hepes	10 min/23°C	10 mM Hepes	60 min/23°C	10 mM Hepes	4 × 3 min/4°C	67, 71
Glucagon-like peptide (GLP-1)	[^{125}I]GLP-1 50 pM ± 0.1 μM cold GLP-1	30 mM Tris–HCl containing 0.1% BSA	15 min/23°C	30 mM Tris–HCl containing 0.1% BSA, 0.5 mg/ml bacitracin and 0.8 mg/ml dithiothreitol	45–180 min/23°C	Same as incubation buffer	4 × 0.5 min/4°C	72
Morphine-modulating peptide (FLFQPQRFamide)	[^{125}I]FLFQPQRF-amide 50 pM ± 1 μM cold peptide	50 mM Tris–HCl + 140 mM NaCl + 0.5% BSA	20 min/23°C	50 mM Tris–HCl + 120 mM NaCl + 0.5% BSA + 0.1 mM bestatin + 1 mM EDTA + 0.5 mM DFP	60 min/25°C	50 mM Tris–HCl	4 × 3 min/4°C	73
Neurokinin1 (NK-1) (Substance P)	[^{125}I]BH-NK-1 0.1 nM ± 1 μM cold NK-1	50 mM Tris–HCl containing 0.005% PEI	10 min/20°C	50 mM Tris–HCl containing 3 mM $MnCl_2$, 200 mg/ml BSA, 2 mg/l chymostatin, 4 mg/l leupeptin + 40 mg/l bacitracin	60 min/20°C	50 mM Tris–HCl	4 × 2 min/4°C	67

Table 2. *Continued*

Receptor subtype	Radioligand and concentration	Preincubation buffer	Preincubation time/ temperature	Incubation buffer	Incubation time/ temperature	Wash buffer	Stop assay time/ temperature	References
Neurokinin-2 (NK-2) (NK-A; Substance k)	[^{125}I]BH-NK-A 0.1 nM ± 1 μM cold NK-A	50 mM Tris–HCl	10 min/19°C	50 mM Tris–HCl containing 3 mM $MnCl_2$, 200 mg/ml BSA, 2 mg/ml chymostatin, 4 mg/ml leupeptin + 40 mg/ml bacitracin	120 min/19°C	50 mM Tris–HCl	4 × 1.5 min/ 4°C	67
Neurokinin-3 (NK-3) (NK-B)	[^{3}H]Senktide 3 nM ± 1 μM cold eledoisin or [^{125}I]eledoisin, 50 pM ± 1 μm cold eledoisin	—	—	50 mM Tris–HCl containing 3 mM $MnCl_2$, 0.02% BSA, 40 μg/ml bacitracin, 2 μg/ml chymostatin, 4 μg/ml leupeptin	90 min/25°C	Same as incubation buffer	4 × 1 min/4°C	74, 75
Neuropeptide Y (NPY)	[^{125}I]NPY 25 pM ± 1 μM cold NPY or [^{3}H]NPY 2.5 nM ± 1 μM cold NPY	50 mM Tris–HCl	20 min/23°C	Krebs-bicarbonate buffer containing 0.1% BSA + 0.05% bacitracin	120 min/23°C	50 mM Tris–HCl	4 × 4 min/4°C	76
Neurotensin (NT)	[^{125}I]Tyr^{3}-NT 0.1 nM ± 0.5 μM cold NT	—	—	50 mM Tris–HCl with 5 mM $MgCl_2$, 0.2% BSA + 50 μM bacitracin	120 min/4°C	50 mM Tris–HCl with 0.25 M sucrose	4 × 2 min/4°C	77
Opioid delta (δ)	[^{3}H]DPDPE 2–10 nM ± 1 μM cold DPDPE or 1 μM naloxone	50 mM Tris–HCl	45 min/37°C	50 mM Tris–HCl containing 5 mM $MgCl_2$, 0.1% BSA + 20 μg/ml bacitracin	90 min/4°C	50 mM Tris–HCl	3 × 5 min/4–23°C	78 Also see Mansour *et al.* (79)
Opioid mu (μ)	[^{3}H]DAMGO 0.5–3 nM ± 1 μM cold DAMGO or naloxone	50 mM Tris–HCl	45 min/37°C	50 mM Tris–HCl	90 min/4–23°C	50 mM Tris–HCl	3 × 5 min/4°C	78 Also see Mansour *et al.* (79)

Opioid kappa (κ)	[^{125}I]Dynorphin (1-8) 0.2 nM containing 0.2 μM cold DAMGO and DPDPE ± 1 μM naloxone	50 mM Tris–HCl	45 min/37°C	50 mM Tris–HCl	90 min/4°C	50 mM Tris–HCl	3 × 5 min/4°C	80, 78 Also see Quirion *et al.* (81) and Jomary *et al.* (82) for use of other radioligands
Opioid kappa-1 (κ_1)	[^{3}H]U69593 5–10 nM ± 1 μM naloxone	50 mM Tris–HCl	45 min/37°C	50 mM Tris–HCl	80 min/23°C	50 mM Tris–HCl	3 × 0.5 min/ 4°C	83 Also see Boyle *et al.* (84) and Clark *et al.* (85) for use of [^{3}H]PD117302 and [^{3}H]CI-977
Oxytocin (OT)	[^{3}H]OT (3 nM) or [^{125}I]OT-antagonist (OTA) (40 pM) ± 1–10 μM cold OT	50 mM Tris–HCl containing 100 mM NaCl, 10 μM GTP followed by rinse for 5 min in 50 mM Tris–HCl	15 min/23°C	50 mM Tris–HCl containing 5–10 mM $MgCl_2$, 0.1 mM bacitracin and 1 mg/ml BSA	60 min/23°C	Same as incubation buffer	2 × 5 min/4°C	86
Peptide YY (PYY)	[^{125}I]PYY 25 pM ± 1 μM cold PYY	50 mM Tris–HCl	20 min/23°C	Krebs-bicarbonate buffer containing 0.1% BSA and 0.05% bacitracin	120 min/23°C	50 mM Tris–HCl	4 × 4 min/4°C	87 Also see Martel *et al.* (76)
Somatostatin (SRIF) (SS-1 and SS-2 subtypes)	[^{125}I]SRIF-analogue 50–125 pM ± 1 μM cold SRIF	150 mM Tris–HCl	60 min/25°C	150 mM Tris–HCl containing 5 mM $MgCl_2$, 0.1% BSA + 0.05% bacitracin	120 min/25°C	150 mM Tris–HCl	4 × 5 min/4°C	88
Thyrotropin-releasing hormone (TRH)	[^{3}H]MeTRH 5–10 nM ± 10 μM TRH	0.9% saline or 50 mM Tris–HCl	15–30 min/4–20°C	20 mM sodium phosphate or 50 mM Tris–HCl buffers containing 0.05–0.1% BSA + 10 μM bacitracin	90 min/4°C	50 mM Tris–HCl	2 × 2.5 min/ 4°C	89, 90
Vasopressin (AVP)	[^{3}H]AVP or [^{3}H]AVP-antagonist 1–3 nM ± 1–10 μM cold AVP	50 mM Tris–HCl containing 100 mM NaCl, 10 μM GTP followed by rinse for 5 min in 50 mM Tris–HCl	15 min/23°C	50 mM Tris–HCl containing 5–10 mM $MgCl_2$, 0.1 mM bacitracin + 1 mg/ ml BSA	60 min/23°C	Same as incubation buffer	2 × 5 min/4°C	86

Table 2. *Continued*

Receptor subtype	Radioligand and concentration	Preincubation buffer	Preincubation time/ temperature	Incubation buffer	Incubation time/ temperature	Wash buffer	Stop assay time/ temperature	References
Vasoactive intestinal peptide (VIP)	$[^{125}I]$VIP 0.1 nM ± 1 μM cold VIP	10 mM Hepes	5 min/20°C	10 mM Hepes containing 130 mM NaCl, 4.7 mM KCl, 5 mM $MnCl_2$, 1 mM EGTA, 1% BSA + 1 mg/ml bacitracin	120 min/20°C	Same as incubation buffer	3 × 5 min/4°C	67 Also see Dietl *et al.* (91)

Table 3. Autoradiographic localization procedures for transmitter transport sites

Transport site	Radioligand and concentration	Preincubation buffer	Preincubation time/ temperature	Incubation buffer	Incubation time/ temperature	Wash buffer	Stop assay time/ temperature	References
Acetylcholine	[^{3}H]HC-3, (hemicholinium) 15 nM ± 10 μM cold HC-3	—	—	50 mM Tris–HCl + 300 mM NaCl	60 min/4°C	50 mM Tris–HCl + 300 mM NaCl	6 × 1 min/4°C	92
	[^{3}H]Vesamicol 40 nM ± 10 μM cold vesamicol	50 mM Tris–HCl + 120 mM NaCl + 5 mM KCl + 2 mM $CaCl_2$ + 1 mM $MgCl_2$	10 min/23°C	Same as preincubation buffer	60 min/23°C	Same as preincubation buffer	2 × 30 sec/4°C	93
Dopamine	[^{3}H]Mazindol 2–15 nM ± 0.3 μM desipramine	50 mM Tris–HCl + 120 mM NaCl + 5 mM KCl	5 min/4°C	50 mM Tris–HCl + 300 mM NaCl + 5 mM KCl	40 min/4°C	Same as incubation buffer	2 × 3 min/4°C	94
	[^{3}H]CFT 5 nM ± 30 μM cold (−)-cocaine	50 mM Tris–HCl + mM NaCl	20 min/4°C	50 mM Tris–HCl + 100 mM NaCl	120 min/4°C	Same as preincubation buffer	2 × 1 min/4°C	95
GABA	[^{3}H]GABA 1 μM ± 1 mM GABA	—	—	Krebs buffer	10–15 min 25°C	Krebs buffer	3 × 2 sec/4°C	96
Glutamate	[^{3}H]Glutamate 3 μM ± 1 mM Glu	—	—	Krebs buffer	10–15 min/ 25°C	Krebs buffer	3 × 2 sec/4°C	96
Norepinephrine (NE)	[^{3}H]desmethyl-imipramine (DMI) 1–10 nM DMI ± 100 μM imipramine [^{3}H]Nisoxetine may be useful in future (Tejani-Butt and Ordway (98))	—	—	50 mM Tris–HCl + 300 mM NaCl	60 min/4°C	50 mM Tris–HCl + 300 mM NaCl	3 × 20 min/4°C	97
Serotonin	[^{3}H]Citalopram 0.5 nM ± 1 μM paroxetine	170 mM Tris–HCl + 120 mM NaCl + 5 mM KCl	15 min/23°C	Same as preincubation buffer	60 min/23°C	Same as preincubation buffer	2 × 10 min/ 23°C	99
	[^{3}H]Paroxetine 0.3 nM ± 10 μM Fluoxetine	50 mM Tris–HCl	15 min/23°C	50 mM Tris–HCl + 120 mM NaCl + 5 mM KCl	60 min/23°C	Same as incubation buffer	2 × 20 min/ 37°C	100

Table 4. Autoradiographic localization procedures for second messenger systems

Acceptor/receptor system	Radioligand and concentration	Preincubation buffer	Preincubation time/temperature	Incubation buffer	Incubation time/temperature	Wash buffer	Stop assay time/temperature	References
Adenylate cyclase	[^{3}H]Forskolin 10 nM ± 10 μM cold forskolin	–	–	50 mM Tris–HCl containing 100 mM NaCl and 5 mM $MgCl_2$	10 min/23°C	Same as incubation buffer	2 × 2 min/4°C	101
Inositol 1,4,5-trisphosphate (IP_3)	[^{3}H]IP_3 50 nM ± 5 μM cold IP_3	–	–	20 mM Tris–HCl containing 20 mM NaCl, 100 mM KCl, 1 mM EDTA and 1 mg/ml BSA	10 min/4°C	Same as incubation buffer	2 × 2 min/4°C	102
Inositol hexakisphosphate (IP_6)	[^{3}H]IP_6 5 nM ± 25 μM cold IP_6	20 mM Tris–HCl containing 20 mM NaCl, 100 mM KCl, 5 mM EDTA and 0.1% BSA	30 min/4°C	Same as preincubation buffer	30 min/4°C	Same as incubation buffer	2 × 2 min/4°C	103
Protein kinase C (PKC)	[^{3}H]Phorbol ester (PDBu) 2.5 nM ± 1 μM cold PDBu	–	–	50 mM Tris–HCl containing 100 mM NaCl and 1 mM $CaCl_2$	60 min/33°C	Same as incubation buffer	1 × 1 min/4°C	101 Also see Thomson *et al.* (104) for [^{3}H]dimethyl-staurosporine binding
Sarcoplasmic reticulum calcium store	[^{3}H]Ryanodine 15 nM ± 10 μM cold ryanodine	20 mM PIPES–KOH containing 1 mM KCl, 550 μM ATP, 100 μM $CaCl_2$ and 1 mM PMSF	15 min/25°C	Same as preincubation buffer	15 min/25°C	Same as incubation buffer	2 × 2.5 min/4°C	105

Table 5. Autoradiographic localization procedures for receptor channels, growth factors, and steroids

Site to be labelled	Radioligand and concentration	Preincubation buffer	Preincubation time/ temperature	Incubation buffer	Incubation time/ temperature	Wash buffer	Stop assay time/ temperature	References
Calcium channels								
1. L-type	[^{35}S]Sadopine 52 pM ± 1 μM (+)PN200-110	–	–	50 mM Tris–HCl + 0.1 mM PMSF	90 min/22°C	50 mM Tris–HCl + 0.1 mM PMSF	3 × 10 min/4°C	106
2. N-Type	[^{125}I]w-conotoxin 0.1–0.5 nM ± 1 μM cold w-conotoxin	5 mM Hepes + 0.32 mM sucrose + 0.01 mg/ml lysozyme + 0.1 mg/ml BSA + 0.1 mg/ml bacitracin	10 min/4°C	Same as preincubation buffer	90 min/4°C	Same as preincubation buffer	3 × 10 min/4°C	107
Potassium channels								
1. ATP-sensitive	[^{125}I]Iodoglibenclamide 100 pM ± 0.1 μM iodoglibenclamide	–	–	20 mM Hepes + 150 mM NaCl	60 min/4°C	20 mM Hepes + 150 mM NaCl	2 × 5 min/4°C	108
2. β-Bungarotoxin (BTX)-sensitive	[^{125}I]BTX 1.5 nM ± 1 μM BTX	–	–	PBS + 1 mM EGTA + 2 mM strontrium chloride + 1 mg/ml BSA	45 min/22°C	Same as preincubation buffer	3 × 5 min/4°C	109
3. Dendrotoxin-sensitive	[^{125}I]dendrotoxin 2 nM ± 0.5 μM DTX	–	–	Krebs–phosphate + 1 mg/ml BSA	45 min/22°C	Same as preincubation buffer	3 × 5 min/4°C	110
4. Apamin-sensitive	[^{125}I]Apamin 25 pM ± 1 μM apamin	–	–	100 mM Tris–HCl + 0.1% BSA	30 min/4°C	100 mM Tris–HCl + 0.1% BSA	4 × 20 sec/4°C	111
NMDA receptor channel	[^{3}H]MK-801 30 nM ± 100 μM cold MK-801 [^{3}H]thienylphencyclidine (TCP) may be a useful probe (Vignon *et al.* (113))	50 mM Tris–HCl + 190 mM sucrose	20 min/23°C	50 mM Tris–HCl	20 min/23°C	50 mM Tris–HCl	2 × 20 sec/4°C	112

Table 5. *Continued*

Site to be labelled	Radioligand and concentration	Preincubation buffer	Preincubation time/ temperature	Incubation buffer	Incubation time/ temperature	Wash buffer	Stop assay time/ temperature	References
Nerve growth factor (NGF)	$[^{125}I]$NGF 75 pM ± 0.1 μM cold NGF	100 mM sodium phosphate + 200 mM NaCl + 0.5 mM $MgCl_2$	3 h/22°C	Same as preincubation + 1 mg/ml cytochrome *c* + 4 μg/ml leupeptin + 0.5 mM PMSF	260 min/22°C	Same as incubation buffer	3 × 1 min/22°C	114
Steroid site	$[^3H]$corticosterone 8 nM ± 0.5 μM steroid	–	–	50 mM Tris–HCl + 5% glycerol + 10 mM DTT + 2 mM EGTA + 5 mM ATP + 6 mM molybdate	20 min/23°C	50 mM Tris–HCl	3 × 5 min/4°C	115

Note added in proof

A specific receptor binding assay has recently been developed for the $5HT_4$ receptor (130). An extension of this procedure resulted in the establishment of a procedure to map the distribution of the $5HT_4$ receptor using autoradiography (130). Tissue sections were incubated with 0.1 nM $[^3H]$GR113808 in HEPES buffer for 30 min at 37°C. Nonspecific binding was defined with 30μM 5HT. The slides were then rinsed in ice-cold buffer for 5 s (×2), dipped in ice-cold water and then rapidly dried. Autoradiograms were generated over 6 weeks (130).

Table 6. Autoradiographic localization procedures for enzymes

Enzyme	Radioligand and concentration	Preincubation buffer	Preincubation time/temperature	Incubation buffer	Incubation time/temperature	Wash buffer	Stop assay time/temperature	References
Angiotensin converting enzyme (ACE)	[^{3}H]Captopril 4 nM ± 1 μM cold captopril	50 mM Tris–HCl	5 min/4°C	50 mM Tris–HCl + 300 mM NaCl + 0.1% BSA	40 min/4°C	Same as incubation buffer	2 × 1 min/4°C	116
Enkephalin convertase	[^{3}H]GEMSA 4 nM ± 10 μM cold GEMSA	50 mM sodium acetate	5 min/4°C	50 mM sodium acetate	30 min/4°C	50 mM sodium acetate	2 × 1 min/4°C	117
Enkephalinase	[^{3}H]HACBO-Gly 3 nM ± 1 μM thiorphan	–	–	50 mM Tris–HCl	60 min/23°C	50 mM Tris–HCl	2 × 5 min/4°C	118
Monoamine oxidase (MAO)	[^{3}H]pargyline 40 nM ± 1 μM cold pargyline	–	–	50 mM Tris–HCl + 300 mM NaCl	30 min/23°C	50 mM Tris–HCl + 300 mM NaCl	2 × 5 min/4°C	119
	[^{3}H]MPTP 20 nM ± 1 μM cold MPTP	–	–	50 mM Tris–HCl + 300 mM NaCl	30 min/23°C	50 mM Tris–HCl + 300 mM NaCl	2 × 5 min/4°C	120
Monoamine oxidase-A	[^{3}H]Ro-41-1049 10–20 nM ± 50 μM clorgyline or L-deprenyl	–	–	50 mM Tris–HCl containing 120 mM NaCl, 1 mM $MgCl_2$, 5 mM KCl, and 0.5 mM EGTA	90 min/20°C	Same as incubation buffer	2 × 1 min/4°C	121
Monoamine oxidase-B	[^{3}H]Ro-19-6327 10–20 nM ± 50 μM clorgyline or L-deprenyl	–	–	50 mM Tris–HCl containing 120 mM NaCl, 1 mM $MgCl_2$, 5 mM KCl, and 0.5 mM EGTA	90 min/20°C	Same as incubation buffer	2 × 1 min/4°C	121
Sodium-potassium ATPase	[^{3}H]Ouabain 45 nM ± 200 mM NaCl or 10 mM ATP	3 mM $MgCl_2$ + 2 mM sodium phosphate + 250 mM sucrose + 0.25 mM EDTA + 1 mM DTT	10 min/37°C	Same as preincubation buffer	195 min/37°C	Same as preincubation buffer	10 min/4°C	122

- sections fall off the slides during the rinsing steps: the gelatin solution used for subbing slides may have been too dilute or the agitation for rinsing may have been too harsh.

3.1 Liquid emulsion autoradiography—the cover-slip technique

While the film autoradiography yields excellent results and is a rapid and reproducible technique, much higher anatomical and cellular resolution can be achieved using the so-called 'cover-slip technique'. However, the latter technique is time-consuming, labour-intensive, and quite precarious. Essentially, either the radiolabelled tissue sections are covered (dipped) with liquid emulsion (11), or emulsion-coated glass coverslips (12) are tightly apposed to the sections to generate the autoradiograms. The emulsion-coating of slides should only be attempted with covalently bound radioligands or ligands that can be irreversibly bound to the receptors using the paraformaldehyde fixation procedure (11) (*Protocol 5*).

Protocol 5. Paraformaldehyde-fixation of radiolabelled tissue sections and emulsion-dipping

1. Ensure the sections are totally dry. Then load the slides in metal slide-holders and place the latter on a platform over a Petri dish containing 10–30 g of paraformaldehyde (PFMD) powder in a desiccator. The PFMD should be equilibrated to room temperature and humidity (50%) for 1–2 days before being put in the desiccator.
2. Place the desiccator (with a good seal) in an oven at 80°C for 2 h to permit the PMFD vapours to fix the sections in. (Wear safety glasses.)
3. The desiccator should be carefully opened in a vented fume-hood and the slide-racks removed immediately in order to prevent the PFMD from condensing on the sections. Leave the slides in the vented hood for 24 h in order to get rid of the PFMD fumes.
4. Defat the sections by rinsing them in 70%, 80%, and 90% ethanol for 1 min each, then for 3–4 min in xylene, then repeat the ethanol steps. Rinse the slides in 2 × 2 min rinses in distilled water and dry the sections at 37°C overnight. Then dip the slides in emulsion as described below.
5. Under safelight conditions (red filter) in a dark-room, melt Kodak NTB-2 emulsion in a small glass beaker in a water-bath (40–45°C), dilute 1:1 with a filtered solution of 0.1% Dreft detergent, and thoroughly mix them together. Alternatively, 0.5% glycerol can be used instead of the detergent to reduce air bubbles. Maintain this mixture in the 40–45°C water-bath for 15–20 min to equilibrate before starting the dipping.
6. Dip the slides in the emulsion (holding the slides from the end), allow to

drain for a few seconds, wipe the back of the slide with tissue paper, and stand them vertically to dry slowly. After ~2 h, transfer the emulsion-coated slides to light-tight slide boxes, seal the latter with black tape, and place them in a −20°C freezer for the appropriate exposure time.

7. Develop and fix the emulsion as described below for the film, *Protocol 7*. These slides can subsequently be stained with cresyl violet, thionin, or Nissl stains.

If the direct emulsion-dipping procedure is deemed unsuitable for your radioligands, then emulsion-coated glass coverslips can be used to generate autoradiograms. Accordingly, detergent-washed coverslips are dipped in the molten emulsion instead of the slides (see above) and allowed to dry. The coverslips are then attached to the slides bearing the radiolabelled tissue sections as shown in *Protocol 6* below.

Protocol 6. Coverslip autoradiography

1. Open the light-tight coverslip box under safe-light conditions in a dark-room (red filter), remove one coverslip, and replace the lid on the box.
2. Apply a small drop of superglue to the frosted end of the slide bearing the PFMD-fixed tissue sections, press the emulsion-coated coverslip on to the slide, and hold it in place for 30–60 sec. Repeat for other slides and also for a set of radiation standards.
3. Place the slides in an X-ray cassette and keep light-tight for the appropriate exposure time.
4. On the day of the autoradiogram production, pour D-19 developer (17°C), distilled water, and Kodak fixer into small slide staining dishes under safe-light dark-room conditions. Open the X-ray cassette, separate the coverslip/slide sandwich with a toothpick, and process this unit carefully through the emulsion developing and fixing procedure described in *Protocol 7*.
5. After development, defat the tissue sections as described in *Protocol 5* and process the slide/coverslip units through the routine staining procedure using thionin, cresyl violet, or Nissl stains.
6. Once the slides and their attached coverslips are dry, add some mounting material to the slides and permanently fix the coverslips to the slides.
7. Once the assemblies are totally dry, examine them microscopically, in light-field to visualize the stained sections and in dark-field to see the silver grains depicting the position of the receptors. Switching between the two fields of view will help pinpoint the cells bearing the receptors. The silver grains can then be counted using a microscope connected to an image analysis machine as described below.

---

4. Generation of autoradiograms

The tight apposition of the labelled tissues sections with the radiation-sensitive sheet-film or emulsion-coated coverslips in the dark for the appropriate exposure time will result in the generation of latent images on the detector material which then needs to be developed and fixed. This is done as follows:

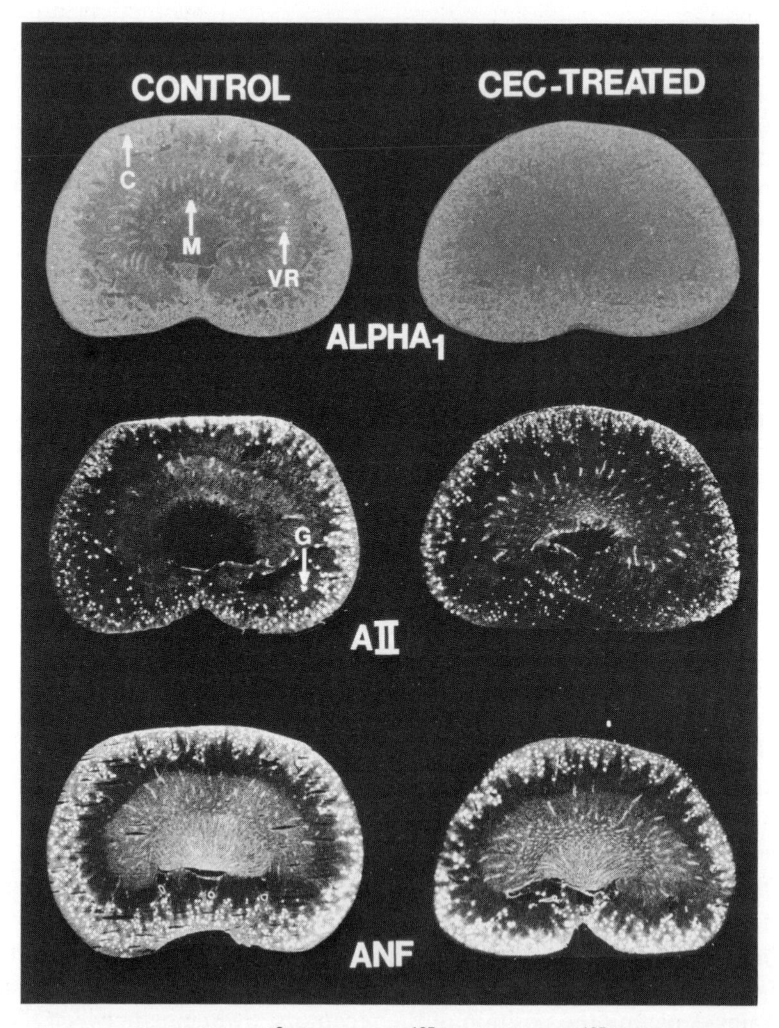

Figure 2. Autoradiography of [^{3}H]prazosin, [^{125}I]Sar-Ile, and [^{125}I]ANF binding to α_1-, angiotensin II-, and ANF-receptors respectively in sections of rat kidneys. Note that only α_1-receptors (α_{1b}) are significantly reduced by chloroethylclonidine (CEC) treatment (see *Table 7*) relative to controls. C = cortex, M = medulla, VR = vasa recta. Dark-field photos are shown—receptors appear as white silver grains. (See ref. 15.)

Protocol 7. Development and fixation of film autoradiograms

The Kodak D-19 developer and fixer should be prepared at least a day before they are needed since both are sold as crystals and have to be dissolved in hot water to make the appropriate strength solutions using the manufacturer's instructions. The solutions then need to be cooled before they can be used.

On the day of autoradiogram development, pour the solutions into separate trays for films (or staining dishes for coverslips, see above) and then perform all the developing and fixing in total darkness in a dark-room to obtain the best results.

1. Equilibrate the X-ray cassettes at room temperature for 1 h if they have been stored at −80°C to reduce chemography. If they have been stored at room temperature then go to step 2.
2. Carefully open the cassette in a dark-room and gently lift the film from the sections by holding it from the edges (should be wearing rubber gloves). Exercise caution in handling the film and try not to scratch it.
3. Hold the film at the top and bottom, again by the edges, with the emulsion side up, and carefully lower it into the developing tray containing Kodak D-19 developer (previously made according to the manufacturer's directions) chilled to 17°C. Ensure that the film is completely covered with the developer and then gently agitate the film for 3 min.
4. Drain excess developer from the film, transfer the film into a tray containing distilled water (22°C), and agitate film for 1 min (stop bath).
5. Drain the water from the film, transfer the film into a tray containing the Kodafix fixer (22°C), and agitate film for 3 min.
6. Drain the fixer from film, transfer the film into a tray containing distilled water (22°C), agitate film for 1–2 min, and then let sit for 5 min. After this time the main lights can be turned on to examine the autoradiograms.
7. The film can now be hung to dry. Care should still be taken to avoid scratching the film since the generation of photographs and image analysis will be adversely affected by the presence of blemishes and scratches on the film.
8. Once the film is dry, carefully identify the various sections with a waterproof pen and then place the film in a transparent plastic sheet-protector with a sheet of white paper as the backing. The film can be held up to the light or placed on a light-box for examination purposes. Subsequently it can be examined and the receptor densities determined in various regions using an image analysis machine.

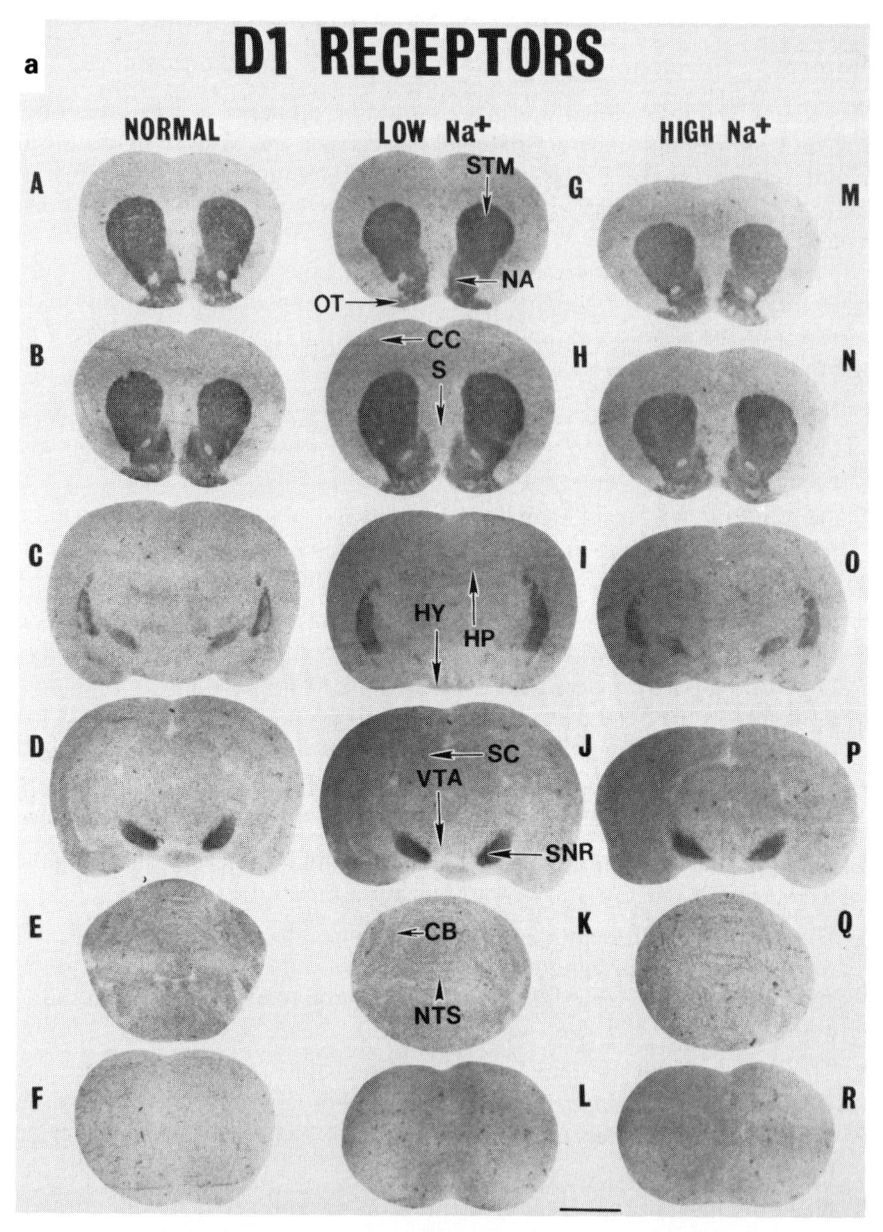

Figure 3. Autoradiography of dopamine (D_1 in **a** and D_2 in **b**) receptor subtypes in rat brain sections is depicted. This study involved chronic manipulation of dietary salt intake and its effects on D_1 and D_2 receptors studied using [^{125}I]SCH23390 and [^{125}I]sulpiride, respectively. The photos are light-field, so receptor localization appears as dark areas. (See refs 128, 129 for more details.)

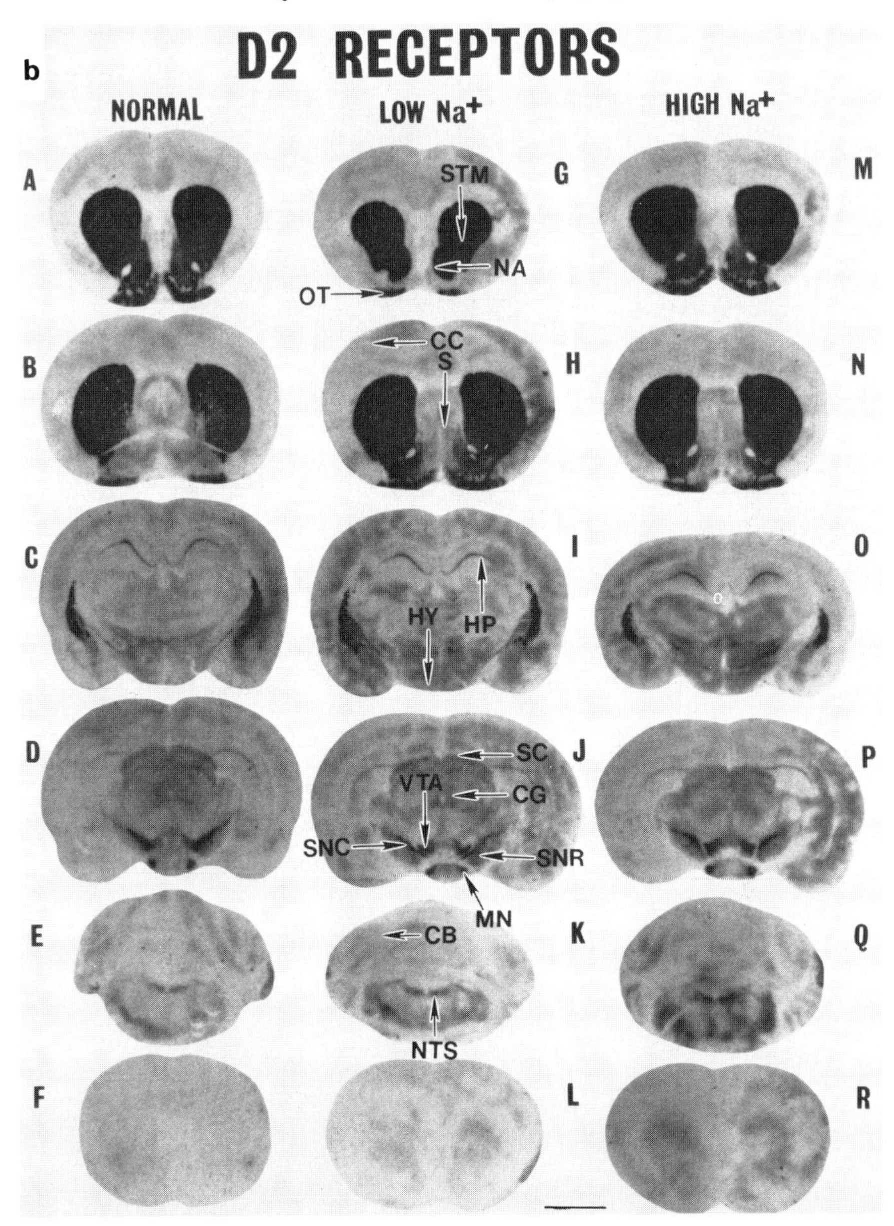
b
D2 RECEPTORS
NORMAL
LOW Na+
HIGH Na+
A
B
C
D
E
F
G
H
I
J
K
L
M
N
O
P
Q
R
STM
NA
OT
CC
S
HY
HP
SC
VTA
CG
SNC
SNR
MN
CB
NTS

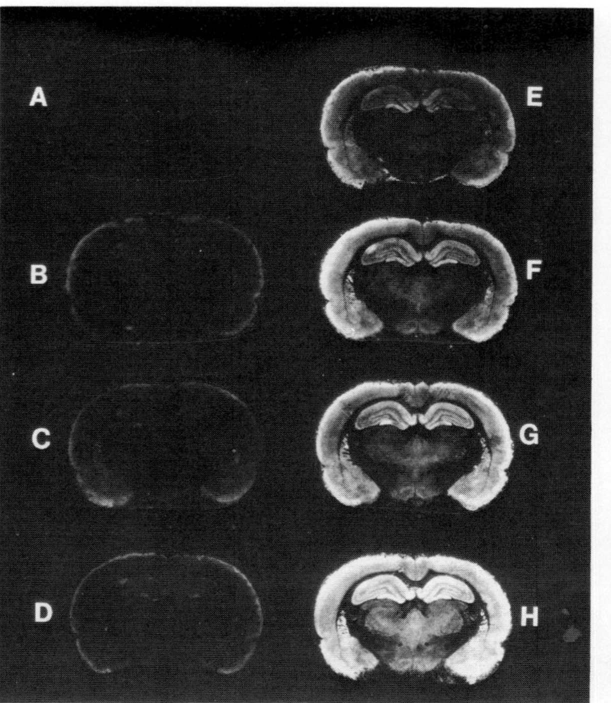

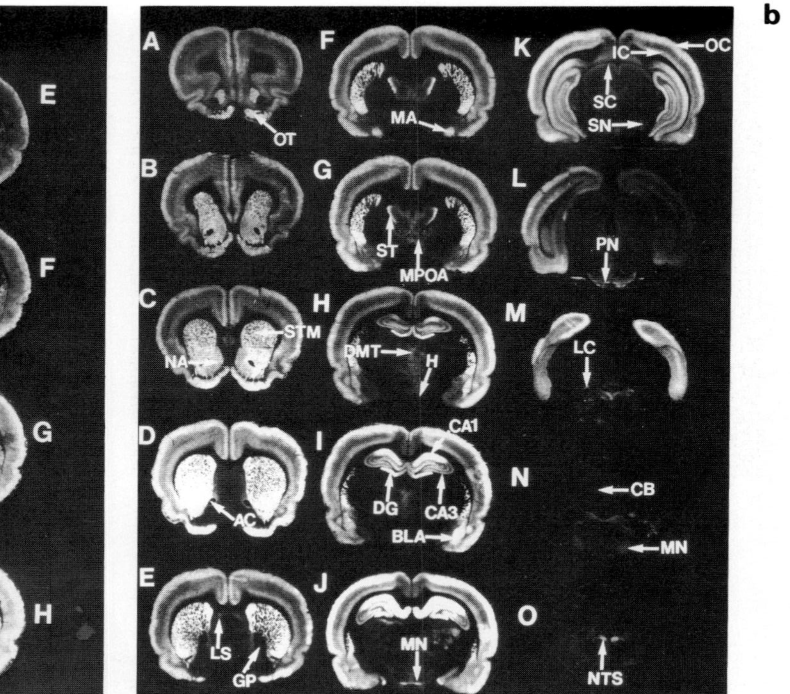

Figure 4. Autoradiography of [^{3}H]4-DAMP binding to M_3/M_1 receptors in adjacent rat brain sections at the hippocampal level. Part **a** shows a saturation experiment starting with low radioligand concentration in section A to the highest concentration in section H. Part **b** shows [^{3}H]4-DAMP (0.6 nM) binding to coronal rat brain sections starting from the forebrain (section A) going through to midbrain and to the hindbrain (sections K–O). Total binding is shown. The level of non-specific binding was close to background and is not shown. AC = anterior commissure; BLA = basolateral nucleus of amygdala; CA1 and CA3 = regions of hippocampus; CB = cerebellum; DG = dentate gyrus of hippocampus; DMT = dorsomedial thalamus; GP = globus pallidus; H = hypothalamus; IC = inner cortex (layers V–VI); LS = lateral

septum; MA = medial nucleus of amygdala; MN = mammillary nucleus; M = motor nucleus; MPOA = medial preoptic area; NA = nucleus accumbens; LC = locus coeruleus; NTS = nucleus of solitary tract; OC = outer cortex (layers I–II); OT = olfactory tubercle; SC = superior colliculus; SN = substantia nigra; ST = stria terminalis; STM = striatum. Bar = 2.76 mm. Part **c** depicts the brain distribution of [^{3}H]pirenzepine binding to rat M_1 muscarinic receptors. See part **b** for identification of brain regions. Part **d** shows the autoradiography of M_1, M_2, and M_3 receptors in coronal sections of the guinea pig ileum. White areas represent the position of receptors. T = total binding (0.5 nM [^{3}H]NMS), N = non-specific binding (with 5 μM atropine); M_1 sites were labelled with 25 nM [^{3}H]pirenzepine; M_2 sites labelled with 0.5 nM [^{3}H]NMS in the presence of 30 nM and 3 nM unlabelled pirenzepine and 4-DAMP respectively; M_3 sites were labelled with 0.8 nM [^{3}H]4-DAMP in the presence of 10 nM pirenzepine. Much of M_1 (11 ± 2% of total), M_2 (64% of total), and M_3 (14 ± 3% of total) sites were localized to the outer longitudinal smooth muscle and the myenteric plexus. (See refs 125, 127 for details.)

c

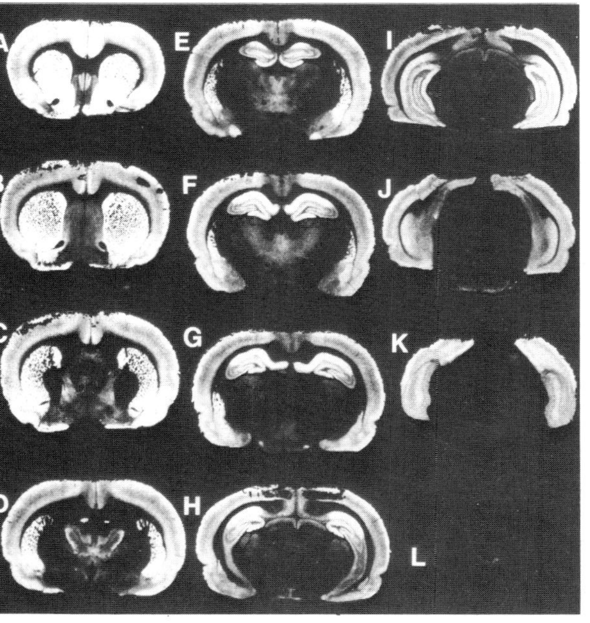

d

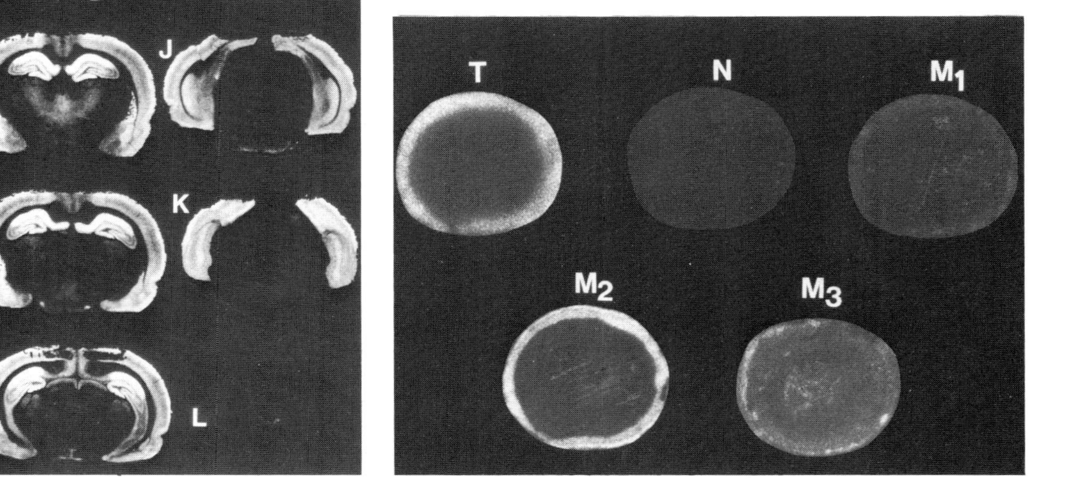

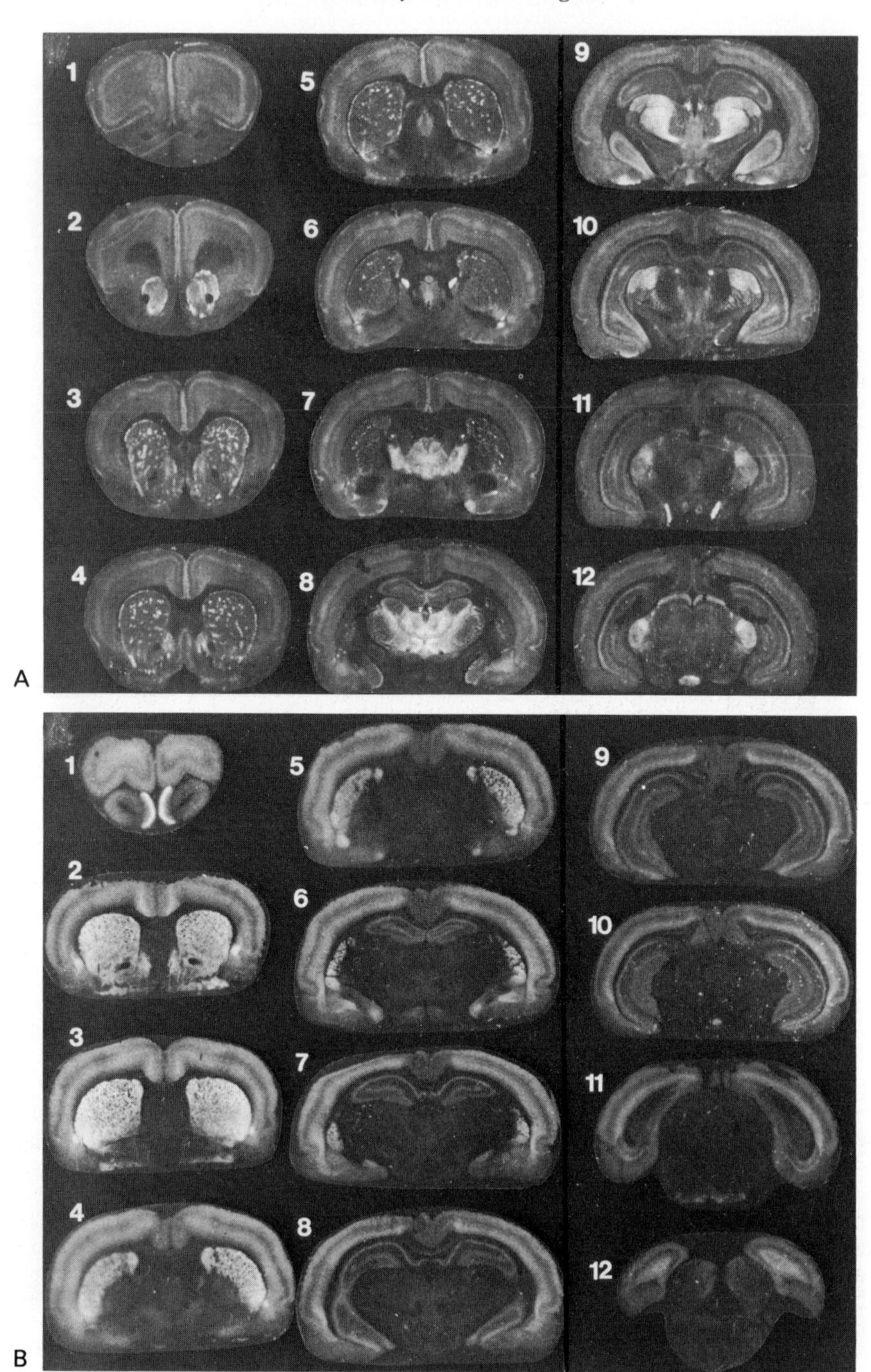
1
5
9
2
6
10
3
7
11
4
8
12
A
1
5
9
2
6
10
3
7
11
4
8
12
B

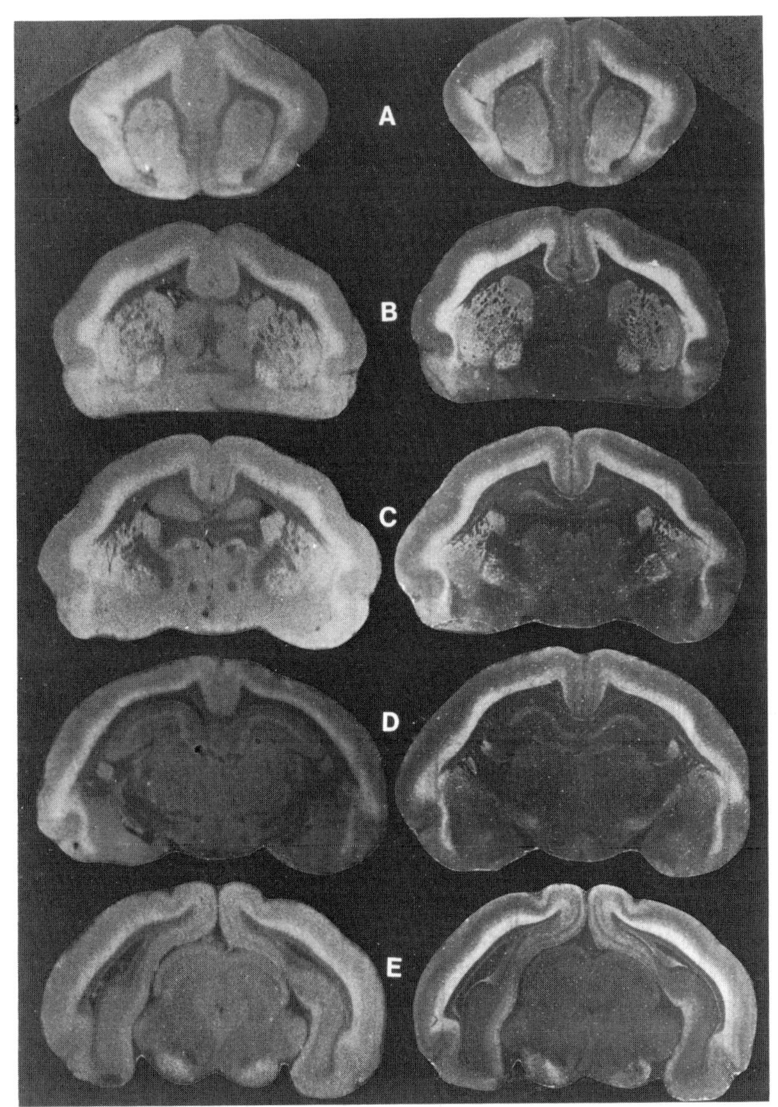

Figure 5. Autoradiography of mu opioid receptors in coronal sections of rat brain using [^{3}H]DAMGO as the peptide radioligand is shown in panel **A**. The brain regional distribution of delta opioid receptors in rat brain is depicted in panel **B** using [^{3}H]DPDPE. In panel **C**, the localization of kappa receptors in guinea pig brain is shown using the non-peptide kappa agonist, [^{3}H]PD117302 (sections on left), and the peptide ligand, [^{3}H]dynorphin (1–8) (sections on right). Note the similarity of the receptor labelling patterns. Panel **D** shows the regional distribution of mu (sections 1–7) and delta (sections 8–11) opioid receptors in horizontal rat brain sections—note the dissimilar localization pattern for these receptors. Panel **E** depicts autoradiography of the mu receptors in parasagittal rat brain sections. (Reproduced with permission from Sharif and Hughes (ref. 9), © 1989 Pergamon Press Ltd, and Sharif and Hughes (ref. 61).)

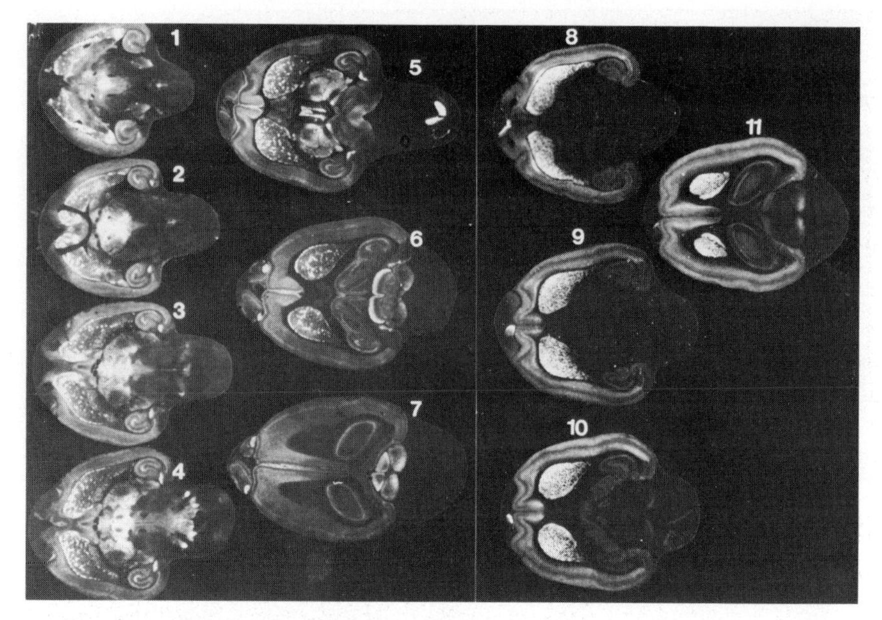

D

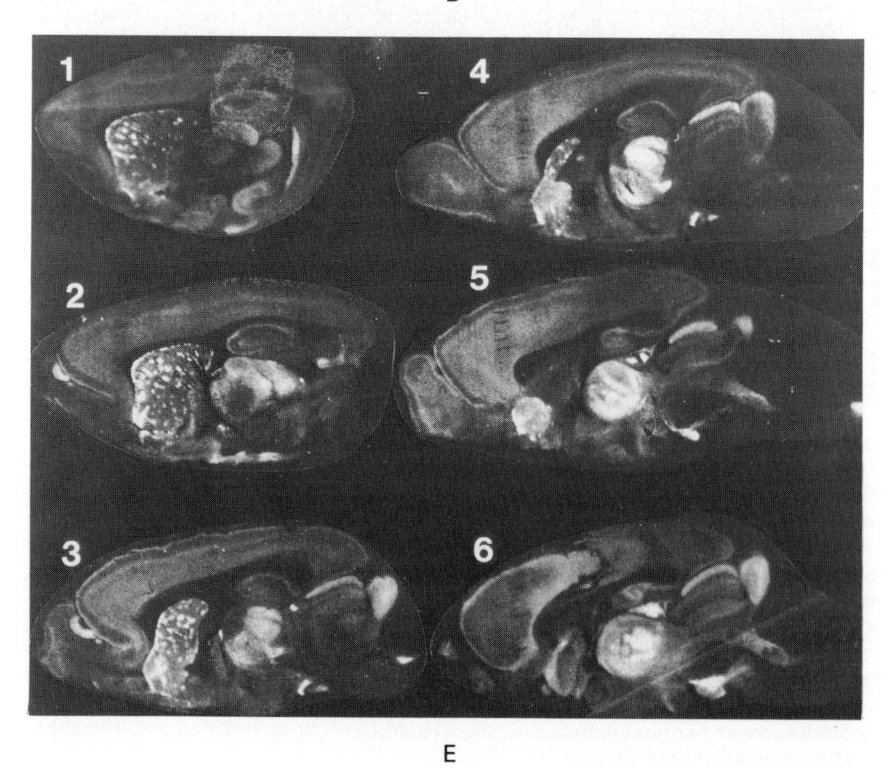

E

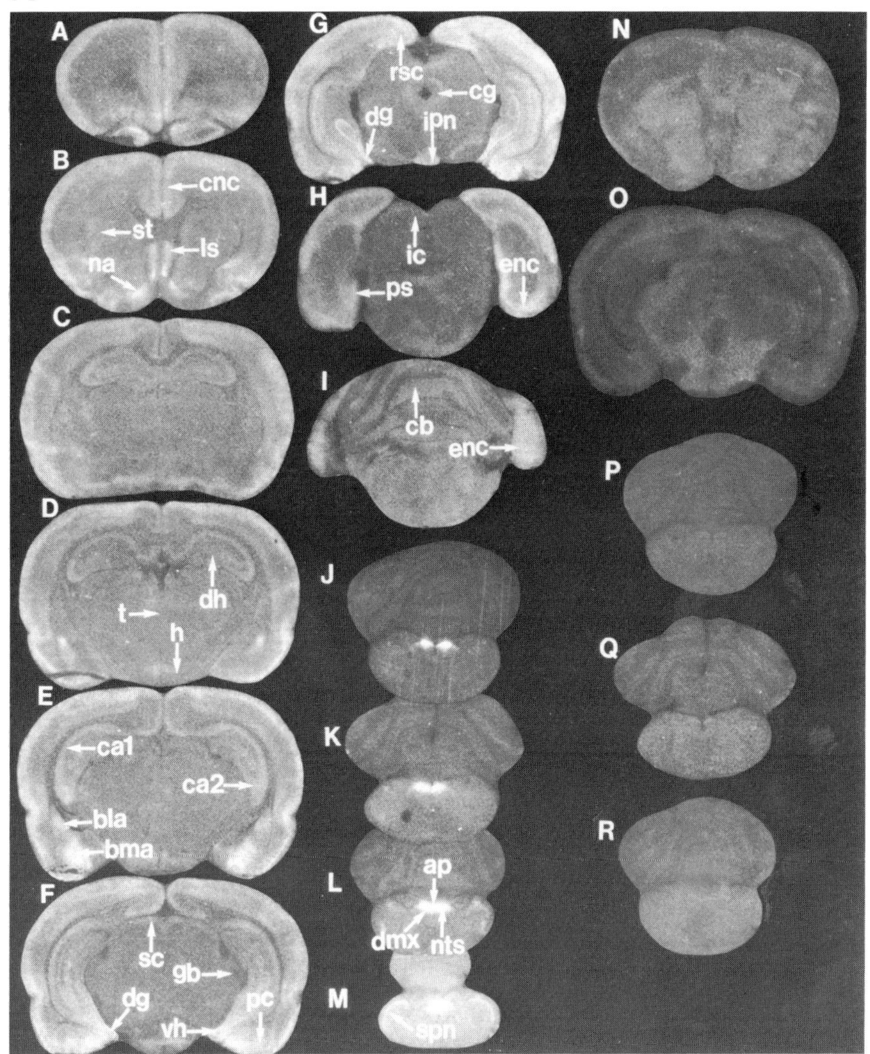

Figure 6. Autoradiography of $5HT_3$ receptors in the CNS of different species using a variety of radioligands. Panel **A** shows the localization pattern in coronal rat brain sections using [^{3}H]quipazine (0.5 nM containing 1 μM paroxetine to prevent interaction at the 5HT transport sites). Sections A–M (rostral to caudal) show total binding while sections N–R are representative sections showing the non-specific binding defined with 1 μM (s)zacopride. Panel **B** shows $5HT_3$ receptors in the brainstem of the rat, ferret, and dog using [^{3}H]GR65630 (top) and [^{3}H]quipazine (bottom). The unlabelled sections under [^{3}H]quipazine in the dog brainstem sections represent binding of [^{3}H]RS-42358. Panel **C** shows a comparison of $5HT_3$ site profiles in rat brain sections using [^{3}H]quipazine (sections A–C), [^{3}H]GR65630 (sections D–F), and [^{3}H]RS-42358 (sections G–I). Panel **D** shows $5HT_3$ receptors in coronal sections of rat brain; panel **E** shows $5HT_3$ receptors in brainstem sections of rat, guinea pig, ferret, and cat brains using [^{3}H]RS-42358 (see *Tables 12–15* and ref. 50 for details).

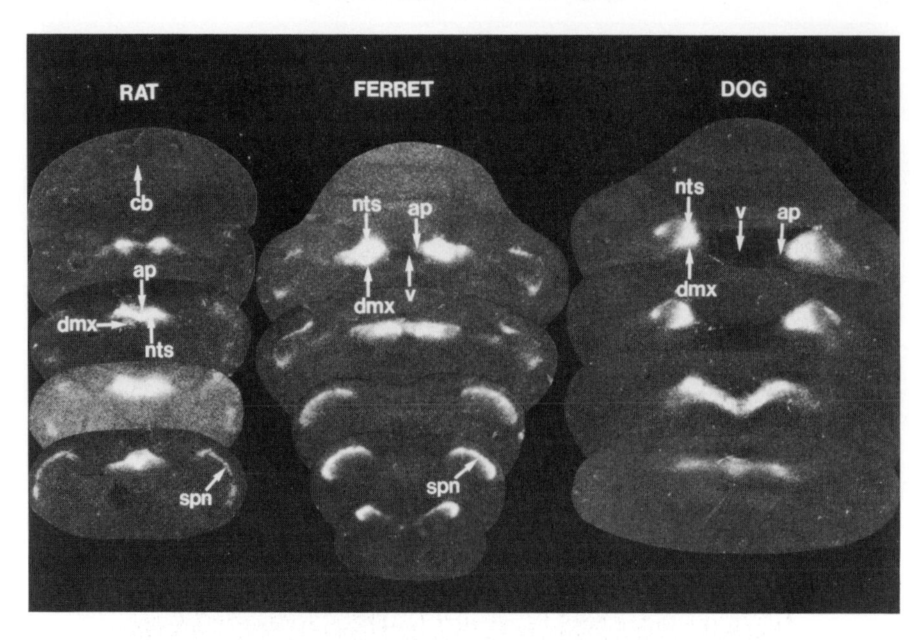
RAT
FERRET
DOG
cb
ap
dmx
nts
spn
nts
ap
dmx
v
spn
nts
v
ap
dmx

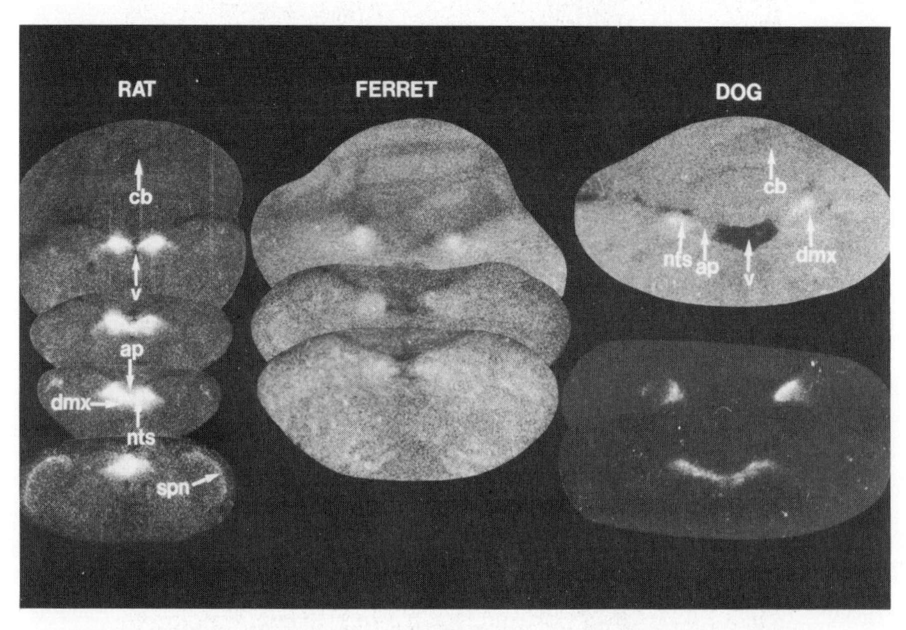
RAT
FERRET
DOG
cb
v
ap
dmx
nts
spn
cb
nts
ap
v
dmx

B

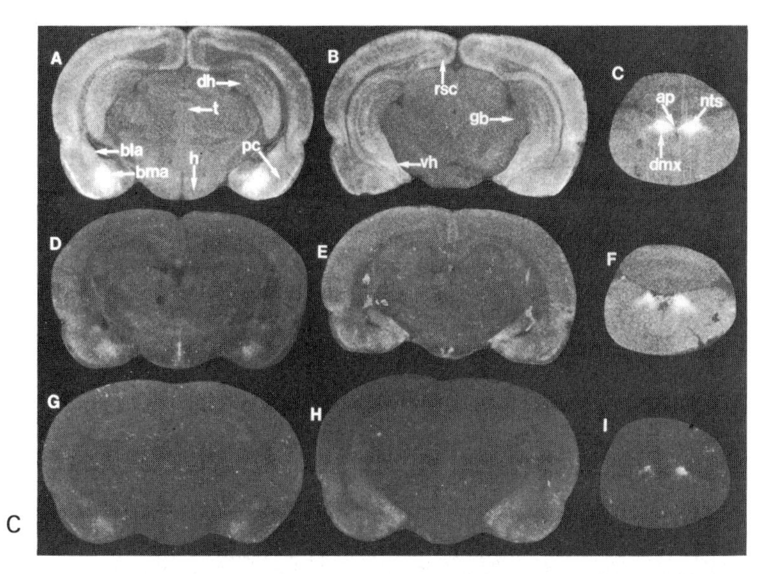
A
dh
t
bla
bma
h
pc
B
rsc
gb
vh
C
ap
nts
dmx
D
E
F
G
H
I

C

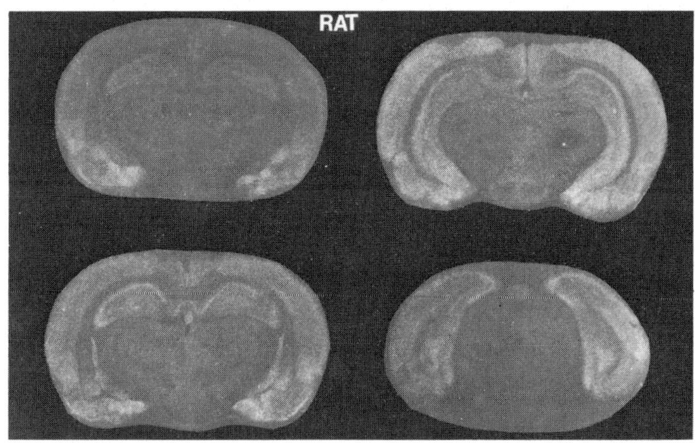
RAT

D

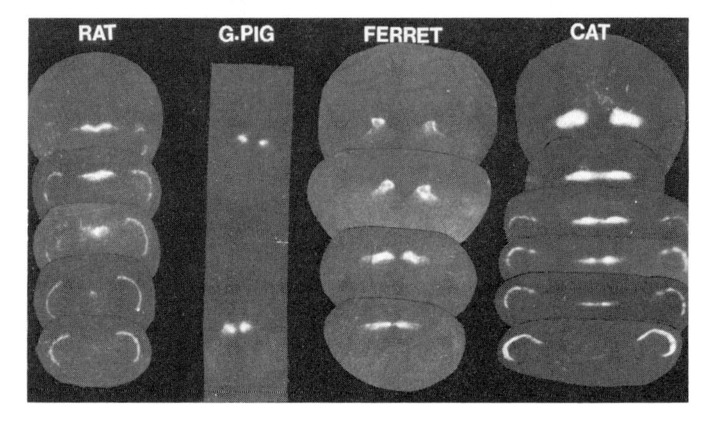
RAT
G.PIG
FERRET
CAT

E

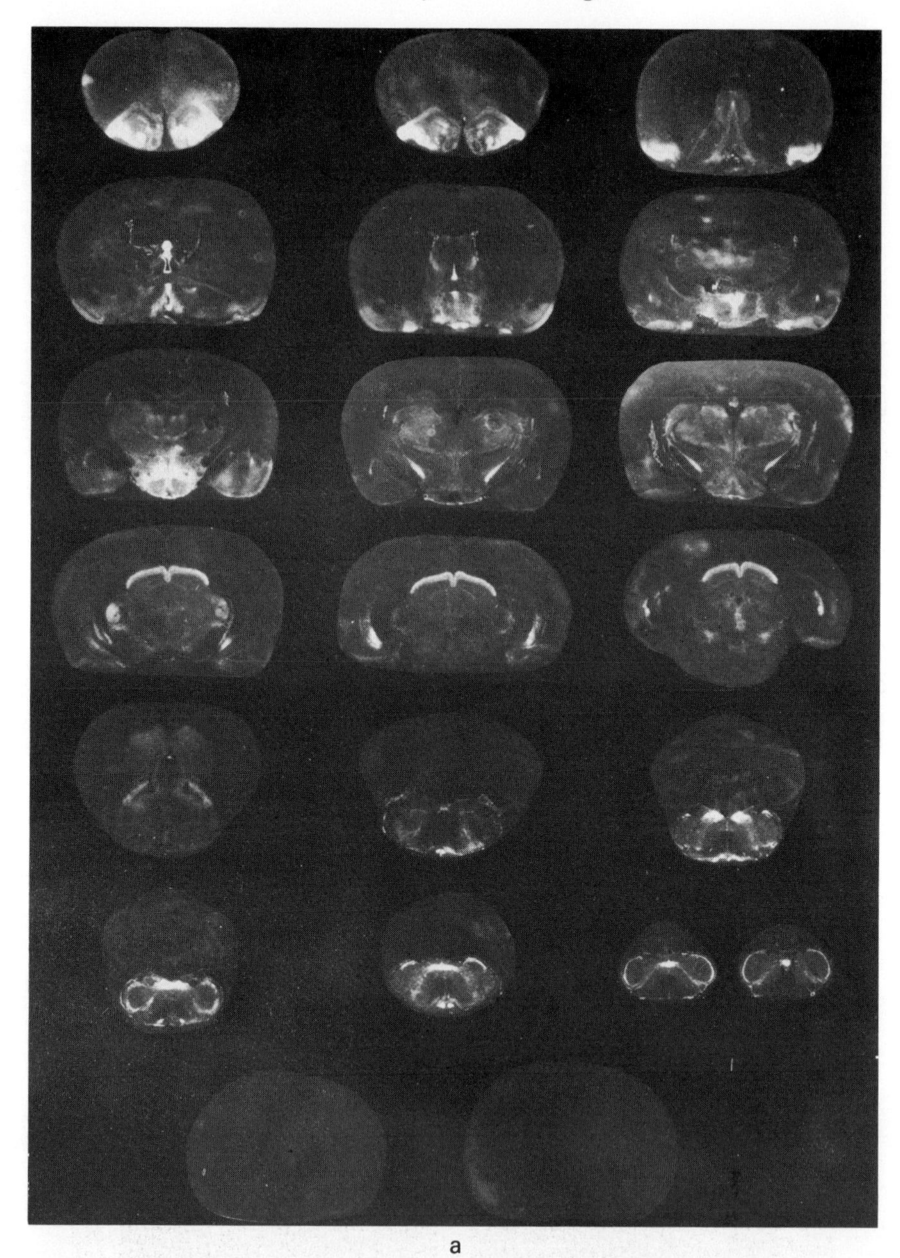

a

Figure 7. Autoradiography of angiotensin II (AII) receptor subtypes in rat central and peripheral tissues using [^{125}I]Sar^1-Ile^8-AII (SARILE) as the radioligand. Panel **a** shows the relative distribution of AII sites in coronal sections of rat brain sections (rostral to caudal sections are shown left to right, top to bottom; non-specific binding is shown in the last two sections at the bottom, almost like background) in dark field format.

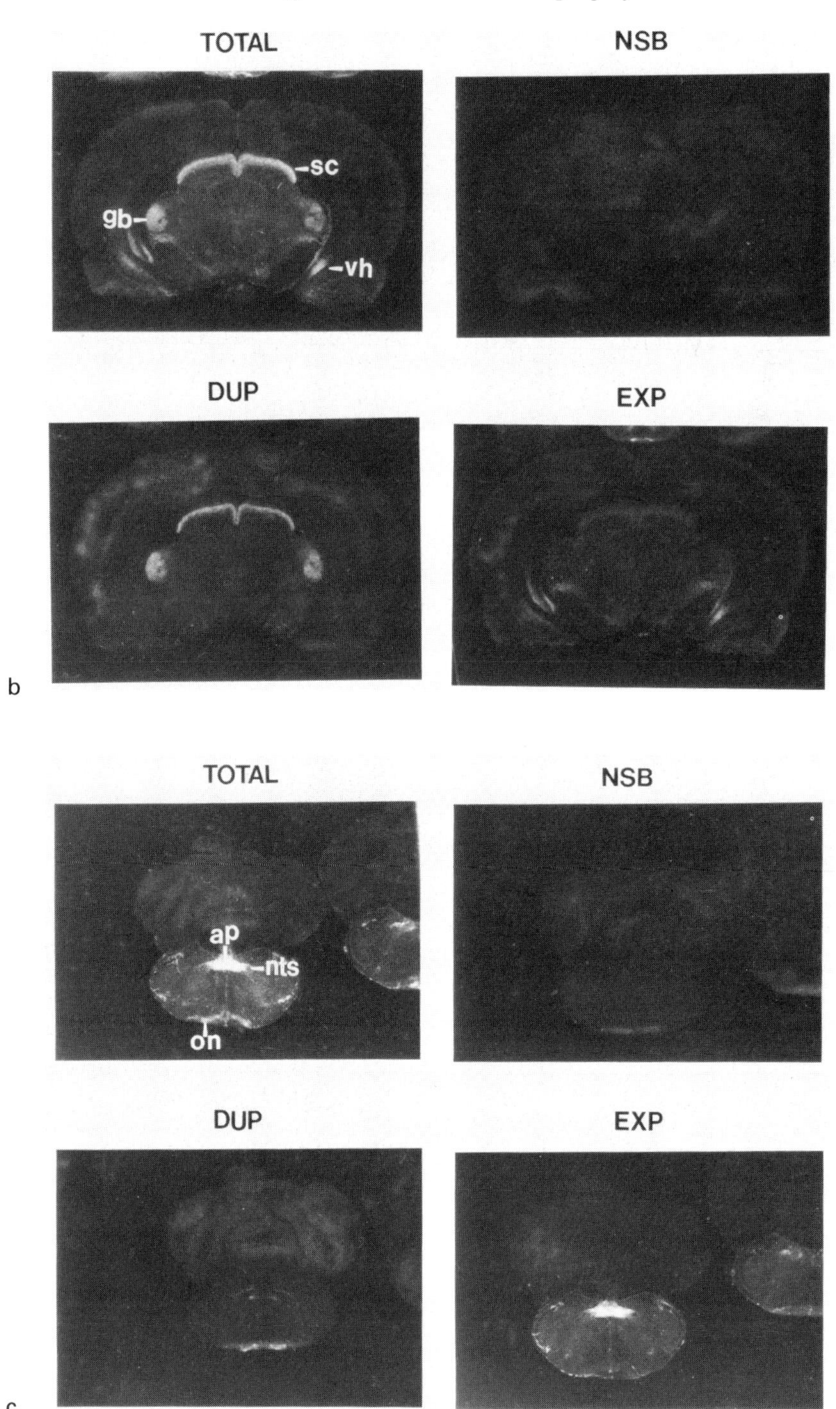
TOTAL
NSB
sc
gb
vh
DUP
EXP
b
TOTAL
NSB
ap
nts
on
DUP
EXP
c

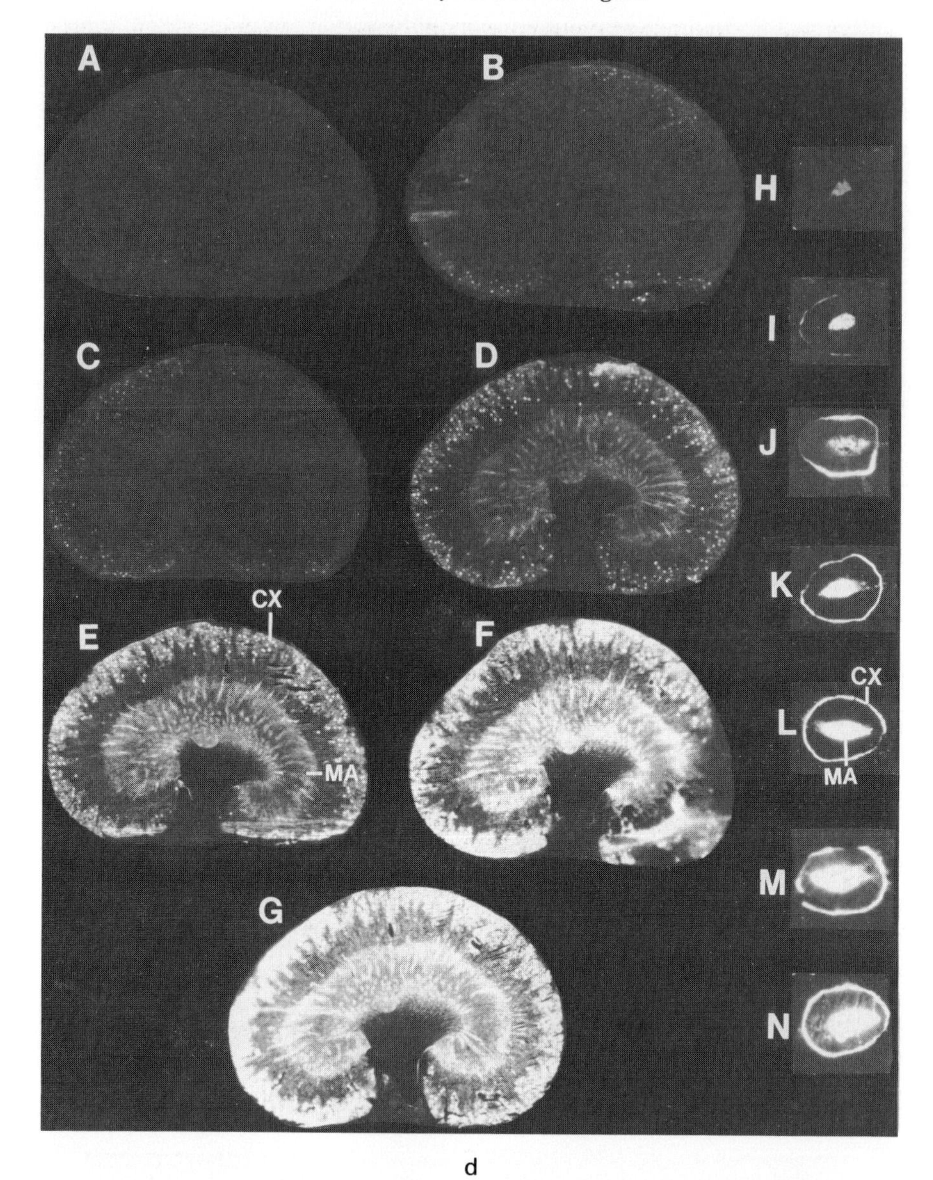

d

Fig. 7 (*cont.*) Panel **b** depicts the total, non-specific (NSB), and AT_1 (defined in the presence of 10 μM EXP655) and AT_2 (defined in the presence of 10 μM DUP753) All receptor subtypes in rat mid-brain and hind-brain (panel **c**). (See *Tables 23–24* for quantitative data from such studies.) Panel **d** shows a typical saturation experiment for [^{125}I]SARILE binding to adjacent sections of rat kidney (sections A–G) and rat adrenal glands (sections H–N) using 25, 50,75, 125, 250, 500, and 750 pM radioligand, respectively. Only total binding is shown, the non-specific binding was relatively low (10–15% of total) and is not shown.

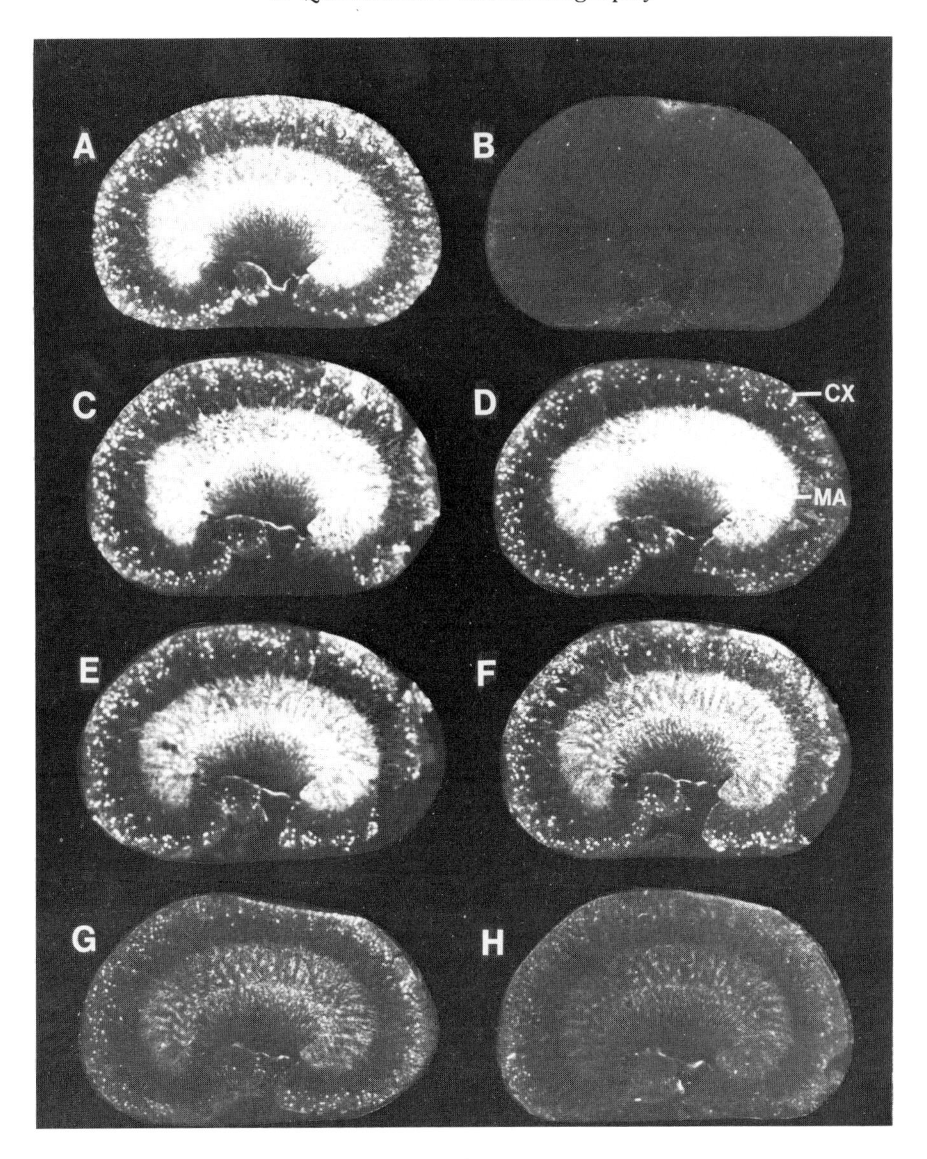

e

Fig. 7 (*cont.*) Panel **e** depicts a typical competition experiment. Sections A and B represent total and NSB binding of [^{125}I]SARILE, respectively, while sections C–H show displacement of [^{125}I]SARILE by 1 nM, 10 nM, 30 nM, 100 nM, 1 μM, and 10 μM DUP753 (AT_1-selective non-peptide antagonist). Data from saturation and competition experiments analysed by image analysis are shown in *Tables 20–22*. CX = cortex; MA = medulla; gb = geniculate body; sc = superior colliculus; vh = ventral hippocampus; ap = area postrema; nts = nuc. solitary tract; on = olivary nucleus (see refs 58, 126 for more details).

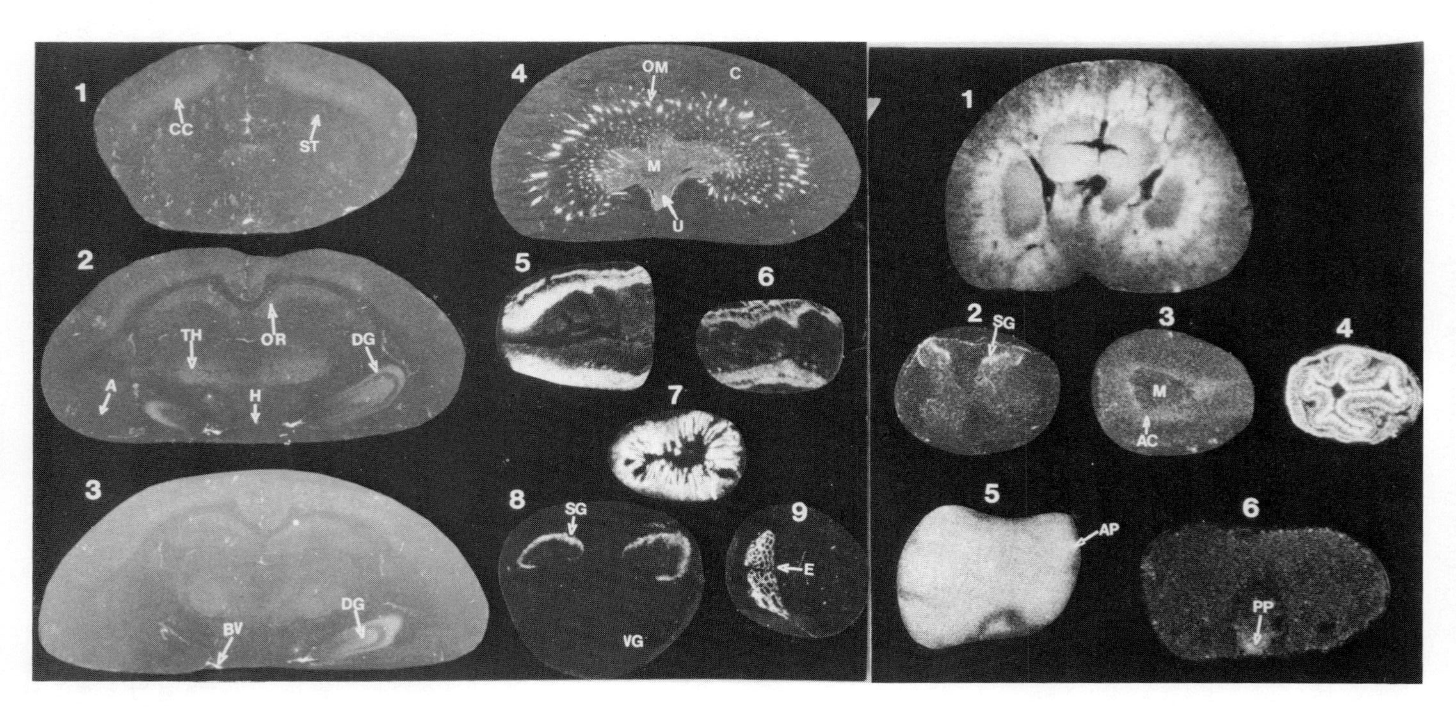

Figure 8. Autoradiography of bradykinin receptors in guinea pig tissues using [^{3}H]bradykinin (left panel, sections 1–9) and using [^{125}I]bradykinin (right panel, sections 1–6). In the left panel, sections 1–3 are coronal brain sections, and sections 4–9 are kidney, uterus, fallopian tubes, duodenum, spinal cord, and epididymus, respectively. Right panel shows sections of kidney, spinal cord, adrenal, ileum, and pituitary in sections 1–5; section 6 shows kappa opioid receptors in the pituitary for comparison with the bradykinin receptors shown in section 5. (Reproduced with permission from Sharif and Hughes (ref. 61).)

5. Quantification of autoradiograms

Autoradiography used to be qualitative in nature until the early 1980s when the first computerized image analysis machines became available. Today, autoradiography is almost totally quantitative, so much so that many of the editors of leading journals expect quantitative data in papers describing autoradiography of receptors or other proteins! Consequently, several image analysis systems are commercially available (e.g. Quantimet 970 (Leica, IL), MCID (Ontario, Canada), RAAS 3000 (Amersham Inc., IL), IBAS (Zeis Kontron, Munich, Germany), and SeeScan (Cambridge, UK)). Although we have used the Quantimet 970 system in our studies, I will not recommend any particular one in order to avoid the inevitable bias, but will outline some of the key features of a good system, and also describe the basic principles and uses of these machines.

5.1 Major components of an image analysis system

The computerized imaging system should possess the following key components:

- a sensitive video camera/scanner (e.g. Newvicon)
- a video/image processor that can interface with a central computer and the scanner
- a central computer interfacing with the video processor, and the monitors, and being capable of controlling the brightness of the lamps/light-box for illuminating the autoradiogram
- monochrome and colour monitors

The main steps in quantification of film-based autoradiograms are described in *Protocol 8*.

Protocol 8. Quantitative image analysis of receptor autoradiograms

1. Switch on the imaging system and place the film on the illuminated macroviewer stage. The images should be illuminated from underneath.
2. Initiate the computer program for checking uneven background illumination, ensuring that the computer-controlled lights are adjusted accordingly and that any shading correction that may be necessary is performed automatically. The detected image should appear on the monitors free from shading, and geometric and photometric distortion.
3. Calibrate the scanner with Kodak filter(s) such that the system can recognize total darkness and full brightness (as of film background) and such that each pixel of the autoradiogram can be assigned an optical density (OD) value on a 0–255 grey level scale using a look-up table in the computer memory.

Protocol 8. *Continued*

4. Now calibrate the scanner and the imaging system with the images of the radiation standards. If data are to be presented as moles radioligand/mm^2, then use a computer routine to calibrate the scanner with a Vernier scale in terms of millimeters. For the standards, use the digitizing tablet ('mouse') to select a uniform area of the radiation standard, get the scanner and computer to digitize the selected area, and assign the radioactivity (moles)/section value for each of the standards in turn.
5. When all the standards have been entered into the system memory, instruct the computer to generate a graph of the ln radioactivity *versus* ln radiation values (*Figure 1*), determine the best fit to the data points, and display this linearized graph along with the correlation coefficient for the best fit. If the correlation coefficient is not >0.95, repeat the calibration of the standards until this is improved upon. Likewise if a mistake was made in selecting the appropriate standard(s) from the film or if an incorrect radiation value was assigned, repeat the calibration as necessary.
6. Now commence analysing the receptor autoradiograms by quantifying the receptor densities in defined sub-regions of the tissue sections using the digitizing tablet. Use the total-binding image for selecting the structures of interest. Use the 'edge-enhancement' routine to highlight the whole of the total-binding section and then retain the outline of this section on the monitor screen.
7. Align and superimpose the non-specific binding image on the outline of the total-binding section image, instruct the computer to subtract the former image from the latter (digital subtraction), and print the data (total, non-specific, and specific binding values) (e.g. *Tables 7, 8, 10–15, 17–19, 23*) for each of the selected structures from the various tissue sections. The names of the structures should be entered after each selected structure.
8. Use specific computer routines to quantify competition and saturation experimental images. If the computer system is capable of direct analysis of the data to yield IC_{50} or K_i (drug concentration producing 50% of the maximal inhibition) (e.g. *Tables 16* and *21*), K_d (radioligand dissociation constant) (e.g. *Tables 9* and *20*) and B_{max} (maximal receptor density) values, then proceed with this kind of curve-fitting program. Alternatively, collect the receptor density values from the image analyser and then use specific computer programs (for example, LIGAND or EBDA) to determine the constants mentioned above.
9. Display selected tissue sections or region(s) on the colour monitor and take a photograph directly from the screen as needed (see book cover or ref. 61).

Table 7. Quantitative autoradiographic distribution of [^{3}H]prazosin binding to α_1-receptors in the rat kidney after treatment with chloroethylclonidine (CEC)

Tissue	Specific [^{3}H]prazosin binding (amol/mm^2)		% decrease in α_{1b}
	Controls	CEC-treated	
Whole kidney	1229 ± 260	351 ± 32 [a]	−71%
Outer cortex	2393 ± 439	704 ± 34 [a]	−71%
Inner cortex	588 ± 92	207 ± 42 [a]	−65%
Outer medulla	855 ± 158	155 ± 82 [a]	−82%
Inner medulla	611 ± 156	196 ± 109 [a]	−68%
Vasa recta	1677 ± 258	Not detected [a]	−100%

Data are means ± SEM (n = 3–4 rats per group).

[a] $P < 0.001$ relative to controls by unpaired Student's t-test. No change in angiotensin II or atrial natriuretic factor binding was observed in the CEC-treated kidney regions relative to controls (see *Figure 2*). These data demonstrate the usefulness of selective inactivation techniques in receptor autoradiography (CEC inactivates α_{1b} receptors) (see ref. 15).

Table 8. Quantitative autoradiographic localization of α_2-adrenoceptors in rat brain using [^{3}H]RS-15385-197

Region	Specific binding (amol/mm^2)
Olfactory system	
Claustrum	2646 ± 226
Olfactory tubercle	2804 ± 308
Anterior olfactory nucleus	3517 ± 249
Septal area	
Lat. septum	3701 ± 696
Medial septum	3763 ± 978
Corpus striatum	
Nucleus accumbens	1841 ± 214
Striatum, head	1663 ± 130
Striatum, body	1212 ± 109
Striatum, tail	3086 ± 282
G. pallidus	399 ± 45
V. palladium	2769 ± 342
Endopiriform nucleus	2781 ± 316
Cortical regions	
Frontal cortex	1565 ± 220
Angular insular cortex	3030 ± 318
Entorhinal cortex	3578 ± 677
Parietal cortex	3319 ± 278
Piriform cortex	4236 ± 276
C. cortex layer 1	2348 ± 195
C. cortex layers 2–3	1861 ± 126
C. cortex layer 4	2164 ± 161
C. cortex layers 5–6	1121 ± 97

Table 8. *contd.*

Region	Specific binding (amol/mm^2)
Hypothalamic regions	
Stria terminalis	2304 ± 387
Med. preoptic area	2095 ± 257
Supraoptic nucleus	2089 ± 446
Dor. med. hypothalamus	3319 ± 277
Ven. med. hypothalamus	2615 ± 508
V. lat. hypothalamus	2277 ± 104
Mammilliary nucleus	3216 ± 791
Arcuate nucleus	2240 ± 366
Amygdaloid nuclei	
Amygdala, basolateral nucleus	4757 ± 411
Amygdala, basomedial nucleus	5106 ± 433
Amygdala, central nucleus	4491 ± 218
Amygdala, cortical nucleus	5218 ± 497
Amygdala, medial nucleus	3344 ± 381
Thalamic nuclei	
Dor. med. thalamic nucleus	2919 ± 297
Vent. lateral thalamic nucleus	421 ± 44
Ventral med. thalamic nucleus	3529 ± 592
Periventricular thalamic nucleus	3058 ± 178
Hippocampal regions	
Hippocampus molecular layer	1686 ± 111
Pyramidal CA_1 layer	496 ± 79
Pyramidal CA_2 layer	657 ± 67
Pyramidal CA_3 layer	565 ± 99
Mid-hindbrain	
Dor. lat. geniculate body	2962 ± 288
Dor. med. geniculate body	1139 ± 219
Superior colliculus	2813 ± 511
Inferior colliculus	1923 ± 203
Central grey	2612 ± 304
Substantia nigra, compacta	2309 ± 480
Ventral tegmental area	2296 ± 668
Dorsal tegmental nucleus	1574 ± 122
Pons/medulla regions	
Locus coeruleus	2961 ± 293
Solitary nucleus (NTS)	4118 ± 341
Area postrema	2294 ± 597
Nucleus of spinal trigeminal nerve	1601 ± 209
Med. vestibular nucleus	1511 ± 319
Facial nucleus	950 ± 86
Dorsal medial vagal nucleus	2760 ± 194
Parabrachial nucleus	3098 ± 210
Cerebellum, molecular layer	442 ± 234
Cerebellum, granular layer	1287 ± 234
White matter	
Internal capsule	190 ± 56
Corpus callosum	154 ± 42

Data are means ± SEM from three to four rats. These data demonstrate the high resolution of the quantitative autoradiography technique (see ref. 21).

Table 9. Equilibrium binding parameters for [^{3}H]4-DAMP binding to M_3/M_1 receptors determined by quantitative autoradiography

Species/tissue	K_d (nM)	B_{max} (amol/mm^2)
Rat brain		
Hypothalamus	0.42 ± 0.06	25 ± 4
Dorsolateral thalamus	0.39 ± 0.05	32 ± 4
Lateral septum	0.42 ± 0.07	36 ± 5
Striatum	0.45 ± 0.08	69 ± 12
Dentate gyrus (U.L.)	0.40 ± 0.06	80 ± 11
CA regions (hippocampus)	0.41 ± 0.07	82 ± 10
Dentate gyrus (L.L.)	0.40 ± 0.06	96 ± 10
Cortex (outer)	0.31 ± 0.04	100 ± 16
Guinea pig brain		
Hypothalamus	0.51 ± 0.07	30 ± 5
Dorsolateral thalamus	0.50 ± 0.11	31 ± 4
Dentate gyrus (U.L.)	0.42 ± 0.06	90 ± 11
Dentate gyrus (L.L.)	0.42 ± 0.08	92 ± 11
Cortex (outer)	0.34 ± 0.08	93 ± 8

Data are means ± SEM from four to five animals per species.
L.L. = lower limb; U.L. = upper limb. These data demonstrate how quantitative autoradiography can be used to obtain kinetic and equilibrium binding parameter estimates for receptor–radioligand interactions. (see Ref. 125).

6. Conclusions

As can be seen from the above discourse, the techniques and procedures for conducting and analysing quantitative autoradiography are not very difficult. With practice and some patience (especially while cutting sections!) excellent results can be obtained. Some examples of quantitative receptor and enzyme mapping are depicted in *Tables 7–23* and *Figures 2–12*. Of course, many more such images are present in the literature cited in this and other chapters in this book.

Acknowledgements

The following colleagues are thanked for their contributions and support over the years: Dr R. L. Whiting, Dr E. Wong, Dr D. Clarke, and Z. To, J. Nunes, and A. Shieh.

a

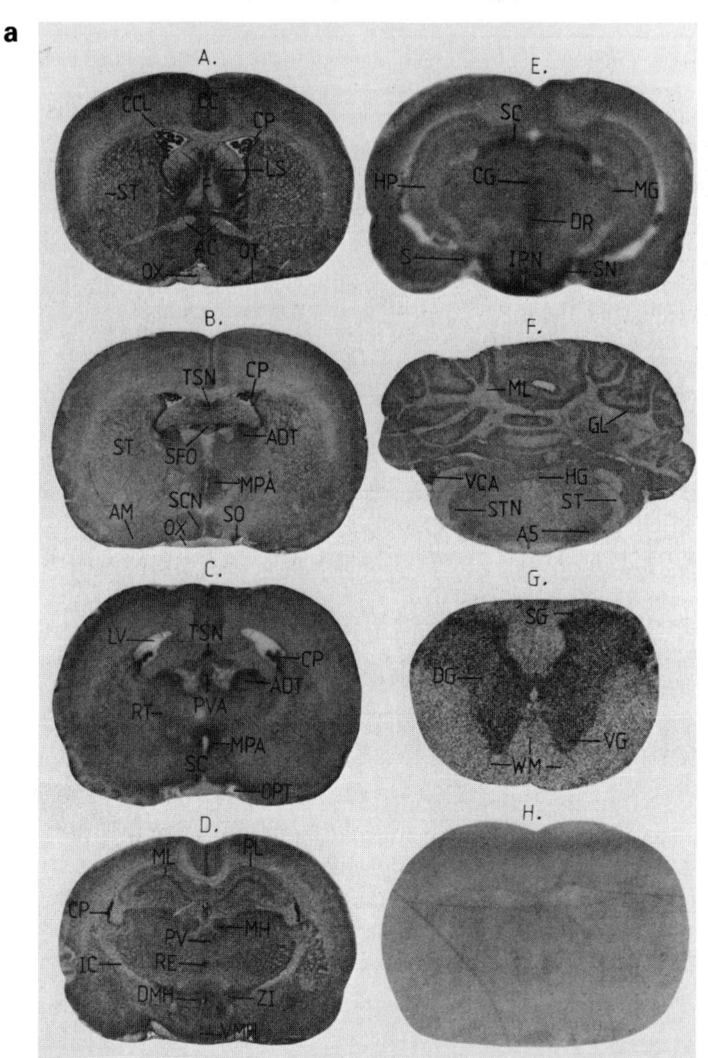

Figure 9. Endothelin (ET) receptor autoradiography is depicted in rodent brain and peripheral tissues. Panel **a** shows the localization pattern in rat brain (dark areas represent the location of the receptors). Section A: AC = anterior commissure, CC = cerebral cortex, CCL = corpus callosum, CP = choroid plexus, F = fornix, LS = lateral septum, OT = olfactory tubercle, OX = optic tract, ST = striatum. Section B: ADT = anterior dorsal thalamic nucleus, AM = amygdala, MPA = medial preoptic area, SCN = suprachiasmatic nucleus, SO = supraoptic nucleus, SFO = subfornical nucleus, TSN = triangular septal nucleus. Section C: LV = lateral ventricle, OPT = optic tract, PVA = anterior paraventricular thalamic nucleus, RT = reticular thalamic nucleus. Section D: DMH = dorsal medial hypothalamic nucleus, IC = internal capsule, MH = medial habenula, ML = molecular hippocampal layer, PL = pyramidal hippocampal layer, PV = paraventricular nucleus, RE = reuniens thalamic nucleus, VMH = ventral medial nucleus, ZI = zona incerta. Section E: CG = central grey, DR = dorsal raphé, HP = hippocampus, MG = medial geniculate body,

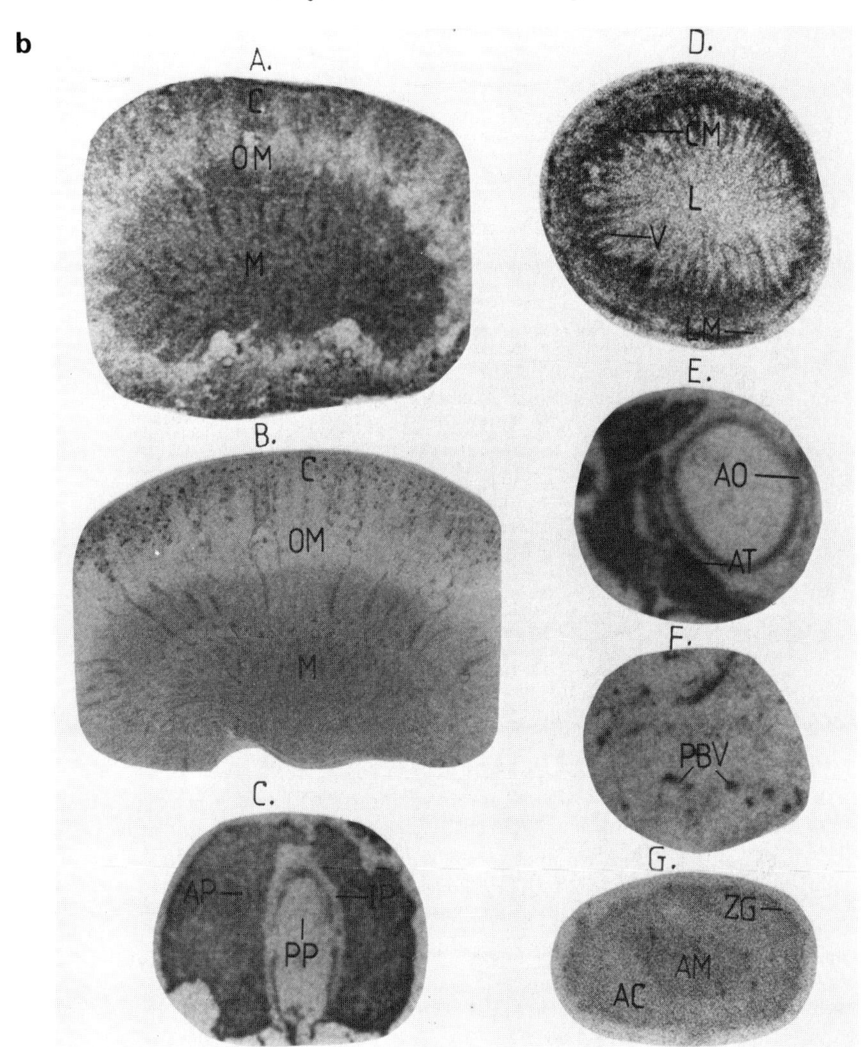

IPN = interpeduncular nucleus, S = subiculum, SC = superior colliculus, SN = substantia nigra. Section F: A5 = A5 noradrenaline cell nuclei, HG = hypoglossal nucleus, GL = cerebellum granular layer, ML = cerebellar molecular layer, ST = spinal tract, STN = spinal trigeminal nucleus, VCA = vestibulocochlear nucleus. Section G: DG = dorsal grey, SG = substantial gelatinosa, VG = ventral grey, WM = white matter. Section H: represents non-specific binding and is an adjacent section to that of Section B. Panel **b** (above) shows [^{125}I]ET binding sites in peripheral tissues. Section A, mouse kidney: C = cortex, M = medulla, OM = outer medulla. Section B, rat kidney: abbreviation as for Section A. Section C, guinea pig pituitary: AP = anterior pituitary, IP = intermediate lobe of pituitary, PP = posterior pituitary. Section D, guinea pig ileum: CM = circular muscle, L = lumen, LM = longitudinal muscle, V = villi. Section E, rat heart: AO = aorta, AT = atrium. Section F, guinea pig pancreas: PBV = pancreatic blood vessels. Section G, rat adrenal: AC = adrenal cortex, AM = adrenal medulla, ZG = zona granulosa (see refs 69, 70).

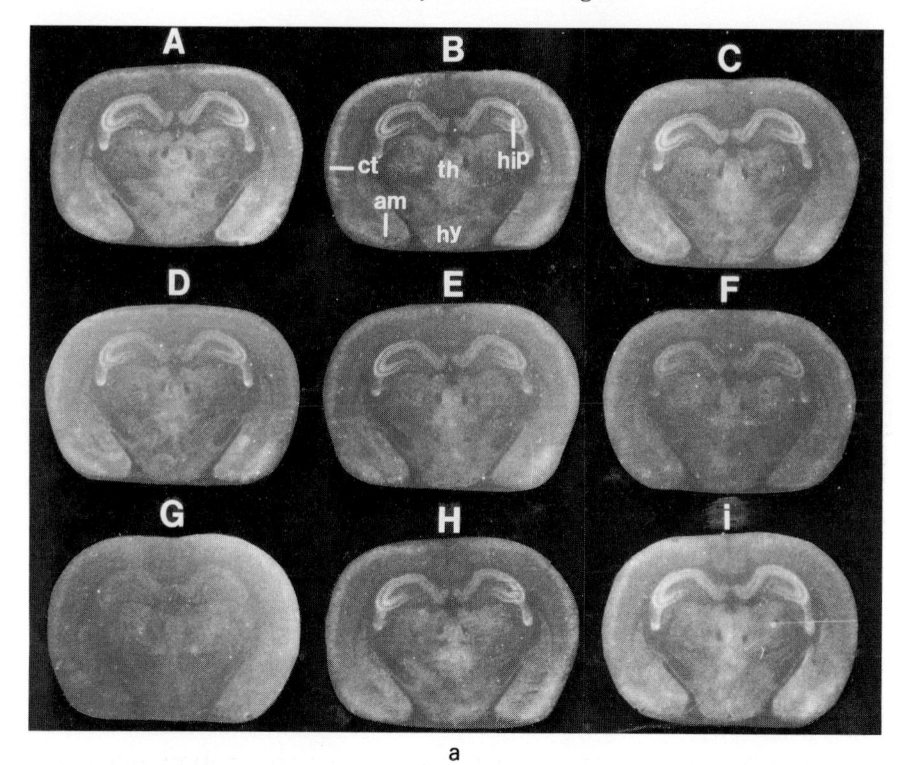

a

Figure 10. Autoradiography of PYY receptors in the brain. Panel **a** shows the autoradiographic visualization of the effects of PYY, and two $5HT_3$ receptor antagonists, ondansetron and zacopride, on [^{125}I]PYY binding in rat brain sections. Section A represents total binding. Increasing concentrations of unlabelled PYY were used to demonstrate the reversible nature of [^{125}I]PYY binding. The concentrations of unlabelled compounds were: section B, 0.01 nM; C, 0.1 nM; D, 0.3 nM; E, 1 nM; F, 3 nM; G, 10 nM. Sections H and I depict the lack of effect of 100 μM ondansetron (H) and 100 μM (s)zacopride (I) on [^{125}I]PYY binding. am = amygdala; ct = cortex; hip = hippocampus; hy = hypothalamus; th = thalamus. Panel **b**: Autoradiographic localization of $5HT_3$ receptors in the dog brainstem and PYY receptors in the cat brainstem. $5HT_3$ receptors in the dog brainstem were labelled with 0.34 nM [^{3}H]GR65630 in the absence (A) or presence (B) of 1 μM (s)zacopride to define total and non-specific binding, respectively. PYY receptors in the cat brainstem were labelled with 25 pM [^{125}I]PYY in the absence (C) or presence (D) of 1 μM PPY to define total and non-specific binding respectively. White areas depict the localization sites of the receptors. ap = area postrema; cb = cerebellum; dms = dorsal motor nucleus of vagus; nts = nucleus of the solitary tract; spn = nucleus of the spinal trigeminal tract; v = fourth ventricle. Panel **c**: Autoradiographic localization of $5HT_3$ and PYY receptors in the brainstem of ferrets. $5HT_3$ receptors were labelled with [^{3}H]GR65630 (0.34 nM) in the absence (A) or presence (B) of 1 μM (S)zacopride to define total and non-specific binding, respectively. PYY receptors were labelled with 25 pM [^{125}I]PYY in the absence (C) or presence (D) of 1 μM PYY to define total and non-specific binding, respectively. White areas depict the sites of the receptors. ap = area postrema; cb = cerebellum; dmx = dorsal motor nucleus of vagus; nts = nucleus of solitary tract; spn = nucleus of spinal trigeminal tract; v = fourth ventricle.

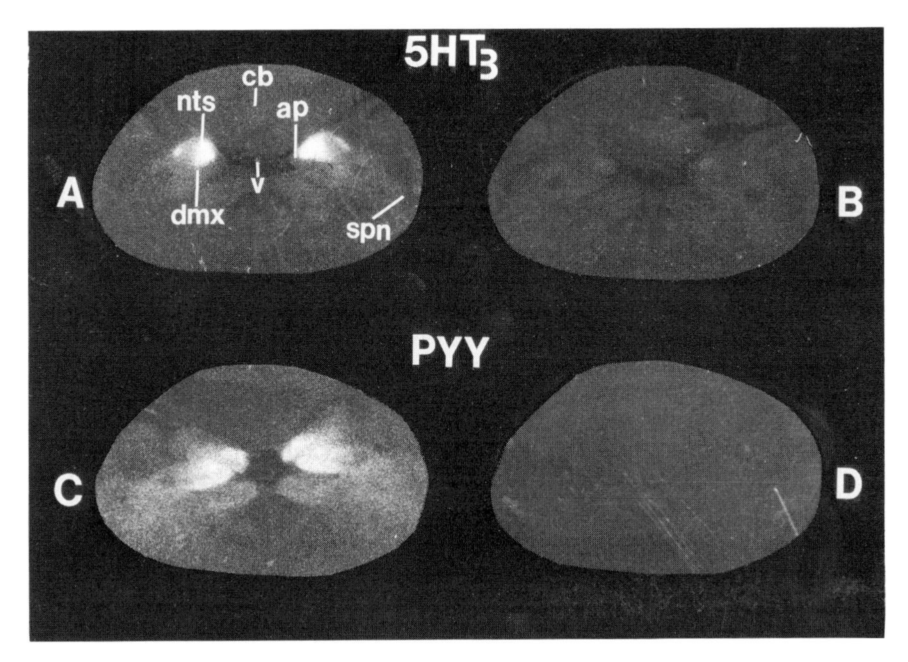
5HT3
cb
nts
ap
A
v
dmx
spn
B
PYY
C
D

b

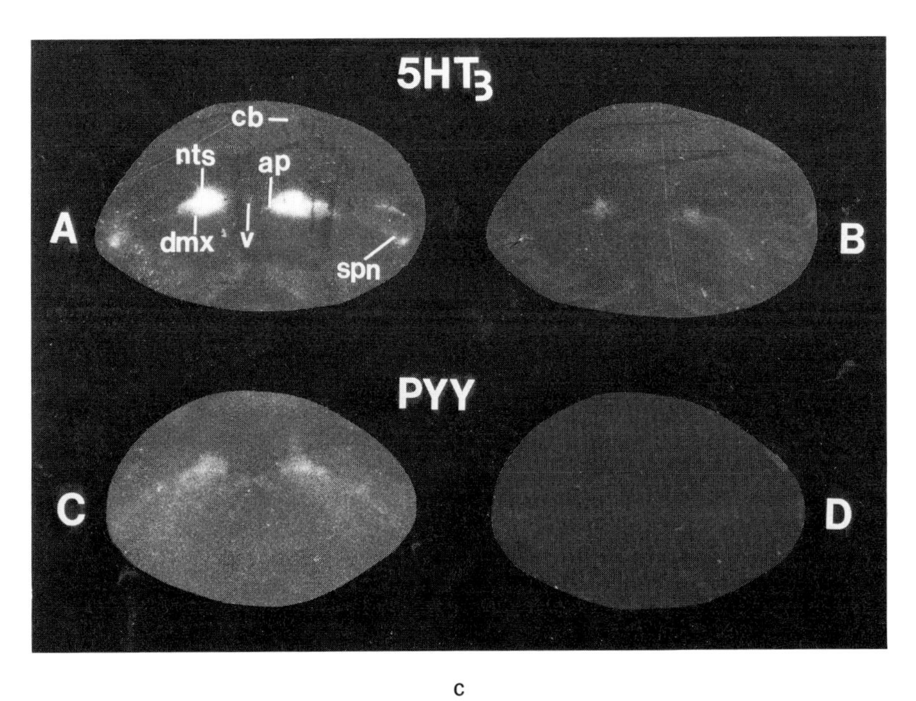
5HT3
cb
nts
ap
A
dmx
v
spn
B
PYY
C
D

c

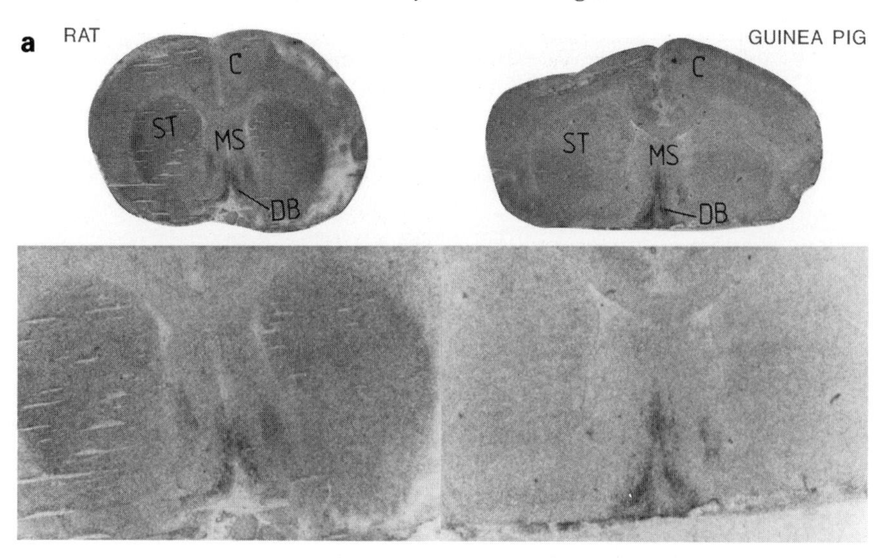

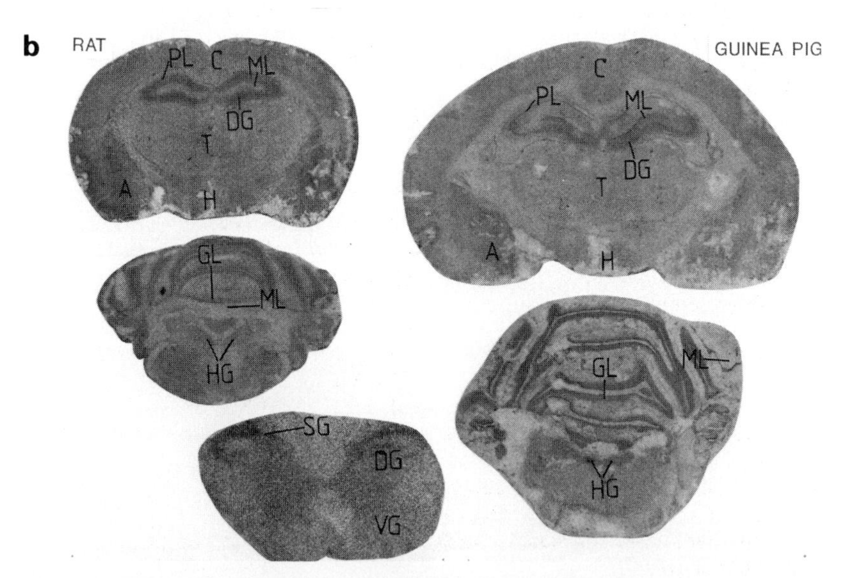

Figure 11. Bright-field photos of nerve growth factor (NGF) receptor autoradiography in rodent brain and spinal cord. Panel **a** shows high densities in the rat and guinea pig diagonal band of Broca (DB), mediolateral septum (MS), and cortex (C). Low power images and high power images are shown at the top and bottom, respectively. Panel **b** shows NGF receptor localization in coronal sections of rat and guinea pig mid- and hind-brain. Heavy labelling can be seen in the hippocampal formation (dentate gyrus (DG), CA of pyramidal cell layer (PL), cortex (C), amygdala (A), granular layer of cerebellum (GL), and hypoglossal nucleus (HG)). The small section at the bottom left depicts the guinea pig spinal cord with a high density of NGF sites in the substantia gelatinosa (SG) and dorsal grey (DG). The concentration of [^{125}I]NGF was 25 pM (see ref. 114).

Table 10. Quantitative autoradiographic distribution of specific [^{3}H]pirenzepine binding sites (M_1) in rat and guinea pig brain

	Specific binding (fmol/mm^2)	
Brain region	**Rat brain**	**Guinea pig brain**
Olfactory tubercle	18.7 ± 0.7	18.3 ± 1.3
Accessory olfactory nucleus	21.2 ± 1.0	ND
Nucleus accumbens	16.4 ± 0.8	20.8 ± 1.9
Endopiriform nucleus	18.1 ± 0.9	17.2 ± 1.8
Lateral septum	4.3 ± 0.6	4.1 ± 0.9
Striatum, head	21.0 ± 1.1	23.4 ± 1.8
Striatum, body	14.5 ± 0.5	13.6 ± 1.1
Striatum, tail	24.9 ± 0.5	23.2 ± 1.2
Globus pallidus	0.9 ± 0.1	1.5 ± 0.2
Stria terminalis	3.3 ± 0.3	3.9 ± 0.3
Medial preoptic area	1.9 ± 0.3	2.3 ± 0.4
Ventral palladium	2.5 ± 0.2	4.1 ± 0.6
Substantia innominata	2.8 ± 0.5	4.1 ± 0.5
Horizontal band of Broca	1.9 ± 0.4	1.8 ± 2.0
Basolateral amygdala	21.3 ± 1.0	20.5 ± 2.3
Cortical amygdala	13.4 ± 2.4	17.9 ± 2.6
Medial amygdala	9.1 ± 0.6	15.8 ± 2.9
Mammilliary nucleus	5.6 ± 0.5	5.3 ± 0.9
Cortex, outer	19.0 ± 1.3	21.8 ± 2.0
Cortex, mid	13.6 ± 0.5	14.6 ± 1.5
Cortex, inner	13.9 ± 0.5	15.2 ± 1.6
Hypothalamus	1.5 ± 0.3	5.4 ± 1.4
Arcuate nucleus	2.2 ± 0.2	ND
Reuniens nucleus	3.3 ± 0.6	3.6 ± 0.7
Rhomboid nucleus	4.9 ± 1.1	4.6 ± 0.9
Dorsomedial thalamus	4.4 ± 0.6	4.1 ± 0.6
Dorsolateral thalamus	3.8 ± 0.2	2.9 ± 0.6
Ventrolateral thalamus	2.4 ± 0.3	2.0 ± 0.3
CA_1 hippocampus	26.9 ± 0.8	25.6 ± 3.1
CA_2 hippocampus	12.4 ± 1.2	11.9 ± 2.1
CA_3 hippocampus	13.7 ± 0.7	11.4 ± 1.7
Mol. hippocampus	24.6 ± 0.9	25.9 ± 3.1
Dentate gyrus	25.1 ± 1.1	22.0 ± 2.7
Substantia nigra	2.3 ± 0.3	3.3 ± 0.4
Interpeduncular nucleus	1.5 ± 0.2	0.9 ± 0.3
Central grey	1.8 ± 0.4	1.5 ± 0.3
Mediolateral geniculate body	3.4 ± 0.4	3.2 ± 0.5
Superior colliculus	2.8 ± 0.2	2.5 ± 0.5
Inferior colliculus	1.5 ± 0.1	2.0 ± 0.5
Raphé	1.4 ± 0.2	0.8 ± 0.1
Dorsal raphé nucleus	1.2 ± 0.1	ND
Pontine nucleus	2.2 ± 0.2	0.8 ± 0.2
Locus coeruleus	1.8 ± 0.4	1.4 ± 0.4
Parabrachial nucleus	1.4 ± 0.1	1.4 ± 0.2
Nucleus tractus solitarius	2.3 ± 0.1	1.5 ± 0.2
Spinal trigeminal nucleus	0.9 ± 0.2	0.7 ± 0.1

Table 10. *contd.*

Brain region	Specific binding (fmol/mm^2)	
	Rat brain	Guinea pig brain
Motor nucleus	0.8 ± 0.1	0.8 ± 0.2
Olivary nucleus	0.7 ± 0.1	0.8 ± 0.1
Spinal cord, gelatinosa	ND	2.1 ± 0.1
Spinal cord, ventral grey	ND	0.7 ± 0.1

Data are means ± SEM from three to four animals per species. The concentration of [^{3}H]pirenzepine (25 nM) accounted for 61% receptor occupancy. No specific binding was detected in rat and guinea pig hearts or kidneys. Note how some 'selectivity' in regional receptor-labelling can be achieved by using [^{3}H]pirenzepine (above) and [^{3}H]4-DAMP (*Table 11*) as radioligands.
ND, not determined. (See refs 125, 127.)

Table 11. Quantitative autoradiographic distribution of specific [^{3}H]4-DAMP binding sites (M_3/M_1) in rat and guinea pig brain

Brain region	Specific binding (fmol/mm^2)	
	Rat brain	Guinea pig brain
Olfactory tubercle	18.9 ± 3.2	18.1 ± 2.6
Accessory olfactory nucleus	24.9 ± 1.2	19.6 ± 5.1
Nucleus accumbens	14.1 ± 0.9	17.8 ± 5.5
Endopiriform nucleus	18.4 ± 0.1	20.9 ± 1.2
Lateral septum	7.6 ± 0.9	7.8 ± 0.5
Striatum, head	24.4 ± 1.8	30.9 ± 2.0
Striatum, body	12.7 ± 2.2	15.4 ± 1.3
Striatum, tail	23.5 ± 2.2	30.1 ± 1.4
Globus pallidus	0.7 ± 0.1	1.6 ± 0.4
Stria terminalis	7.8 ± 1.3	16.3 ± 2.1
Medial preoptic area	4.3 ± 0.6	5.4 ± 0.5
Ventral palladium	4.1 ± 0.5	6.9 ± 0.9
Substantia innominata	3.4 ± 0.4	5.6 ± 0.9
Basolateral amygdala	14.3 ± 3.0	15.2 ± 0.9
Cortical amygdala	15.1 ± 2.5	16.5 ± 1.0
Medial amygdala	12.7 ± 1.6	11.3 ± 0.7
Mammilliary nucleus	11.9 ± 1.9	4.6 ± 0.5
Horizontal band of Broca	3.0 ± 0.3	3.9 ± 0.5
Cortex, outer	22.5 ± 2.6	22.4 ± 1.7
Cortex, mid	9.7 ± 1.6	12.8 ± 3.5
Cortex, inner	13.2 ± 0.9	16.2 ± 0.9
Hypothalamus	3.3 ± 0.4	7.6 ± 0.8
Arcuate nucleus	2.1 ± 0.4	9.4 ± 0.9
Reuniens nucleus	6.7 ± 0.7	6.5 ± 0.4
Rhomboid nucleus	8.0 ± 0.4	7.7 ± 0.6
Periventricular nucleus	8.5 ± 0.8	4.6 ± 0.6
Dorsomedial thalamus	5.2 ± 0.7	8.0 ± 0.5
Dorsolateral thalamus	5.8 ± 0.6	7.9 ± 0.6

Table 11. *contd.*

Brain region	Specific binding (fmol/mm^2)	
	Rat brain	**Guinea pig brain**
Ventrolateral thalamus	4.7 ± 0.6	3.2 ± 0.1
CA_1 hippocampus	26.0 ± 1.8	25.1 ± 2.9
CA_2 hippocampus	11.7 ± 1.5	16.1 ± 0.7
CA_3 hippocampus	13.2 ± 1.5	16.3 ± 2.2
Molecular hippocampus	22.9 ± 1.9	25.6 ± 3.3
Dentate gyrus	25.5 ± 1.3	24.6 ± 2.6
Substantia nigra	2.6 ± 0.4	4.8 ± 0.3
Interpeduncular nucleus	3.0 ± 0.9	3.2 ± 0.7
Central grey	2.9 ± 0.9	2.9 ± 0.6
Medio-lateral geniculate body	3.2 ± 0.7	6.4 ± 1.3
Superior colliculus	9.2 ± 1.3	7.5 ± 1.8
Inferior colliculus	2.5 ± 0.3	5.3 ± 1.2
Raphé	3.9 ± 0.3	ND
Dorsal raphé nucleus	1.7 ± 0.4	ND
Pontine nucleus	6.0 ± 1.3	7.1 ± 2.5
Locus coeruleus	5.5 ± 0.7	5.1 ± 0.9
Parabrachial nucleus	2.1 ± 0.3	3.7 ± 0.9
Nucleus tractus solitarius	2.4 ± 0.2	3.1 ± 0.8
Spinal trigeminal nucleus	0.6 ± 0.1	1.9 ± 0.5
Motor nucleus	1.3 ± 0.1	1.5 ± 0.2
Olivary nucleus	1.1 ± 0.4	0.8 ± 0.1
Molecular cerebellum	0.2 ± 0.06	1.2 ± 0.3
Granular cerebellum	0.9 ± 0.2	2.1 ± 0.5

Data are means ± SEM from three to four animals per species. The concentration of [^{3}H]4-DAMP (0.6 nM) used accounted for 63% receptor occupancy. ND = not determined. No specific binding was detected in rat and guinea pig hearts or kidneys. [^{3}H]4-DAMP is therefore not selective for M_3 receptors but also labels M_1 receptors (*Table 10*) (see refs 125, 127).

Table 12. Quantitative autoradiographic localization of $5HT_3$ receptors in the brainstem nuclei of three species using [^{3}H]GR65630 as the radioligand

Brain region	Specific binding (amol/mm^2)		
	Rat	**Ferret**	**Dog**
Nuc. solitary tract	8000 ± 551	7167 ± 491	8495 ± 207
Area postrema	2449 ± 221	3176 ± 454	2208 ± 204
Dorsal nucleus of Vagus	3594 ± 238	2436 ± 383	2757 ± 537
Spinal trigeminal nucleus (SPN5)	1238 ± 163	3687 ± 498	ND

Data are means ± SEM from 3–5 animals of each species. Note the high resolution of the quantitative autoradiography technique. ND = not detected. (See ref. 50.)

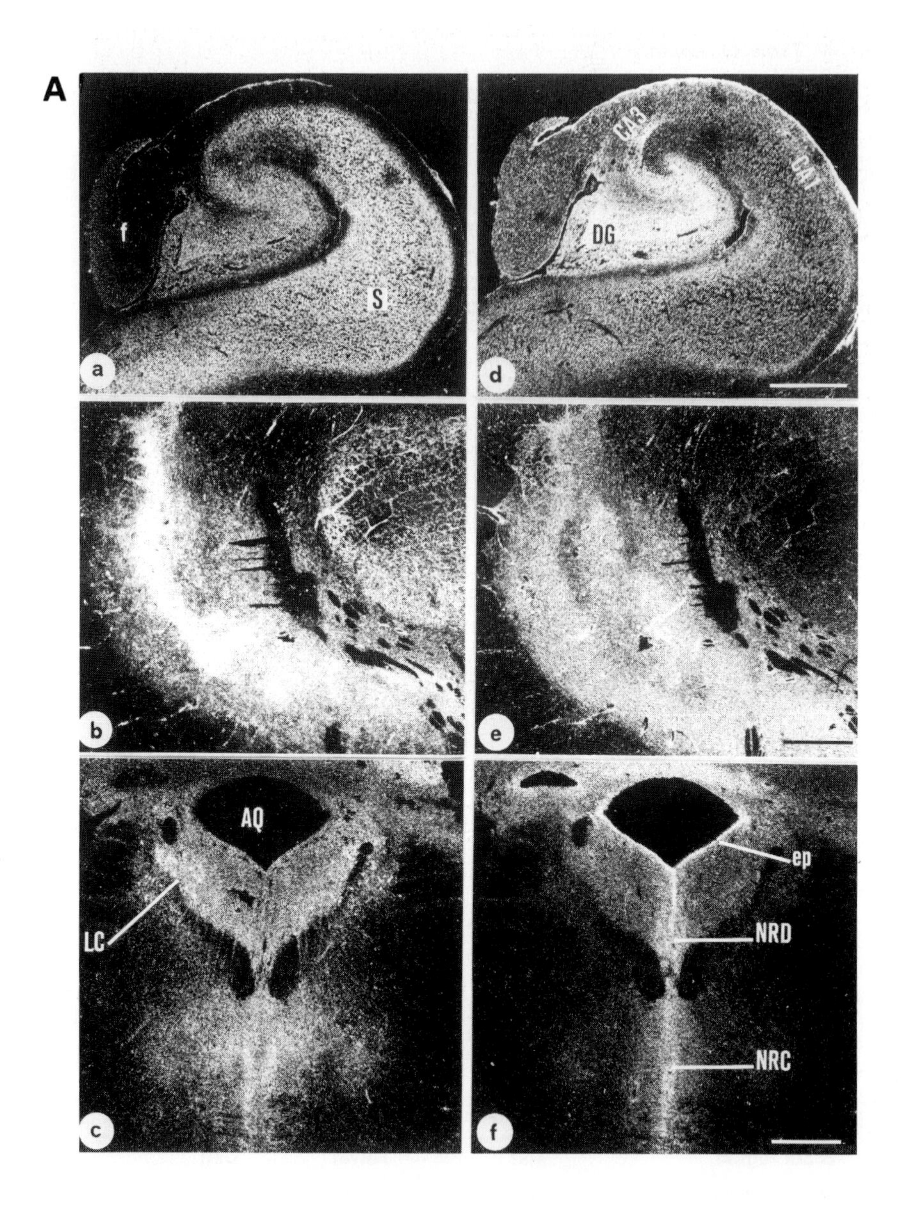
A
f
S
a
CA3
CA1
DG
d
b
e
AQ
LC
c
ep
NRD
NRC
f

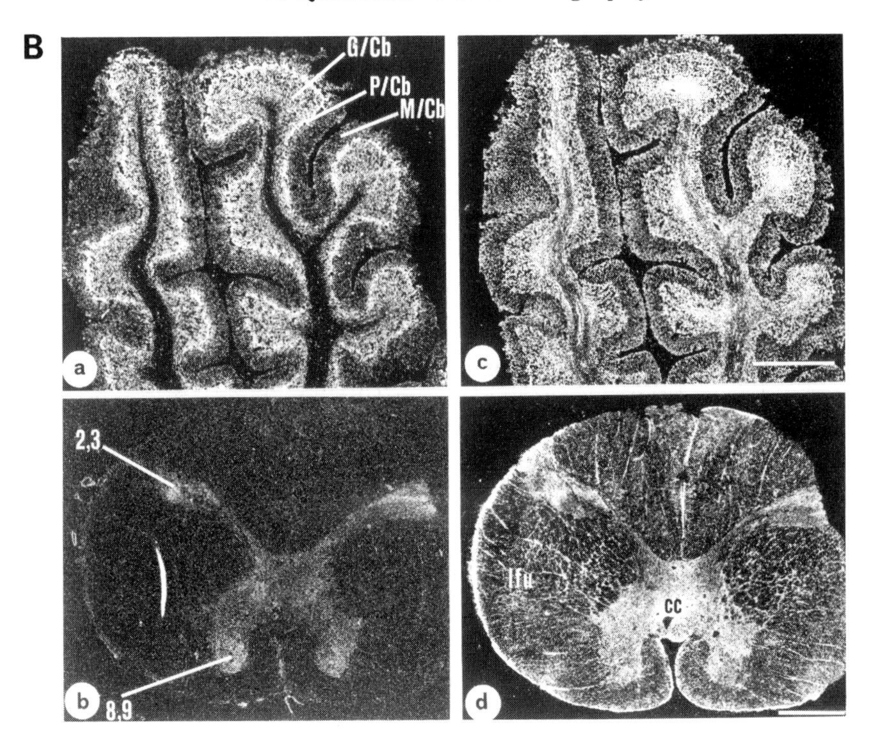

Figure 12. Distribution of MOA-A (sections a–c) and MAO-B (sections d–f) binding sites in adjacent sections of human post-mortem brain (panel A) and spinal cord (panel B). Note: the abundance of MAO-B in the intragranular layer of the hilus in the dentate gyrus (DG) and in the CA3 region of the hippocampus (panel A, section d); the presence of both MAO-A and MAO-B (the former more discrete) in the substantia nigra (SN); the presence of both enzymes (MAO-B more discrete) in the dorsal and central raphé nuclei (NRD, NRC) and the abundance of MAO-B in the ependyma (ep) lining the aqueduct (AQ) and of MAO-A in the locus coeruleus (LC) (panel A, sections c and f); the presence of MAO-A in the cerebellar Purkinje cell layer (P/Cb, panel A, section a) and the presence of both enzymes (MAO-A more discrete) in layers of the dorsal (2, 3) and ventral (8, 9) horns of the spinal cord (panel B, sections b and d). MAO-B is also abundant around the central canal (cc). Further abbreviations: f, fimbria; S, subiculum; CA1, hippocampal region; cp, cerebral peduncle; G/Cb, cerebellar granular layer; M/Cb, cerebellar molecular layer; lfu, lateral funiculus. (Courtesy of Dr J. G. Richards, see ref. 121. Reproduced with permission from Oxford University Press.)

Table 13. Quantitative autoradiography of $5HT_3$ receptors in rat brain regions using three different radioligands

Brain regions	Specific binding (amol/mm^2)		
	[^{3}H]GR65630	[^{3}H]Quipazine	[^{3}H]RS-42358
Nuc. solitary tract	8612 ± 593	19207 ± 126	3746 ± 358
Dorsal n. vagus	3868 ± 256	8926 ± 1066	1516 ± 146
Area postrema	2637 ± 238	9576 ± 254	1373 ± 151
Spinal trigeminal nucleus (SPN5)	1332 ± 176	3816 ± 548	370 ± 49
Amygdala (basomedial n)	1067 ± 212	4752 ± 365	1090 ± 261
Hippocampus (ventral dentate gyrus)	1066 ± 238	4916 ± 443	878 ± 139
Entorhinal cortex	823 ± 90	3443 ± 326	380 ± 55
Amygdala (basolateral nucleus)	769 ± 84	2267 ± 348	753 ± 184
Piriform cortex	741 ± 112	2951 ± 249	642 ± 130
Hippocampus, CA_2	726 ± 53	1601 ± 176	296 ± 82
Hippocampus, CA_1	641 ± 53	705 ± 139	299 ± 75
Hippocampus, CA_3	489 ± 35	720 ± 135	284 ± 53
Temporal cortex	ND	ND	222 ± 44
Occipital cortex	ND	ND	265 ± 58
Inferior colliculus	306 ± 90	564 ± 157	40 ± 5
Central grey	68 ± 24	569 ± 206	102 ± 30

Data are means ± SEM from 3–6 rats for [^{3}H]GR65630, 3–6 rats for [^{3}H]quipazine (in the presence of 1 μM paroxetine), and 3–14 rats for [^{3}H]RS-42358. ND = not determined. (See ref. 50.)

Table 14. Quantitative autoradiographic localization of $5HT_3$ receptors in the rat cerebral cortex using [^{3}H]RS-42358 as the radioligand

Cortex	Specific binding (amol/mm^2)		
	Layers I–II	Layers III–IV	Layers V–VI
Perirhinal	439 ± 52	399 ± 52	399 ± 51
Retrospinal	369 ± 34 [a]	224 ± 19	174 ± 51
Occipital	312 ± 58 [a]	138 ± 24	154 ± 25
Parietal	339 ± 23	261 ± 47	215 ± 63
Temporal	403 ± 89	268 ± 46	273 ± 62
Entorhinal*	380 ± 55*	–	–
Piriform*	642 ± 130*	–	–

Data are means ± SEM from 6–14 rats.
[a] $P < 0.05$ compared with data for layers V–VI. * Data represent all layers since these cortical structures are not well differentiated.

Table 15. Quantitative autoradiographic localization of $5HT_3$ binding sites in mammalian brainstem structures using [^{3}H]RS-42358

Brain structures	Specific radioligand binding (amol/mm^2; means ± SEM)					
	Rat	**Mouse**	**Guinea pig**	**Ferret**	**Cat**	**Dog**
Area postrema	1373 ± 151	3488 ± 617	261 ± 46	2126 ± 251	4233 ± 715	1578 ± 217
Nuc. solitary tract (medial)	3746 ± 358	6934 ± 601	3267 ± 726	3319 ± 360	5423 ± 893	3647 ± 291
Nuc. solitary tract (central)	3295 ± 227	6079 ± 436	3717 ± 812	4583 ± 576	7444 ± 1103	–
Nuc. solitary tract (lateral)	1786 ± 229	2043 ± 199	1911 ± 315	2087 ± 397	–	–
Dorsal motor nuc. vagus	1516 ± 146	1423 ± 174	1033 ± 240	2428 ± 200	1562 ± 336	2574 ± 348
Spinal trigeminal tract (SPN5)	370 ± 49	2041 ± 250	–	297 ± 59	1775 ± 125	ND

The data are means ± SEM of 3–15 animals of each species.
ND = not detected.
The density of [^{3}H]GR65630-labelled $5HT_3$ receptors in the dog brainstem was (amol/mm^2): area postrema = 1991 ± 226; NTS (medial) = 7416 ± 251; DMX = 2104 ± 494 (n = 3). Note the power of the quantitative autoradiography technique for making quantitative assessment of species differences in receptor labelling and detecting receptors in very small brain nuclei.

Table 16. Affinity estimates (pK_i s) for RS-42358-197 and ondansetron competing for [^{3}H]RS-42358-197 binding to $5HT_3$ receptors determined by quantitative autoradiography

	RS-42358-197		**Ondansetron**	
Rat brain region	**pK_i**	**nH**	**pK_i**	**nH**
Area postrema	10.9 ± 0.28	1.1 ± 0.3	9.3 ± 0.1	0.7 ± 0.2
Nuc. solitary tract	9.90 ± 0.10	1.6 ± 0.4	9.2 ± 0.1	1.0 ± 0.2
Entorhinal cortex	10.0 ± 0.10	2.1 ± 0.5	8.7 ± 0.1	1.0 ± 0.2
Amygdaloid complex	10.5 ± 0.10	1.3 ± 0.2	9.5 ± 0.3	0.8 ± 0.1
Hippocampal v. dentate gyrus	10.5 ± 0.10	1.3 ± 0.3	8.6 ± 0.3	1.1 ± 0.3

Data are means ± SEM of 3–10 rats. Each competition curve for each compound was iteratively fitted to at least seven data points. These data demonstrate the power of quantitative autoradiography for permitting determination of drug affinities in very small brain regions, structures that would be difficult to study using conventional homogenate-based assays. pK_i = apparent-log of inhibition constant; nH = Hill coefficient.

Table 17. Quantitative autoradiography of peptide YY (PYY) receptors in the brainstem of different species

	Specific autoradiographic binding (amol/mm^2)				
Brain region	**Rat**	**Ferret**	**Mouse**	**Guinea pig**	**Cat**
Area postrema	19.2 ± 4.7	5.4 ± 1.2	34.8 ± 9.1	17.7 ± 2.8	47.8 ± 3.5
Nuc. solitary tract (M)	31.6 ± 5.5	6.7 ± 1.1	39.2 ± 10.6	14.9 ± 2.4	59.8 ± 4.5
Nuc. solitary tract (C)	33.1 ± 3.9	–	54.4 ± 5.5	12.1 ± 1.5	61.8 ± 1.1
Nuc. solitary tract (L)	24.5 ± 3.2	–	44.0 ± 3.3	10.5 ± 2.0	–
Dorsal motor nuc. vagus	15.2 ± 1.2	6.0 ± 1.3	40.0 ± 6.2	9.8 ± 2.0	47.3 ± 6.2
Spinal trigeminal tract nuc.	6.2 ± 1.1	7.3 ± 2.5	20.0 ± 3.8	—	14.8 ± 4.2

Data are means ± SEM of 3–7 animals of each species and represent specific binding. [^{125}I]PYY concentration was 22–25 pM and non-specific binding was defined with 1 μM unlabelled PYY. The specific binding component generally represented (as a percentage of total binding) about 70% in the area postrema, 66–93% in the nucleus of solitary tract, 50–80% in the motor nucleus of vagus, and 50–70% in the spinal trigeminal tract nucleus.

The dog area postrema, nucleus of solitary tract, and dorsal motor nucleus of vagus also contained a high density of PYY binding sites (Harding and McDonald (123)).

Table 18. Quantitative autoradiography of $5HT_3$ and peptide YY (Y_1 and Y_2) receptors in the brainstem

	Specific binding (amol/mm^2)		
Brain region	**$5HT_3$ in dog**	**PYY in dog***	**PYY in cat**
Nuc. solitary tract	7416 ± 251	0.35 ± 0.02	78 ± 2
Dorsal nuc. vagus	2104 ± 494	0.37 ± 0.02	28 ± 2
Area postrema	1991 ± 226	0.29 ± 0.02	26 ± 1
Spinal trigeminal tract nuc.	Not detected	ND	13 ± 3
Cerebellum	152 ± 68	ND	14 ± 2

Data are means ± SEM from 3–4 animals.
* From ref. 124; data are in optical density values. ND = not determined.

Table 19. Quantitative autoradiography of $5HT_3$ and PYY receptors in the ferret brainstem

	Specific binding (amol/mm^2)	
Brain region	**$5HT_3$ receptors**	**PYY receptors**
Nuc. solitary tract	6166 ± 981	34 ± 5
Nuc. spinal trigeminal tract	3453 ± 596	13 ± 3
Area postrema	3221 ± 533	26 ± 1
Dorsal motor nuc. vagus	2193 ± 386	28 ± 2
Cerebellum	95 ± 17	14 ± 2

Data are means ± SEM from 3–4 ferrets.

Table 20. Saturation analysis of [^{125}I]Sar^1-Ile^8-AII binding to angiotensin II (AII) receptors determined by quantitative autoradiography

Tissue	**Apparent affinity (K_d, nM)**	**Apparent maximal receptor density (B_{max})**
Rat brain		
Piriform cortex	0.26 ± 0.02	149.0 ± 8.5
Nuc. olfactory tract	0.16 ± 0.03	147.3 ± 4.6
Superior colliculus	0.34 ± 0.03	135.7 ± 13.8
Nuc. spinal trigeminal tract	0.23 ± 0.02	115.0 ± 9.8
Nuc. solitary tract	0.14 ± 0.06	141.3 ± 2.0
Bovine cerebellum	0.02 ± 0.01	147.0 ± 3.1
Rat peripheral tissues		
Kidney cortex	0.19 ± 0.01	149.0 ± 8.5
Kidney medulla	0.22 ± 0.02	162.7 ± 7.8
Adrenal medulla	0.014 ± 0.001	165.7 ± 3.2

Data are means ± SEM from three animals. (See refs 58, 126.)

Table 21. Affinity estimates for Sar[1]-Ile[8]-AII and DUP753 at the AT_1 angiotensin receptors in different tissues determined by quantitative autoradiography

	Sar[1]-Ile[8]-AII		DUP753	
	pK_i	nH	pK_i	nH
Rat brain				
Piriform cortex	ND		7.01 ± 0.05 (6)	0.95 ± 0.10
N. olfactory tract	ND		7.20 ± 0.04 (6)	0.88 ± 0.09
V. hippocampus (dentate gyrus)	8.92 ± 0.05 (5)	1.76 ± 0.18	7.40 ± 0.08 (6)	1.10 ± 0.18
Nuc. spinal trigeminal	9.11 ± 0.09 (4)	1.19 ± 0.24	7.99 ± 0.09 (12)	0.85 ± 0.07
Nuc. solitary tract	8.77 ± 0.08 (5)	1.94 ± 0.42	7.31 ± 0.07 (12)	1.08 ± 0.08
Superior colliculus	9.00 ± 0.10 (7)	0.95 ± 0.16	7.06 ± 0.22 (4)	0.75 ± 0.03*
Rat kidney				
Kidney cortex	8.55 ± 0.11 (5)	1.12 ± 0.28	6.79 ± 0.09 (10)	1.19 ± 0.16
Kidney medulla	8.49 ± 0.15 (5)	1.10 ± 0.22	6.87 ± 0.04 (11)	1.26 ± 0.08
Bovine cerebellum	9.77 ± 0.11 (5)	1.15 ± 0.15	<4	

Data are means ± SEM from 3–12 animals. * $P < 0.05$, significantly less than unity. ND = not determined. nH = Hill coefficient. (See ref. 58.)

Table 22. Affinity estimates for EXP655 and CGP42112A at AT_2 angiotensin receptors determined by quantitative autoradiography

	EXP655		CGP42112A	
	pK_i	nH[a]	pK_i	nH[a]
Rat superior colliculus	6.87 ± 0.11 (13)	0.77 ± 0.06*	8.34 ± 0.22 (5)	1.32 ± 0.32
Bovine cerebellum	7.24 ± 0.09 (11)	1.08 ± 0.14	9.48 ± 0.08 (9)	1.31 ± 0.10
Rat adrenal medulla	6.49 ± 0.11 (13)	0.93 ± 0.08	8.83 ± 0.17 (8)	0.99 ± 0.10
Rat kidney cortex	<4		<5	

Data are means ± SEM from 5–13 animals. * $P < 0.05$, significantly less than unity.
[a] nH = Hill coefficient. (See refs 58, 126.)

Table 23. Quantitative autoradiography of angiotensin II receptor subtypes in the rat brain and peripheral tissues

Tissue	Specific binding (amol/mm^2)	% AT_1	% AT_2
Brain			
Area postrema	63.7 ± 7.4 (6)	96 ± 1	11 ± 6
Subfornical organ	62.2 ± 11.7 (6)	100 ± 1	–
Nuc. solit. tract	56.4 ± 5.0 (22)	96 ± 1	11 ± 2
Nuc. olf. tract	40.4 ± 11.5 (3)	98 ± 1	–
Piriform cortex	34.3 ± 5.8 (9)	99 ± 1	–
Olfact. cortex	31.6 ± 2.8 (5)	99 ± 1	–
Olivary nucleus	31.0 ± 3.9 (3)	41 ± 9	78 ± 6
Suprachiasm. nuc.	30.2 ± 3.4 (5)	99 ± 1	–
V. hippo. d. gyrus	25.7 ± 5.4 (5)	81 ± 19	–
Lat. hypothalamus	24.4 ± 2.1 (5)	99 ± 1	–
Med. amygdala nuc.	23.7 ± 4.4 (4)	70 ± 13	32 ± 20
Periventric. nuc.	23.2 ± 1.7 (3)	98 ± 2	–
Subthalamic nuc.	22.4 ± 2.2 (6)	23 ± 12	88 ± 5
Spinal trigem. nuc.	20.2 ± 2.3 (21)	94 ± 1	–
Lat. preoptic area	19.3 ± 1.9 (4)	99 ± 1	–
Lat. genicu. body	17.9 ± 3.6 (5)	6 ± 1	84 ± 7
Sup. colliculus	16.2 ± 1.6 (20)	35 ± 4	72 ± 2
Dorsal raphé	15.9 ± 2.7 (5)	94 ± 2	–
Inf. colliculus	14.4 ± 2.2 (5)	44 ± 11	79 ± 4
Dr. med. thalamus	12.1 ± 2.3 (5)	–	89 ± 2
Ven. med. thalamus	10.0 ± 2.0 (6)	–	89 ± 2
Lat. septum	9.3 ± 3.6 (6)	31 ± 9	73 ± 6
V. diag. b. Broca	8.7 ± 2.4 (3)	77 ± 14	31 ± 21
Choroid plexus	8.7 ± 1.2 (6)	99 ± 1	–
(Bovine cerebellum)	80.6 ± 5.7 (9)	23 ± 1	87 ± 4
Peripheral tissues			
Adrenal medulla	129.4 ± 3.3 (8)	5 ± 3	65 ± 4
Ant. pituitary	87.5 ± 9.0 (10)	98 ± 1	
Kidney glomeruli	72.7 ± 3.2 (6)	99 ± 1	
Kidney vasa recta	71.3 ± 4.5 (6)	99 ± 1	
Uterine muscle	67.3 ± 6.1 (12)	28 ± 6	36 ± 9
Kidney cortex	58.6 ± 6.1 (13)	99 ± 1	
Kidney medulla	51.5 ± 4.0 (13)	99 ± 1	
Adrenal cortex	40.2 ± 5.6 (8)	22 ± 8	46 ± 8
Spleen	13.7 ± 2.9 (6)	92 ± 2	

Data are means ± SEM from 3–22 animals. The subtypes, AT_1 and AT_2, were defined with 10 μM DUP753 and EXP655, respectively. [^{125}I]SARILE concentration was 0.1–0.2 nM. (See refs 58, 126.)

References

1. Rogers, A. W. (1979). *Techniques of Autoradiography* (3rd Edn). Elsevier Press, Amsterdam.
2. Baker, J. R. J. (1989). *Autoradiography: A comprehensive overview.* Microscopy Handbooks, Vol. 18, Royal Microscopical Society, Oxford Science Publications.
3. Kuhar, M. J. (1981). *Trends Neurosci.,* **4,** 60.
4. Palacios, J. M., Probst, A., and Cortes, R. (1986). *Trends Neurosci.,* **9,** 284.
5. Sokoloff, L. (1989). In *Brain Imaging: Techniques and Applications* (ed. N. A. Sharif and M. E. Lewis), pp. 230–61. Ellis Horwood Ltd., Chichester.
6. Beaudet, A. and Szigethy, E. (1989). In *Brain Imaging: Techniques and Applications* (ed. N. A. Sharif and M. E. Lewis), pp. 186–208. Ellis Horwood Ltd., Chichester.
7. Ulberg, S. (1977). *The technique of whole body autoradiography,* Science Tools (special Issue) LKB-Produkter AB, Stockholm.
8. Kuhar, M. J., De Souza, E. B., and Unnerstall, J. R. (1986). *Annu. Rev. Neurosci.,* **9,** 27.
9. Sharif, N. A. and Hughes, J. (1989). *Peptides,* **10,** 499.
10. Palacios, J. M. and Dietl, M. M. (1987). *Experentia,* **43,** 750.
11. Herkenham, M. (1988). In *Molecular Neuroanatomy* (ed. F. W. van Leeuwen, R. M. Buijs, C. W. Pool, and O. Pach). Chapter 11, pp. 111–20. Elsevier Science Publishers, Amsterdam.
12. Young, W. S. and Kuhar, M. J. (1979). *Brain Res.,* **179,** 255.
13. Fastbom, J., Pazos, A., Probst, A., and Palacios, J. M. (1987). *Neuroscience,* **22,** 827.
14. Jarvis, M. F. and Williams, M. (1989). *Eur. J. Pharmacol.,* **168,** 243.
15. Sharif, N. A., Shieh, A., Blue, D., and Clarke, D. E. (1991). ASPET (American Society for Pharmacology & Experimental Therapeutics), Abst. no. 386.
16. Dashwood, M. R. (1983). *Eur. J. Pharmacol.,* **86,** 51.
17. Feng, F., Pettinger, W. A., Abel, P. W., and Jeffries, W. B. (1991). *J. Pharmacol. Expt. Therapeutics,* **258,** 263.
18. Laporte, A.-M., Schechter, L. E., Bolanos, F. J., Vergé, D., Hamon, M., and Gozlan, H. (1991). *Eur. J. Pharmacol.,* **198,** 59.
19. Young, W. S. and Kuhar, M. J. (1980). *Proc. Natl Acad. Sci. USA,* **77,** 1696.
20. Stephenson, J. A. and Summers, R. J. (1985). *Eur. J. Pharmacol.,* **116,** 271.
21. MacKinnon, A. C., Brown, C. M., and Sharif, N. A. (1993). *Eur. J. Pharmacol.,* in prep.
22. Boyajian, C. L., Loughlin, S. E., and Leslie, F. M. (1987). *J. Pharmacol. Exp. Ther.,* **241,** 1079.
23. Hamid, Q. A., Mak, J. C. W., Sheppard, M. N., Corrin, B., Venter, J. C., and Barnes, P. J. (1991). *Eur. J. Pharmacol.,* **206,** 13.
24. Buxton, B. F., Jones, C. R., Molenaar, P., and Summers, R. J. (1987). *Brit. J. Pharmacol.,* **92,** 299.
25. Altar, C. A. and Marien, M. R. (1987). *J. Neurosci.,* **7,** 213.
26. Dawson, T. M., Barone, P., Sidhu, A., Wamsley, J. K., and Chase, T. N. (1988). *Neuroscience,* **26,** 8.
27. Bouthenet, M.-L., Martres, M.-P., Sales, N., and Schwartz, J. C. (1987). *Neuroscience,* **20,** 117.

28. Olsen, R. W., McCabe, R. T., and Wamsley, J. K. (1990). *J. Chem. Neuroanat.*, **3,** 59.
29. Knight, A. T. and Bowery, N. G. (1992). *J. Neurol. Transn.* (suppl.), **35,** 189.
30. Herkenham, M., Lynn, A. B., Little, M. D., Johnson, M. R., Melvin, L. S., De Costa, B. R., and Rice, K. C. (1990). *Proc. Natl Acad. Sci. USA,* **87,** 1932.
31. Monagham, D. T., Holets, V. R., Toy, D. W., and Cotman, C. W. (1983). *Nature,* **306,** 176.
32. Monagham, D. T. and Cotman, C. W. (1982). *Brain Res.,* **252,** 91.
33. Monagham, D. T. and Cotman, C. W. (1985). *J. Neurosci.,* **5,** 2909.
34. Morgan, R. C., Mercer, L. D., Cincotta, M., and Beart, P. M. (1991). *Neurochem. Int.,* **18,** 75.
35. Ruat, M., Traiffort, E., Bouthenet, M. L., Souil, E., *et al.* (1991). *Agents Action* (suppl.), **33,** 123.
36. Traiffort, E., Pollard, H., Moreau, J., Ruat, M., Schwartz, J.-C., Martinez-Mir, M. I., and Palacios, J. M. (1992). *J. Neurochem.,* **59,** 290.
37. Arrang, J.-M., Garbarg, M., Lancelot, J.-C., Lecomte, J.-M., Pollard, H., Robba, M., Schunack, W., and Scwartz, J.-C. (1987). *Nature,* **327,** 117.
38. Wamsley, J. K., Gehlert, D. R., Roeske, W. R., and Yamamura, H. I. (1984). *Life Sci.,* **34,** 1395.
39. Cortes, R., Probst, A., Tobler, H.-J., and Palacios, J. M. (1986). *Brain Res.,* **362,** 239.
40. Regenold, W., Araujo, D., and Quirion, R. (1987). *Eur. J. Pharmacol.,* **144,** 417.
41. Wang, J.-X., Roeske, W. R., Hawkins, K. N., Gehlert, D. R., and Yamamura, H. I. (1989). *Brain Res.,* **477,** 322.
42. Araujo, D. M., Lapchak, P. A., and Quirion, R. (1991). *Synapse,* **9,** 165.
43. Clarke, P. B. S., Schwartz, R. D., Paul, S. M., Pert, C. B., and Pert, A. (1985). *J. Neurosci.,* **5,** 1307.
44. Hoyer, D., Pazos, A., Probst, A., and Palacios, J. M. (1986). *Brain Res.,* **376,** 85.
45. Vergé, D., Daval, G., Marcinkiewicz, M., Patey, A., El Mestikawy, S., Gozlan, H., and Hamon, M. (1986). *J. Neurosci.,* **6,** 3474.
46. Radja, F., Laporte, A.-M., Daval, G., Verge, D., Gozlan, H., and Hamon, M. (1991). *Neurochem. Int.,* **18,** 1.
47. Pazos, A., Hoyer, D., and Palacios, J. M. (1984). *Eur. J. Pharmacol.,* **106,** 531. Academic Press, New York.
48. Waeber, C., Dietl, M., Hoyer, D., Probst, A., and Palacios, J. M. (1988). *Neurosci. Lett.,* **88,** 11.
49. Pazos, A., Gonzalez, A. M., Waeber, C., and Palacios, J. M. (1991). In *Receptors in the Human Central Nervous System* (ed. F. A. O. Mendelsohn and G. Paxinos), Chapter 4, pp. 71–101. Academic Press, New York.
50. Sharif, N. A., Nunes, J. L., To, Z. P., Wong, E. H. F., Eglen, R. M., and Whiting, R. L. (1991). *Brit. J. Pharmacol.,* **102,** 142.
51. Pratt, G. D. and Bowery, N. G. (1989). *Neuropharmacology,* **12,** 1367.
52. Barnes, J. M., Barnes, N. M., Champaneria, S., Costall, B., and Naylor, R. J. (1990). *Neuropharmacology,* **29,** 1037.
53. Koscielniak, T., Ponchant, M., Laporte, A.-M., Guminski, Y., Verge, D., Hamon, M., and Gozlan, H. (1990). *C.R. Acad. Sci. Paris,* **311** (series 3), 231.

54. Waeber, C., Hoyer, D., and Palacios, J. M. (1989). *Neuroscience,* **31,** 393.
55. Sharif, N. A., Nunes, J. L., To, Z. P., Jakeman, L., Wong, E. F. H., Eglen, R. M., and Whiting, R. L. (1993). *J. Neurochem.*, in prep.
56. Palacios, J. M., Waeber, C., Bruinvels, A. T., and Hoyer, D. (1992). *Mol. Brain Res.,* **13,** 175.
57. Quirion, R., Dalpe, M., De Lean, A., Gutkowska, J., Cantin, M., and Genest, J. (1984). *Peptides,* **5,** 1167.
58. Sharif, N. A., To, Z. P., Gazvini, S., Wong, E. F. H., Eglen, R. M., and Whiting, R. L. (1991). *Proc. IBRO Meeeting*, Abst. no. 59.6.
59. Tsutsumi, K. and Saavedra, J. M. (1991). *J. Neurochem.,* **56,** 348.
60. Harding, J. W., Cook, V. I., Miller-Wing, A. V., Hanesworth, J. M., Sardinia, M. F., Hall, K. L., Stobb, J. W., Swanson, G. N., Coleman, J. K. M., Wroight, J. W., and Harding, E. C. (1992). *Brain Res.,* **583,** 340.
61. Sharif, N. A. and Hughes, J. (1989). In *Brain Imaging: Techniques and Applications* (ed. N. A. Sharif and M. E. Lewis). Chapter 3, p. 36. Ellis Horwood Ltd., Chichester.
62. Privitera, P. J., Daum, P. R., Hill, D. R., and Hiley, C. R. (1991). *Brain Res.,* **577,** 73.
63. Burch, R. M., Kyle, D. J., Martin, J. A., Hiner, R. N., Connor, J., and Wan, Y. P. (1992). *Du Pont Biotech. Update,* **7,** 127. (Also see Burch, R. M. and Kyle, D. J. (1992). *Life Sci.,* **50,** 829.)
64. Zarbin, M. A., Kuhar, M. J., O'Donohue, T. L., Wolfe, S. S., and Moody, T. W. (1985). *J. Neurosci.,* **5,** 429.
65. Beresford, I. J. M., Hall, M. D., Clark, C. R., Hill, R. G., Hughes, J., and Sirinathsinghji, D. J. S. (1987). *Neuropeptides,* **10,** 109.
66. Hill, D. R. and Woodruff, G. N. (1990). *Brain Res.,* **526,** 276.
67. Mantyh, P., Catton, M., Magio, J. E., and Vigna, S. R. (1991). In *Sensory Nerves and Neuropeptides in Gastroenterology* (ed. M. Costa *et al.*), *Adv. Exp. Med. Biol.*, p. 253. Plenum Press, New York.
68. De Souza, E. B., Perrin, M. H., Insel, T. R., Rivier, J., Vale, W. W., and Kuhar, M. J. (1984). *Science,* **224,** 1449.
69. Hoyer, D., Waeber, C., and Palacios, J. M. (1989). *J. Cardiovas. Pharmacol.,* **13** (suppl. 5), S162.
70. Davenport, A. P. and Morton, A. J. (1991). *Brain Res.,* **554,** 278.
71. Melander, T., Kohler, C., Nilsson, S., Hokfelt, T., Brodin, E., Theodorsson, E., and Bartfai, T. (1988). *J. Chem. Neuroanatom.,* **1,** 213.
72. Uttenthal, L. O., Toledano, A., and Blazquez, E. (1992). *Neuropeptides,* **21,** 143.
73. Allard, M., Zajac, J.-M., and Simonnet, G. (1992). *Neuroscience,* **49,** 101.
74. Dam, T.-V., Escher, E., and Quirion, R. (1990). *Brain Res.,* **506,** 175.
75. Yashpal, K., Dam, T.-V., and Quirion, R. (1990). *Brain Res.,* **506,** 259.
76. Martel, J.-C., Fourness, A., St. Pierre, S., and Quirion, R. (1990). *Neuroscience,* **36,** 255.
77. Moyse, E., Rosténe, W., Via, M., Leonard, K., Mazelle, J., Kitabgi, P., Vincent, J.-P., and Beaudet, A. (1987). *Neuroscience,* **22,** 525.
78. Sharif, N. A. and Hughes, J. (1989). *Peptides,* **10,** 499.
79. Mansour, A., Kachaturian, H., Lewis, M. E., Akil, H., and Watson, S. J. (1987). *J. Neurosci.,* **7,** 2445.

80. Sharif, N. A., Hunter, J. C., Hill, R. G., and Hughes, J. (1988). *Neurosci. Lett.*, **86,** 272.
81. Quirion, R., Dam, T.-V., Sarrieau, A., and Rosténe, W. (1989). In *Brain Imaging: Techniques and Applications* (ed. N. A. Sharif and M. E. Lewis), Chapter 4, p. 77. Ellis Horwood Ltd., Chichester.
82. Jomary, C., Gairin, J. E., and Beaudet, A. (1992). *Proc. Natl Acad. Sci. USA,* **89,** 564.
83. Hunter, J. C., Birchmore, B., Woodruff, R., and Hughes, J. (1989). *Neuroscience,* **31,** 735.
84. Boyle, S. J., Meecham, K. G., Hunter, J. C., and Hughes, J. (1990). *Mol. Neuropharmacol.,* **1,** 23.
85. Clark, C. R., Birchmore, B., Sharif, N. A., Hunter, J. C., Hill, R. G., and Hughes, J. (1988). *Brit. J. Pharmacol.,* **93,** 618.
86. Loup, F., Tribollet, E., Dubois-Dauphin, M., and Dreifuss, J. J. (1991). *Brain Res.,* **555,** 220.
87. Sharif, N. A. and Nunes, J. L. (1993). *Neurochem. Int.,* in prep.
88. Krantic, S., Martel, J.-C., Weissmann, D., Pujol, J.-F., and Quirion, R. (1990). *Neuroscience,* **39,** 127.
89. Sharif, N. A. (1989). *Ann. New York Acad. Sci.,* **553,** 147.
90. Sharif, N. A. and Burt, D. R. (1985). *Neurochem. Int.,* **7,** 525.
91. Dietl, M. M., Hof, P. R., Martin, J.-L., Magistretti, P. J., and Palacios, J. M. (1990). *Brain Res.,* **520,** 14.
92. Quirion, R. (1987). *Synapse,* **1,** 293.
93. Altar, C. A. and Marien, M. R. (1988). *Synapse,* **2,** 486.
94. Graybiel, A. M. and Moratalla, R. (1989). *Proc. Natl Acad. Sci. USA,* **86,** 9020.
95. Kaufman, M. J., Spealman, R. D., and Madras, B. K. (1991). *Synapse,* **9,** 177.
96. Taxt, T. and Storm-Mathisen, J. (1984). *Neuroscience,* **11,** 79.
97. Biegon, A. and Rainbow, T. C. (1983). *J. Neurosci.,* **3,** 1069.
98. Tejani-Butt, S. M. and Ordway, G. A. (1992). *Brain Res.,* **583,** 312.
99. Lidow, M. S., Goldman-Rakic, P. S., Gallager, D. W., and Rakic, P. (1989). *J. Comp. Neurol.,* **280,** 27.
100. Hridina, P. D., Foy, B., Hepner, A., and Summers, R. J. (1990). *J. Pharmacol. Exp. Ther.,* **252,** 410.
101. Worley, P. F., Baraban, J. M., De Souza, E. B., and Snyder, S. H. (1986). *Proc. Natl Acad. Sci. USA,* **83,** 4053.
102. Worley, P. F., Baraban, J. M., Colvin, J. S., and Snyder, S. H. (1987). *Nature,* **325,** 159.
103. Hawkins, P. T., Reynolds, D. J. M., Poyner, D. R., and Hanley, M. R. (1990). *Biochem. Biophys. Res. Commun.,* **167,** 819.
104. Thomson, F. J., Mitchell, R., MacEwna, D. J., Harvey, J., and Johnson, M. S. (1991). *Brit. J. Pharmacol.,* **104,** 273.
105. Padua, R. A., Wan, W., Nagy, J. I., and Geiger, J. D. (1991). *Brain Res.,* **542,** 135.
106. Knaus, H.-G., Striessnig, J., Hering, S., Marrer, S., Schwenner, E., Holtje, H.-D., and Glossman, H. (1992). *Mol. Pharmacol.,* **41,** 298.
107. Takemura, M., Kiyama, H., Fukui, H., Toyhama, M., and Wada, H. (1988). *Brain Res.,* **451,** 386.
108. Gehlert, D. R., Mais, D. E., Gackenheimer, S. L., Krushinski, J. H., and Robertson, D. W. (1990). *Eur. J. Pharmacol.,* **186,** 373.

109. Awan, K. A. and Dolly, J. O. (1991). *Neuroscience,* **40,** 29.
110. Pechin-Mathews, A. and Dolly, J. O. (1989). *Neuroscience,* **29,** 347.
111. Mourre, C., Hugues, M., and Lazdunski, M. (1986). *Brain Res.,* **382,** 239.
112. Bowery, N. G., Wong, E. H. F., and Hudson, A. L. (1988). *Brit. J. Pharmacol.,* **93,** 944.
113. Vignon, J., Privat, A., Chaudieu, I., Thierry, A., Kamenka, J.-M., and Chicheportiche, R. (1986). *Brain Res.,* **378,** 133.
114. Altar, C. A., Burton, L. E., Benett, G. L., and Gugich-Djordjevic, M. (1991). *Proc. Natl Acad. Sci. USA,* **88,** 281.
115. Sarrieau, A., Vial, M., Philibert, D., and Rostene, W. (1984). *Eur. J. Pharmacol.,* **98,** 151.
116. Stritmatter, S. M., Lo, M. M. S., Javitch, J. A., and Snyder, S. H. (1984). *Proc. Natl Acad. Sci. USA,* **81,** 1599.
117. Lynch, D., Strittmatter, S. M., and Snyder, S. H. (1984). *Proc. Natl Acad. Sci. USA,* **81,** 653.
118. Waksman, G., Hamel, E., Fournie-Zaluski, M.-C., and Roques, B. P. (1986). *Proc. Natl Acad. Sci. USA,* **83,** 1523.
119. Rainbow, T. C., Parsons, B., Wieczorek, C., and Manaker, S. (1985). *Brain Res.,* **330,** 337.
120. Parsons, B. and Rainbow, T. C. (1984). *Eur. J. Pharmacol.,* **102,** 375.
121. Saura, J., Kettler, R., Da Prada, M., and Richards, J. G. (1992). *J. Neurosci.,* **12,** 1977.
122. Antonelli, M. C., Baskin, D. G., Garland, M., and Stahl, W. L. (1989). *J. Neurochem.,* **52,** 193.
123. Harding, R. K. and McDonald, T. J. (1989). *Peptides,* **10,** 21.
124. Leslie, R. A., McDonald, T. J., and Robertson, H. A. (1988). **9,** 1071.
125. Sharif, N. A., To, Z. P., Eglen, R. M., and Whiting, R. L. (1992). *Proc. IBRO Meeting, Abst.,* No. 54.32.
126. Jakeman, L. B., Sharif, N. A., To, Z. P., Eglen, R. M., Bonhaus, D. W., and Wong, E. H. F. (1994). *Brain Res.* (in press).
127. Whiting, R. L., Ford, A. P. D. W., Baxter, G. S., Harris, G. C., Sharif, N. A., and Eglen, R. M. (1991). *Brit. J. Pharmacol.*, **102,** 191.
128. Nunes, J. L., Lake, K., Sharif, N. A., *et al.* (1993). *J. Pharmacol. Expt. Ther.* (in press).
129. Sharif, N. A., Nunes, J. L., Rapp, J. M., *et al.* (1990). Dopamine '90, Satellite Meeting of XI IUPHAR, Como, Italy. Abst. #90.
130. Grossman, C. J., Kilpatrick, G. J., and Bunce, K. T. (1993). *Brit. J. Pharmacol.*, **109,** 618.

5

Use of antibodies to visualize neurotransmitter receptor subtypes

ALLAN I. LEVEY

1. Introduction

Methods for visualization and quantification of receptors have provided information that is fundamental for understanding the roles of neurotransmission and specific receptors in development, synaptic plasticity, learning, memory, control of movement, and a variety of other neural processes. Receptor autoradiography is one of the most widely used methods for this purpose because it is a relatively easy and sensitive technique. However, this technique is limited by the specificity of available radioligands and by its relatively poor spatial resolution. These factors have become more problematic in the past several years as a result of the molecular cloning of large gene families of neurotransmitter receptors. Many receptor subtypes for a given neurotransmitter are genetically and pharmacologically so highly related that ligands are not able to discriminate amongst the subtypes. For example, whereas there are only two or three pharmacologically distinct muscarinic acetylcholine receptors, five muscarinic acetylcholine receptors (m_1–m_5) have been identified (1–4). Similarly, D_1 and D_2 dopamine binding sites have been well characterized with selective ligands, but at least five dopamine receptor gene products (D_1–D_5) share these binding properties (5–9). The same problem exists for other G-protein coupled receptors, all of which are related members of the same gene superfamily (10). Moreover, because several mRNAs encoding related subtypes are often present in the same brain region (11), it is important to determine the cellular and subcellular localizations of the individual receptors for clues to their potential functional relevance. Conventional autoradiographic methods, even if subtype-selective ligands are available, are not useful for such precise anatomical localization. For these reasons, other methods for quantification and localization of specific receptor subtypes are much needed (12).

Immunological methods provide an alternative method for studying receptor subtypes that circumvent several limitations of conventional pharmacological techniques. Antibodies have been widely applied for biochemical and immunological characterization of many receptors (13). Antibodies have also been valuable for immunocytochemical localization of many neurotransmitter receptors, including muscarinic (14) and nicotinic (15) acetylcholine, glutamate (16), GABA (17, 18), glycine (19), serotonin (20), and β-adrenergic (21, 22). However, it has been difficult to develop antibodies selective for receptor subtypes. Only recently have receptor gene families and superfamilies been identified; most antibodies developed previously are not subtype-specific and very few antibodies have been tested for cross-reactivity to other members of the particular gene family (14, 15, 23). Traditional methods for immunizations with purified proteins are likely to induce antibodies to several receptors, because most tissues express several subtypes with such similar primary and secondary structures that biophysical techniques, including affinity purification, are frequently incapable of separating them. In addition, analysis of the genes and the predicted protein sequences reveals regions of the receptors that are highly conserved among family members and which may induce cross-reactive antibodies. For instance, many antibodies to short synthetic peptides corresponding to sequences in the β_2 adrenergic receptor also recognize the β_1 subtype (24). Some epitopes are conserved even among other families of receptors; anti-peptide antibodies cross-react with the first inner cytoplasmic loop of β_2 adrenergic receptor and rhodopsin (25). Hybridoma production is one method for selection of monospecific antibodies, although some monoclonal antibodies are likely to react with several receptor subtypes or subunits (23, 26, 27). For the purpose of studying highly related receptors selectively, it is most efficient to target antibodies directly to regions of the proteins that are unique to a single subtype.

Antibodies can be targeted to specific regions of proteins by immunization with either synthetic peptides or fusion proteins that have the desired sequence (28–30). Synthetic peptide antigens have been useful for producing excellent antibodies reactive with several receptor subtypes (16, 22, 31). This approach is most practical when there are only short stretches (10–20 amino acids) of divergence among the individual proteins. The advantages of this method are that it requires few skills (the entire procedure can be performed commercially) and that antibodies are targeted to highly defined sites. One major drawback of anti-peptide antibodies is that many antibodies will show little or no binding to the native receptor. The terminal regions of proteins are frequently targeted because these regions are highly flexible and often react with the corresponding anti-peptide antibodies (28–30); this strategy has worked well for glutamate receptor subunit proteins (16, 31). However, many receptors have N- and C-terminal regions that either are not unique or have modifications which limit anti-peptide antibody binding (2, 5–9). For example, we found that anti-peptide antibodies to these sites on muscarinic acetylcholine receptors were not suitable for many purposes, including visualization of the subtypes by immunocytochemistry (32). For this reason, we used fusion proteins derived from unique regions of the five muscarinic acetylcholine receptors for producing antibodies to the subtypes (14, 33). The large size of the fusion proteins (at least 10-fold larger than typical synthetic peptides) has the advantage of producing a wide variety of antibodies to many epitopes. Moreover, because fusion proteins may recreate the structure and function of the native proteins such as receptors (34), the antibodies may also preferentially recognize the receptors. In practice, we found that anti-fusion antibodies were much better than anti-peptide antibodies for visualization and quantification of muscarinic acetylcholine receptors. Once a few basic molecular and immunological techniques have been mastered, excellent subtype-specific antibodies can be efficiently and economically developed against more genetically distinct receptors.

The purpose of this chapter is to describe the methods that we have developed for visualization and quantification of receptor subtypes with antibodies raised against fusion proteins. A flow chart illustrating the overall strategy is shown in *Figure 1*. Divergent regions of the receptor genes (cDNAs) are subcloned into bacterial expression vectors that, when transfected into *Escherichia coli*, will result in the synthesis of large amounts of receptor polypeptide fused to other carrier proteins encoded by the vector. These fusion proteins are then purified and separately injected into rabbits to induce polyclonal antibody responses. The antibodies are affinity purified using the respective fusion proteins immobilized on a matrix. Antibodies are characterized by immunoblotting and immunoprecipitation with the fusion proteins and cloned receptors to verify subtype-specificity. Simple immunoprecipitation and immunocytochemical assays with subtype-specific antibodies can be used to identify, quantify, and localize the native receptor proteins in tissues. These methods complement conventional ligand binding techniques, and in addition, offer much better specificity and sensitivity than pharmacological agents for characterizing individual receptor subtypes.

2. Receptor fusion proteins

2.1 Selection of subtype-specific sequences

The first step in developing antibodies selective for neurotransmitter receptor subtypes is choosing the best region for targeting the antibodies. The amino acid sequences of the receptor family must be available for this purpose; if some members of the family have not been cloned, predictions can be made about the regions likely to maintain divergence. All G-protein coupled receptors share the general structure of the muscarinic acetylcholine receptor family (2), as illustrated in *Figure 2*. Each muscarinic receptor comprises a single protein of 460–590 amino acids, and seven transmembrane regions are

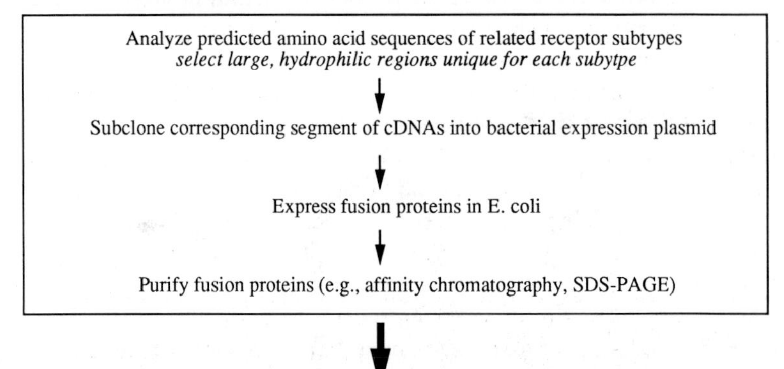

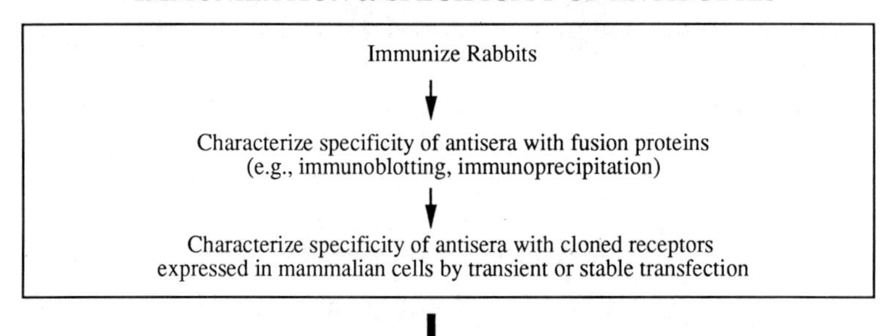

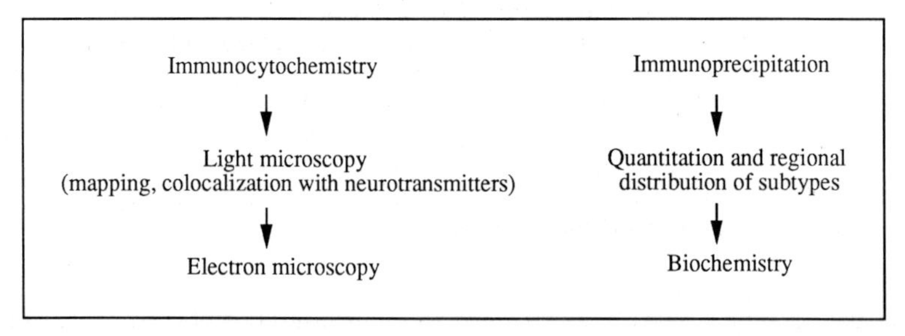

Figure 1. Overview of molecular and immunological approaches used to develop anti-fusion protein antibodies for visualization and quantification of muscarinic acetylcholine and other neurotransmitter receptor subtypes.

highly conserved among the five subtypes. These domains are believed to be an important part of the ligand binding sites, which may in part explain the sequence conservation of this region and the pharmacological similarities among subtypes. These regions should be avoided for antibody production because hydrophobic regions are poor immunogens, and even if antibodies

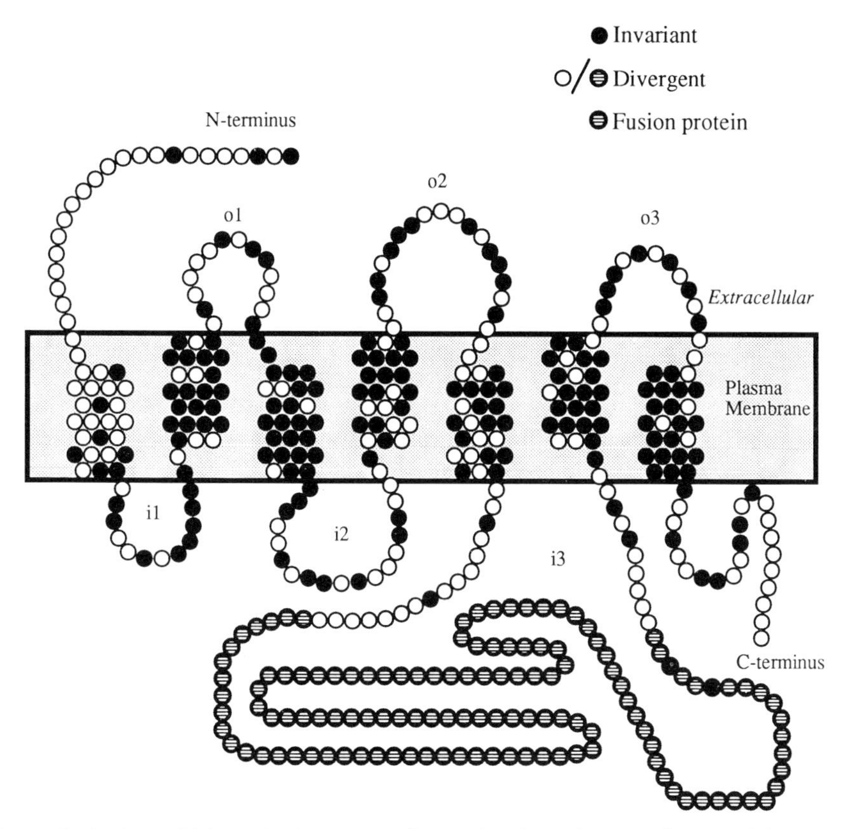

Figure 2. Amino acid homologies among five molecular subtypes of muscarinic acetylcholine receptors. The m_1 receptor is represented: m_2–m_5 have additional amino acid residues in the i3 loops. The structure shown is based on that predicted for all G-protein coupled receptors. Each receptor consists of one protein that spans the membrane seven times. The protein sequences are most related (i.e. identical amino acids in all five receptors) in the transmembrane domains. Divergent regions are in the N- and C-termini and the large intracellular i3 loop. Subtype-specific antibodies are targeted to fusion protein corresponding to the i3 loops as shown.

are induced, they are likely to cross-react. Although the N-termini are one of the least conserved regions of G-protein coupled receptors, they are not good targets for antibodies because they are short and contain one to three asparagine residues that are consensus sites for N-linked glycosylation (2). Thus, antibodies directed against epitopes on the protein core are likely to be masked by carbohydrate. Indeed, we found that although a panel of anti-peptide antibodies to this region of the five muscarinic receptors bound each peptide selectively, they did not bind to the native receptors (32). The C-termini are better targets for selective antibodies, although palmitoylation and phosphorylation may mask the native epitopes since potential sites for these modifications are present in most G-protein coupled receptors. Moreover, the C-termini are all extremely short and may have a limited spectrum of epitopes. In practice, we found that anti-peptide antibodies raised to the C-termini were able to bind native receptor subtypes but with limited efficiency (32).

The third intracellular (i3) loops are excellent targets for selective antibodies to muscarinic acetylcholine receptors and many other G-protein coupled receptors. These regions are highly divergent, very large (157–203 amino acids), contain many basic residues, and are hydrophilic, all features that may promote antigenicity (28–30). However, relatively small segments of the i3 loops are conserved among functionally related subtypes (35). For example, the regions immediately adjacent to transmembrane regions V and VI are similar in m_2 and m_4, but divergent from the other muscarinic receptor subtypes (m_1, m_3, and m_5) which themselves share sequence homologies in these regions. Antibodies must avoid these conserved regions to ensure subtype-specificity. The large size of the i3 loops makes them well suited for fusion protein expression and development of antibodies against the engineered proteins.

2.2 Construction of bacterial expression vectors

A variety of efficient protein expression systems are commercially available and potentially useful for developing good antibodies. We have found the pGEX vectors developed by Smith and Johnson (36) (available from Pharmacia) to offer several advantages. In the pGEX system, receptor (or other) gene segments are fused to the *Schistosoma japonicum* gene encoding glutathione S-transferase (GST). The vast majority of proteins can be expressed with this system, and importantly, most are soluble after non-denaturing lysis of the bacteria. GST is a 26 kd protein that binds with high affinity to glutathione and permits simple affinity purification of the fusion proteins from bacterial lysates with glutathione–agarose beads. Typical yields are 10–20 mg of purified protein per litre of culture. The pGEX vectors incorporate recognition sites for either thrombin (pGEX-2T) or factor X (pGEX-3X) at the fusion site, enabling simple proteolytic cleavage of the recombinant proteins and isolation of the receptor fragments without GST. When constructing the genes, we usually exclude sequences that encode hydrophobic regions because the translated proteins may be insoluble and/or difficult to isolate from bacterial inclusion bodies. Polymerase chain reaction, ligation, and identification of the recombinant vectors are all performed using standard molecular biological techniques (37).

Protocol 1. Construction of recombinant pGEX-2T plasmids

1. Synthesize or purchase oligonucleotie primers (48-mers) complementary to sense and antisense strands of the cDNA sequence to be amplified by polymerase chain reaction (PCR). Incorporate restriction sites for *Bam*HI

and *Eco*RI into the primers at the 5′ and 3′ ends of the gene fragments, respectively (make sure these sites are not present in the cDNA).

2. Amplify receptor DNA fragments by PCR.
3. Digest receptor DNA and pGEX-2T with *Bam*HI and *Eco*RI endonucleases.
4. Gel purify the digested DNAs.
5. Ligate the receptor DNA and pGEX-2T and transform *E. coli*.
6. Analyse transformants by miniprep DNA purification and digestion with *Bam*HI and *Eco*RI.
7. Verify recombinant plasmids by double stranded dideoxy sequencing (38).

2.3 Bacterial expression and purification of receptor fusion proteins

Recombinant pGEX-2T plasmids can be expressed in most *E. coli* strains using *Protocol 2*, as modified from Smith and Johnson (36). Smaller cultures (e.g. 5–15 ml) can be grown for analytical purposes. The plasmid-encoded *lacI^q^* allele represses the efficient tac promoter until transcription is induced with isopropylthiogalactopyranoside (IPTG). The soluble fusion proteins expressed with the pGEX system can be purified easily from *E. coli* lysates by single-step affinity chromatography on immobilized glutathione (glutathione–agarose, Sigma Chemicals). Although GST can be proteolytically cleaved from the GST–receptor fusion as described above, we have found that immunizations with the entire protein are more convenient. GST might act as a carrier protein to enhance immunogenicity of the receptor moiety, and antibodies reactive with GST can be removed simply by absorption on GST if necessary.

Protocol 2. Expression and purification of GST–receptor fusion proteins

1. Transform *E. coli* with either a recombinant or parent pGEX-2T vector, grow in 400 ml LB broth containing 100 μg/ml ampicillin to an optical density of 0.4, and then induce fusion proteins by adding 400 μl of 1.0 M IPTG stock for 3–4 h.
2. Check induction by SDS–PAGE analysis of 25 μl samples of cultures taken immediately before and 4 h after IPTG addition. With good results, the fusion protein is the most abundant protein in culture.
3. Harvest cultures by centrifugation at 5000 *g* for 10 min, wash cell pellets with 10 ml of buffer A (50 mM Tris, pH 8.0, 10 mM EDTA, 25% sucrose), resuspend, and digest cell walls with 12 ml buffer A containing 40 mg of lysozyme on ice for 1 h.

Protocol 2. *Continued*

4. Wash cells and resuspend in 10 ml buffer B (10 mM Tris, 1.0 mM EDTA, 1.0 mM DTT) containing a protease inhibitor mixture (1.0 mM PMSF, 5.0 μg/ml aprotinin, 1.0 μg/ml leupeptin, and 1 μg/ml pepstatin).
5. Freeze–thaw three to four times on dry ice (may be frozen overnight at this step).
6. Add DNase (20 μg/ml) or sonicate gently on ice to reduce viscosity.
7. Add 0.1 vol 10% Triton X-100.
8. Pellet by centrifugation at 5000 *g*, remove, and save both supernatant and pellet. (Check both fractions by SDS–PAGE for fusion protein; only soluble protein in the supernatant can be further purified. The above steps can be repeated on the pellet if a significant amount of protein is in this fraction.)
9. To the supernatant add 15 ml buffer C (20 mM Hepes, pH 7.6, 100 mM KCl, 0.2 mM EDTA, 20% glycerol) containing fresh 1 mM DTT and protease inhibitor mixture.
10. Add 4 ml of 1:1 slurry of glutathione–agarose beads (Sigma Chemicals); preabsorb beads in buffer C overnight.
11. Shake at 4°C for 1–2 h.
12. Pellet at 1000 *g* for 5 min, and batch wash beads twice with 40 ml buffer C.
13. Elute protein with 6 × 2 ml of 5.0 mM reduced glutathione (Sigma Chemicals) in buffer C.
14. Collect elutions and centrifuge at 2500 *g* for 2 min to remove contaminating beads.
15. Analyse fractions by SDS–PAGE, pool, and dialyse into phosphate buffered saline using Centriprep chambers (Millipore) or tubing. Check protein concentration (glutathione interferes with most assays if not completely dialysed), aliquot, and freeze at −80°C.

Many other vectors, including the pET series (39), can be used to produce high levels of receptor fusion proteins (33), although often the fusion proteins are recovered in insoluble bacterial inclusion bodies. In these cases low speed centrifugation and washing of the insoluble material will remove soluble bacterial proteins. Further purification to near homogeneity can be obtained by preparative SDS–PAGE. Gel slices containing the fusion proteins can be minced by repeated passage through successively smaller bore needles and directly injected for immunization. Although it is desirable to obtain highly purified fusion protein for immunizations, either purification method is likely to result in small amounts of contaminating *E. coli* proteins. We have not been concerned with these contaminants because preimmune sera from rabbits and other species used for immunizations already harbour natural antibodies to enteric *E. coli* proteins; such antibodies can be removed by absorption on bacterial lysates.

3. Antibody production and characterization

3.1 Immunization

Rabbits are used for immunization to produce polyclonal antibodies to receptors because they are relatively inexpensive, services are commercially available, and sufficient quantities of sera can be obtained readily. Monoclonal antibodies can also be generated against fusion proteins if unlimited quantities or more precisely targeted antibodies are necessary. However, the usefulness of the antibodies for a wide range of applications may be enhanced by the presence of many antibody populations available in polyclonal antisera. Although many immunization scheme are available, little is known about the conditions necessary for producing the 'best' antibodies for a novel antigen. It is likely that the conformations of the fusion protein (and processed fragments) at time of antigen presentation are critical in determining the reactivities of the antibodies. For example, antibodies raised to non-denatured immunogens (e.g. soluble fusion proteins) may bind to native receptor proteins with higher affinities than to denatured receptor proteins. Similarly, antibodies raised to denatured immunogens (e.g. by SDS) may bind to denatured receptor proteins better than native receptor proteins. If the primary goal is to develop an antibody reactive with the receptors on immunoblots, it may be useful either to gel-purify or to denature using SDS soluble fusion proteins prior to immunization. We have found that antibodies raised to soluble fusion proteins are better than antibodies to SDS–PAGE purified proteins for both immunocytochemistry and immunoprecipitation (14, 33, 40). We give rabbits monthly injections of ~200 μg of soluble purified fusion protein emulsified in Freund's adjuvant. Bleeds are obtained 3 and 4 weeks after boost, clotted overnight at 4°C, clarified, and the sera aliquoted and frozen at −80°C. For repeated use, 1–2 ml aliquots are stored at 4°C for several months using aseptic techniques.

3.2 Specificity of antibodies

Characterization of antibody specificity is critical for validation and interpretation of studies using the antibodies, including visualization and quantification of native receptors in tissues. Since gene families are identified by sequence homologies, it is likely that the most highly related and immunologically cross-reactive proteins in nature are the members of a given receptor family. Therefore, much effort can be saved by careful comparisons (preferably by computer algorithms) of amino acid sequence homologies among receptor subtypes and avoiding conserved regions during construction of the fusion proteins. None the less, it is essential that antibody specificity be determined directly with the related proteins.

As a first step it is convenient to rapidly determine any cross-reactive epitopes by testing antisera binding against all related fusion proteins. For this purpose, we test crude antisera on immunoblots of *E. coli* cultures expressing the various fusion proteins (33). Since all sera, preimmune and immune, react with many *E. coli* proteins, background immunoreactivity can be eliminated easily by a 30 min preincubation of sera with *E. coli* lysates (100-fold excess of concentrated culture lysates prepared by boiling in 1% SDS works well). Most immune sera will be reactive with epitopes on GST, the fragment common to each immunogen (if not cleaved previously), and will bind to these sites on all receptor fusion proteins. To determine the reactivities of antibodies selectively using receptor epitopes, GST-reactive antibodies can be removed (e.g. by preabsorption of antisera with 50 μg/ml of purified GST or lysates of *E. coli* cultures transfected with pGEX-2T), or alternatively, the fusion proteins can be cleaved with thrombin prior to immunoblotting in order to separate physically the GST and receptor proteins.

Antibody reactivities must also be tested with the cloned receptors directly to determine whether native receptors are recognized and to establish subtype-specificity. Transfected cells (either permanently or transiently) expressing the individual receptor cDNAs are the most rigorous targets; tissues (and cell lines derived from tissues) are not good substitutes because they frequently express mutiple receptor subtypes. Shown in *Protocol 3* is a double antibody immunoprecipitation assay that we developed for characterization of antibody binding to muscarinic receptor subtypes, including quantification of any cross-reactivities. The assay takes advantage of the high specific activity of [^{3}H]NMS, a ligand that binds to all five muscarinic receptor subtypes with high affinity (41). The principle is the same as in other receptor binding assays with measurement of bound versus free radioligand. However, in this assay soluble receptor–radioligand complexes are bound by the rabbit anti-receptor antibody and then precipitated with a goat anti-rabbit antibody. 'Bound' antibody–receptor–[^{3}H]NMS complexes are measured in washed precipitates, and 'free' receptor–[^{3}H]NMS sites remain soluble.

We have found using the immunoprecipitation assay that 5 μl of crude antisera raised against i3 loop fusion proteins can quantitatively precipitate at least 20–40 fmol of muscarinic receptor. Moreover, tests of over 20 antisera on the entire panel of five cloned muscarinic receptor subtypes have revealed virtually complete subtype-specificity. The assay is fast, simple, and sensitive (as little as 1–2 fmol receptor can be detected reliably). Once the antisera have been characterized, the assay can be used to determine the composition of receptor subtypes in a tissue (see Section 4.1). This method can also be adapted to other families of receptors for which solubilization conditions have been established and ligands with high specific/non-specific binding ratios are available.

Protocol 3. Immunoprecipitation assay for soluble muscarinic receptors

1. Harvest transfected cells expressing muscarinic receptor cDNAs with a rubber policeman, wash, resuspend in cold TE buffer (10.0 mM Tris, 1.0 mM EDTA, pH 7.4) containing a mixture of protease inhibitors (5 μg/ml aprotinin, 1–2 μg/ml leupeptin, 100 μg/ml PMSF) and homogenize (low setting) on ice with a polytron. Animal tissues can be substituted for determining subtype composition.
2. Centrifuge at 500 *g* to remove nuclei and cell debris and then recentrifuge supernatant at 30 000 *g* for 20 min. Wash membrane pellet by repeating centrifugation. Resuspend in TE (plus inhibitors) and determine protein concentration. Save aliquot for membrane filtration binding and protein assays to assess total number of receptors/mg membrane protein.
3. Resuspend membranes in TE containing 1.0% digitonin (Wako) and 0.1% cholate (Sigma) to give a protein concentration of 1 mg/ml. The optimal concentration of detergents varies with the source of digitonin and should be titred separately. Concentrations as low as 0.4% digitonin are typically effective.
4. Polytron and leave on ice for 60 min to solubilize the receptors.
5. Centrifuge at 30 000 *g* for 30 min at 4°C.
6. Transfer supernatant to new tube and add [^{3}H]NMS to give a final concentration of 10 nM.
7. Pipette 200 μl of receptor–[^{3}H]NMS mixture into 96 well microtitre plates. Save aliquots of the mixture for gel filtration on G25 Sephadex to determine total specific binding to soluble receptors (calculated from the difference in binding in the presence and absence of 1.0 μM atropine).
8. Add 5 μl of rabbit antiserum to each well and incubate for 4 h at 4°C. Control experiments substitute non-immune serum and serum pre-absorbed with fusion proteins.
9. Add 50 μl of goat anti-rabbit serum (diluted 1:1 in TE) to each well and incubate overnight at 4°C. The optimal concentration of goat anti-rabbit sera varies slightly depending on the batch.
10. Centrifuge microtitre plate at 2400 *g* for 5 min at 4°C. Rapidly aspirate supernatants, resuspend pellets in 250 μl ice-cold TE containing 0.1% digitonin/cholate, and recentrifuge.
11. Repeat aspiration, resuspension, and pelleting.

Protocol 3. *Continued*

12. Resuspend pellets with 100 μl of 10% SDS and count radioactivity in each sample.

13. Percentage bound by each antibody is calculated by formula:

$$\% \text{ Bound} = \frac{\text{Specific binding in immunoprecipitates}}{\text{Total specific binding of soluble receptors (gel filtration)}}$$

$$= \frac{([^3\text{H}]\text{NMS in immune pellets}) - ([^3\text{H}]\text{NMS in non-immune pellets})}{[^3\text{H}]\text{NMS-labelled soluble receptors} - \text{atropine}}$$

A variety of other methods can be used to determine subtype-specificity, including immunoblotting, immunocytochemistry, enzyme-linked immunosorbent assays, or immunoprecipitation of metabolically labelled receptor proteins. The critical factor with any method is that each antibody is tested against every genetically identified member of the receptor family directly. For this purpose, transfected cells that express a single receptor cDNA are excellent targets. Preabsorption on the respective fusion proteins is an important control with any assay to verify immunological binding of the antibodies (versus non-specific protein interactions). Assuming the antibodies are selective in these simple test systems, the 'gold standard' is to demonstrate a single band (of the appropriate size) on immunoblots of brain proteins. However, many antibodies cannot be evaluated in this manner because they will not recognize SDS-denatured receptors. Even with this method one can never be absolutely certain that antibodies are monospecific for all applications, because potentially cross-reactive proteins may also be denatured. Therefore, the most prudent approach is to characterize the antibodies by several independent methods. In addition, the distributions of the receptor proteins in tissues should be the same using independent methods (e.g. immunoprecipitation and immunocytochemistry), and these results should be generally consistent with the distributions of the mRNAs and binding sites.

4. Characterization of receptor proteins in tissues

4.1 Quantification of receptor subtypes by immunoprecipitation

Monospecific antibodies reactive with muscarinic acetylcholine or other receptor subtypes enables quantification of individual proteins by a variety of methods. The major advantage of immunological methods over conventional pharmacological binding assays is that some antibodies are much more specific for a single subtype. For example, pirenzepine, one of the most highly selective drugs for M_1 binding sites, has less than 10-fold differences in affinity between m_1 and m_4 receptors (41). Therefore, even pirenzepine cannot be used to establish the relative abundance of m_1 and m_4 in a complex mixture of receptors as occurs in tissues. The immunoprecipitation assay described above is one approach we have used to determine the composition of individual receptors in tissues. By incubating non-selective ligands, such as *N*-methyl scopolamine (NMS) or (QNB), with receptors solubilized from tissues, the majority of binding sites are saturated. The panel of subtype-specific antibodies is then applied to samples of mixture and each receptor is independently measured in the respective immunoprecipitates. We have identified m_1–m_4 receptors in brain and peripheral tissues with this method, in distributions that agree well with the distributions of mRNAs (14, 40). The assay is extremely sensitive as a result of the high specific activity of the radioligands. Limitations of this approach are that the receptors must be solubilized under non-denaturing conditions in order to retain ligand binding capacity, and the yield of receptors is significantly lower than the number of sites present in membranes. Interpretations of data should reflect the possibilities that some subtypes may solubilize less efficiently, have altered binding affinities, or even be preferentially degraded. A variety of other immunoassays can be designed to quantify individual receptor proteins. For example, enzyme-linked immunosorbent assays and immunoblotting can be standardized with known quantities of cloned receptor subtypes and used to measure receptors present in unknown samples. These assays might have the advantage of being independent of ligand binding parameters. Regardless of the particular method, well characterized immunoassays with monospecific antibodies are extremely useful for identification of the receptor proteins in tissues and determination of alterations in the relative levels of each receptor as a result of experimental manipulations (e.g. drug treatments or lesions) and natural conditions (e.g. ageing or diseases of the nervous system).

4.2 Visualization of receptor subtypes by immunocytochemistry

One of the major advantages of antibodies to receptors is the ability to localize individual subtypes in tissues with cellular and subcellular resolution. Like other applications of anti-receptor antibodies, the validity of the findings rests upon use of rigorously characterized antibodies. Therefore, we do not attempt immunocytochemical localization of receptors until antibodies have been tested by immunoblotting and immunoprecipitation as described above (Section 3.2). Moreover, we have found it essential to affinity purify antibodies using the respective fusion proteins immobilized on Affi-Gel (Sigma) using standard procedures (42). Since fusion proteins used for immunizations contain both receptor and GST (or other) polypeptide fragments, it is necessary to remove antibody populations reactive with GST since these might cross-react with irrelevant brain proteins. This can be achieved by absorption on immobilized GST columns either before or after affinity purification with the fusion proteins. Yields of ~0.1–0.5 mg of purified antibody are obtained from 1.0 ml of antiserum.

A critical factor for immunocytochemistry is tissue fixation. Overfixation typically leads to a marked loss of immunoreactivity and little, if any, ability to detect receptor localization. On the other hand, underfixation may result in loss of receptor proteins from the tissues and even tissue disintegration. Because it is impractical to test many fixation parameters independently, we have adapted one commonly used procedure. *Protocol 4* is our method for perfusion fixation of rat brains with 3% paraformaldehyde; this works extremely well for light microscopic visualization of almost all receptors and other neural antigens tested. However, if the quality of staining is poor, a wide variety of other fixatives and conditions may be tried.

Protocol 4. Perfusion fixation of rats with paraformaldehyde

1. Deeply anaesthetize rats with 50 mg/kg pentobarbital administered intraperitoneally.
2. Incise abdomen and expose thoracic cavity widely after incision of diaphragm and retraction of thoracic cage.
3. Insert 16 gauge intravenous catheter into left ventricle, remove stylette, and thread catheter sleeve into aortic root. Incise right atrium for outflow.
4. Perfuse 20–30 ml of ice-cold 0.9% saline at 20 ml/min flow rate with peristaltic pump.
5. Clamp descending aorta with haemostat immediately after starting saline rinse.
6. Perfuse 200 ml of freshly prepared 3% paraformaldehyde in 0.1 M phosphate buffer, pH 7.6.
7. Perfuse 200 ml of ice-cold 10% sucrose in 0.1 M phosphate buffer, pH 7.6.
8. Remove brain and place in 30% sucrose in 0.1 M phosphate buffer (pH 7.6 overnight).

Many immunocytochemical procedures can be used to localize antibodies in tissue sections for light microscopy. We use unlabelled antibody enzyme methods as described in *Protocol 5*. Tissues are first treated with Triton and normal goat serum to facilitate antibody penetration and prevent non-specific binding of immunoglobulin, respectively. Although every antibody should be titred for use in immunocytochemistry, 0.5 μg/ml is a standard final concentration for most affinity purified anti-receptor antibodies that results in an excellent signal to noise ratio. Tissue-bound rabbit antibodies are bound by goat anti-rabbit immunoglobulin, and this complex is localized with rabbit peroxidase anti-peroxidase complexes followed by enzymatic detection of peroxidase using diaminobenzidine. Avidin–biotin peroxidase complex methods (available commercially in kits) also work well, although that increased sensitivity is balanced by higher background staining.

Protocol 5. Receptor immunocytochemistry

1. Section brains at 40 μm on a freezing sliding microtome and collect sections in Tris-buffered saline (TBS) on ice.
2. Treat sections in 4% normal goat serum (NGS)/0.4% Triton X-100/TBS for at least 1 h at 4°C on a rotary shaker table.
3. Incubate sections in 2% NGS, 0.1% Triton X-100, TBS containing affinity purified antibody (final concentration of ~0.5 μg/ml) or control for 12–48 h at 4°C.
4. Rinse sections for 3 × 10 min in cold TBS.
5. Incubate sections in 2% NGS, 0.1% Triton X-100, TBS containing 1% goat anti-rabbit serum for 1 h at 4°C.
6. Rinse sections for 3 × 10 min in cold TBS.
7. Incubate sections in 2% NGS/TBS (no Triton) containing 0.5% rabbit peroxidase anti-peroxidase complexes for 1 h at 4°C.
8. Rinse sections for 3 × 10 min in cold TBS.
9. Develop peroxidase reaction by incubating sections in 0.05% diaminobenzidine/0.01% hydrogen peroxide for 5–10 min (by visual inspection of brown precipitate).
10. Rinse sections for 3 × 10 min in cold TBS.
11. Mount sections on to gel-alum subbed slides, air dry, dehydrate, and cover with coverslips.

Examples of immunocytochemical localization of muscarinic acetylcholine receptor subtypes in rat forebrain are shown in *Figure 3*. As with any immunological assay, control experiments must be performed to verify the specificity of the reaction. One control performed with all experiments is omission of the primary antibody (or use of non-specific affinity purified rabbit immunoglobulin at the same final concentration as the primary antibody). Staining of tissue with this control is a consequence of non-specific binding of secondary antibodies or PAP complexes, or possibly endogenous peroxidase activity. Demonstration of staining inhibition by preabsorption of the antibody with purified antigen (fusion protein) is also critical to demonstrate immunological binding of the primary antibody (*Figure 3D′–F′*).

Successful localization of receptors by immunocytochemistry complements and extends autoradiographic ligand techniques. In many cases antibodies achieve a higher degree of subtype-selectivity than is possible with radioligands. Moreover, immunocytochemistry offers much greater spatial resolution than autoradiographic techniques, including localization of receptors at cellular (*Figure 4*) and ultrastructural levels.

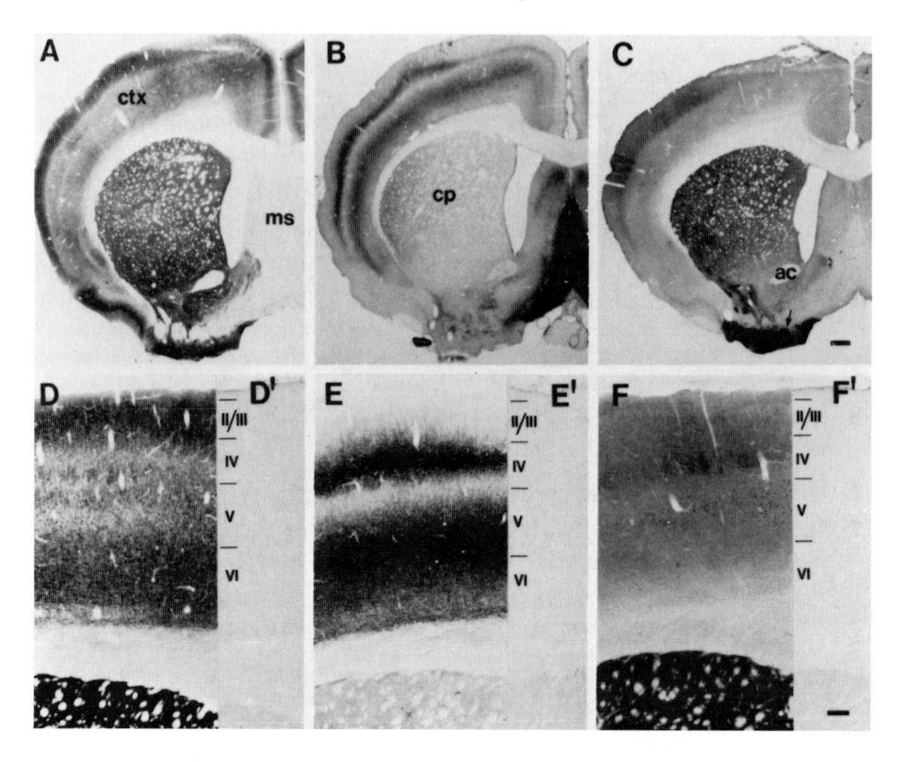

Figure 3. Example of immunocytochemical localization of m_1 (**A, D**), m_2 (**B, E**), and m_4 (**C, F**) muscarinic acetylcholine receptor proteins in rat forebrain. Note the complementary distributions of the receptors in cortex (ctx), caudate-putamen (cp), and medial septum (ms). The cortical laminar distributions are also different, as shown in panels D–F. Immunoreactivity is completely abolished by preabsorption of the antibodies with the respective fusion proteins (**D′–F′**). Scale bars: A–C, 500 μm; D–F, 220 μm. Reprinted by permission from Levey *et al.* (14).

Acknowledgements

This work was supported by NIH NS 01387, NIH NS30454, and a Faculty Scholar Award from the Alzheimer's Disease and Related Disorders Association.

References

1. Hulme, E. C., Birdsall, N. J. M., and Buckley, N. J. (1990). *Annu. Rev. Pharmacol. Toxicol.,* **30,** 633.
2. Bonner, T. I., Buckley, N. J., Young, A. C., and Brann, M. R. (1987). *Science,* **237,** 527.
3. Bonner, T. I., Young, A. C., Brann, M. R., and Buckley, N. J. (1988). *Neuron,* **1,** 403.

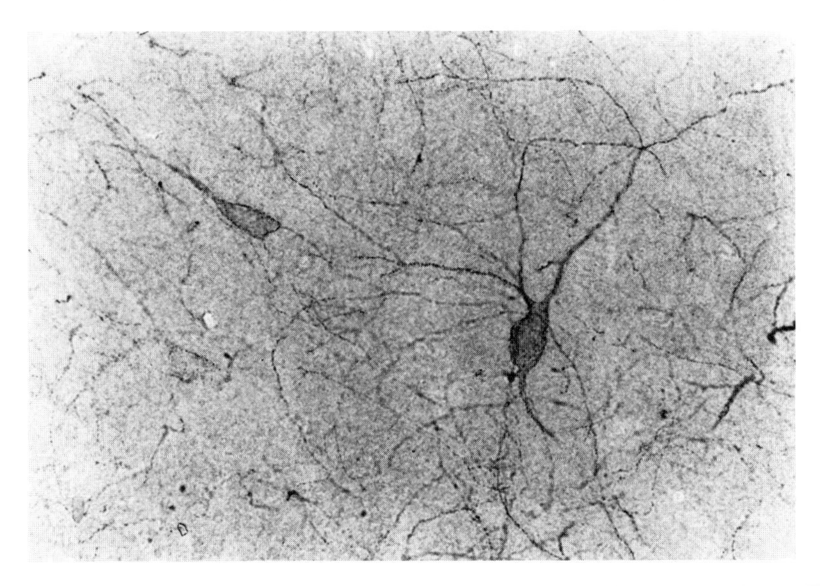

Figure 4. Immunocytochemical localization of the m_2 receptor in rat neostriatum. This method provides high resolution, including identification of the cellular and subcellular distribution of the receptors. Note the distribution of the receptor protein on the surface of scattered cell bodies and dendrites.

4. Buckley, N. J., Bonner, T. I., Buckley, C. M., and Brann, M. R. (1989). *Mol. Pharmacol.*, **35,** 469.
5. Sunahara, R. K., Niznik, H. B., Weiner, D. M., *et al.* (1990). *Nature,* **347,** 80.
6. Stormann, T. M., Gdula, D. C., Weiner, D. M., and Brann, M. R. (1990). *Mol. Pharmacol.*, **37,** 1.
7. Sokoloff, P., Giros, B., Martres, M.-P., Bouthenet, M.-L., and Schwartz, J.-C. (1990). *Nature,* **347,** 146.
8. Sunahara, R. K., Guan, H.-C., O'Dowd, B. F., *et al.* (1991). *Nature,* **350,** 614.
9. Van Tol, H. H. M., Bunzow, J. R., Guan, H.-C., *et al.* (1991). *Nature,* **350,** 610.
10. Hall, Z. W. (1987). *Trends Neurosci.*, **10,** 99.
11. Buckley, N. J., Bonner, T. I., and Brann, M. R. (1988). *J. Neurosci.*, **8,** 4646.
12. Kuhar, M. J., De Souza, E. B., and Unnerstall, J. R. (1986). *Annu. Rev. Neurosci.*, **9,** 27.
13. Bahouth, S. W., Wang, H.-Y., and Malbon, C. C. (1991). *Trends Pharmacol. Sci.*, **12,** 338.
14. Levey, A. I., Kitt, C. A., Simonds, W. F., Price, D. L., and Brann, M. R. (1991). *J. Neurosci.*, **11,** 3218.
15. Swanson, L. W., Simmons, D. M., Whiting, P. J., and Lindstrom, J. (1987). *J. Neurosci.*, **7,** 3334.
16. Blackstone, C. D., Moss, S. J., Martin, L. J., Levey, A. I., Price, D. L., and Huganir, R. L. (1992). *J. Neurochem.*, in press.
17. Richards, J. G., Schoch, P., Haring, P., Takacs, B., and Mohler, H. (1987). *J. Neurosci.*, **7,** 1866.

18. Somogyi, P., Takagi, H., Richards, J. G., and Mohler, H. (1989). *J. Neurosci.,* **9,** 2197.
19. Araki, T., Yamano, M., Murakami, T., Wanaka, A., Betz, H., and Tohyama, M. (1988). *Neuroscience,* **25,** 613.
20. Sotelo, C., Cholley, B., Mestikawy, S. E., Gozlan, H., and Hamon, M. (1990). *Eur. J. Neurosci.,* **2,** 1144.
21. Wanaka, A., Kiyama, H., Murakami, T., *et al.* (1989). *Brain Res.,* **485,** 125.
22. Aoki, C., Zemcik, B. A., Strader, C. D., and Pickel, V. M. (1989). *Brain Res.,* **493,** 331.
23. Ewert, M., Shivers, B. D., Luddens, H., Mohler, H., and Seeburg, P. H. (1990). *J. Cell Biol.,* **110,** 2043.
24. Wang, H., Lipfert, L., Malbon, C. C., and Bahouth, S. (1992). *J. Biol. Chem.,* **264,** 14424.
25. Weiss, E. R., Hadcock, J. R., Johnson, G. L., and Malbon, C. C. (1987). *J. Biol. Chem.,* **262,** 4319.
26. Venter, J. C., Eddy, B., Hall, L. M., and Fraser, C. M. (1984). *Proc. Natl Acad. Sci. USA,* **81,** 272.
27. Lindstrom, J., Schoepfer, R., and Whiting, P. (1987). *Mol. Neurobiol.,* **1,** 281.
28. Walter, G. (1986). *J. Immunol. Methods,* **88,** 149.
29. Lerner, R. A. (1984). *Adv. Immunol.,* **36,** 1.
30. Stern, P. (1991). *TIBTECH,* **9,** 163.
31. Wenthold, R. J., Yokotani, N., Doi, K., and Wada, K. (1992). *J. Biol. Chem.,* **267,** 1.
32. Levey, A. I., Simonds, W. F., Spiegel, A. M., and Brann, M. R. (1989). *Soc. Neurosci. Abst.,* **15,** 64.
33. Levey, A. I., Stormann, T. M., and Brann, M. R. (1990). *FEBS Lett.,* **275,** 65.
34. Freissmuth, M., Selzer, E., Marullo, S., Schutz, W., and Strosberg, A. D. (1991). *Proc. Natl Acad. Sci. USA,* **88,** 8548.
35. Wess, J., Bonner, T. I., Dorje, F., and Brann, M. R. (1990). *Mol. Pharmacol.,* **38,** 517.
36. Smith, D. B. and Johnson, K. S. (1988). *Gene,* **67,** 31.
37. Sambrook, J., Fritsch, E. F., and Maniatis, T. (1989). *Molecular cloning: a laboratory manual*, 2nd edition. Cold Spring Harbor Laboratory Press, Cold Spring Harbor, NY.
38. Kraft, R. J., Tardoff, K. S., Krauter, K. S., and Leinwand, L. A. (1988). *Biotechniques,* **6,** 544.
39. Rosenberg, A. H., Lade, B. N., Chui, D.-S., Lin, S.-W., Dunn, J. J., and Studier, F. W. (1987). *Gene,* **56,** 125.
40. Dorje, F., Levey, A. I., and Brann, M. R. (1991). *Mol. Pharmacol.,* **40,** 459.
41. Dorje, F., Wess, J., Lambrecht, G., Tacke, R., Mutschler, E., and Brann, M. R. (1991). *J. Pharmacol. Exp. Ther.,* **256,** 727.
42. Harlow, D. and Lane, D. (1988). *Antibodies: a laboratory manual.* Cold Spring Harbor Laboratory Press, Cold Spring Harbor, NY.

6

Immunofluorescent and immunogold methods to localize G proteins

J. M. LEWIS

1. Introduction

GTP-binding regulatory proteins (G proteins) underlie pathways of signal transduction that include inhibition of adenylyl cyclase, stimulation of K^+ channels, and stimulation of phospholipase C. Members of the family of G proteins—G_i, G_s, G_o, G_z, or G_i—may also regulate events such as fusion of endosomes, and there are probably other, as yet unknown receptors and effectors modulated by G proteins. While the participation of G proteins in the regulation of events at the cell surface has been well characterized, less is known about the intracellular control of such signal transduction pathways. One way of modulating G protein function, and thereby co-ordinating different paths of information transfer, involves spatial and temporal controls over G protein function in intact cells. Through these controls a cell can modulate the ability of its complement of different G proteins to interact directly with particular receptors or effectors. Spatial and temporal parameters governing G protein function within intact cells possibly include the compartmentalization of different G protein subtypes with selected receptors and effectors within the plasma membrane. In addition, the redistribution of G proteins within cells may take place in response to specific agonists, and qualitative or quantitative changes in G protein subunits may occur throughout the cell cycle.

Methods to investigate the co-localization of G proteins with particular receptors and effectors, or with other proteins in previously unknown associations, include immunocytochemistry and subcellular fractionation. Subcellular fractionation enables the study of stable interactions between proteins within an averaged cell population, but is limited by confidence in the purity or recovery of a particular organelle. Indeed, the distribution of proteins may change upon disruption of the cell. Immunocytochemistry circumvents many of these issues, preserves spatial relationships, and permits one to assess directly the distribution of G proteins within intact, individual cells (1).

The primary drawback of studies involving immunocytochemistry is the potential for non-specific interaction of antibody. To minimize the potential that questions of antibody specificity might compromise results, the following experimental strategy is recommended:

- use of a panel of antibodies directed against peptides corresponding to regions of primary structure of G proteins, in addition to antibodies directed against intact protein, rather than reliance upon antisera from one or two rabbits
- demonstration of the specificity of each antibody for specific G protein subunits (e.g. $G\alpha_{i1}$ or $G\alpha_{i2}$) by a method such as immunotransfer blotting
- specific peptide inhibition and other controls at the level of immunocytochemistry

This chapter will therefore focus on the preparation and purification of antibodies and the design of control experiments such that one can ascertain that the presence of antibody represents the interaction of only G protein subunits with antibodies.

2. Production of antibodies

Antibodies are capable of detecting peptide sequences as short as three or four amino acids in length; the chances that at least several proteins within a cell contain one such short, specific sequence are quite high. However, the chances that unrelated proteins would contain two epitopes in common with a G protein α subunit are small, and it is even less likely that proteins unrelated to the α subunit would share three or more epitopes. In this regard, the use of several peptide-directed antibodies is a powerful means to ensure that the protein visualized in fluorescently labelled cells or gold-labelled sections is indeed the protein of interest.

2.1 Conjugation of peptides with carrier proteins for use as antigens

Keyhole limpet haemocyanin (KLH) will greatly enhance the immunogenicity of a peptide and is widely used. The use of alternative carrier proteins such as bovine serum albumin (BSA) provides an important control for potential KLH-directed antibodies. It is important to note that antibodies directed against BSA will often cross-react with serum components deposited within the extracellular matrix surrounding cultured cells, thereby increasing non-specific background of immunofluorescent samples.

Protocol 1. Conjugation of peptides with keyhole limpet haemocyanin

1. Obtain purified peptides of 10–15 amino acids in length. These should correspond to regions of primary structure of the protein to be studied,

with an additional N-terminal cysteine to which KLH can be cross-linked. For peptides without an N-terminal cysteine, see *Protocol 2*.

2. Activate KLH by reaction with *m*-maleimidobenzoyl-*N*-hydroxysuccinimide ester (MBS). Dissolve 10 mg KLH (Sigma) in 625 μl of 10 mM $NaPO_4$ buffer, pH 7.2. Add 1.75 mg MBS (Pierce) in 200 μl dimethylformamide drop-wise to the KLH solution, with constant stirring. Stir the mixture for 30 min at room temperature.
3. Purify the KLH-MB conjugate by gel filtration chromatography using a 20 ml Sephadex G-25 column equilibrated with 50 mM $NaPO_4$ buffer, pH 6.0; combine fractions containing the activated KLH, which will have a pale yellow colour. Adjust the pH of the KLH-MB to 7.2 by drop-wise addition of 1 M NaOH.
4. Determine the amount of free sulfhydryl groups on the peptide by the method of Ellman (2). Assuming that 100% of the sulfhydryl groups remain free, dissolve the equivalent of 12.5 mg of peptide in 1 ml of 20 mM $NaPO_4$, pH 7.2. Adjust the amount of peptide accordingly. Add the peptide solution to all of the KLH-MB. Stir the mixture gently for 3 h at room temperature; store in aliquots at −70°C.

Protocol 2. Conjugation of peptides with bovine serum albumin

1. Obtain purified peptides of 10–15 amino acids in length corresponding to regions of primary structure of the protein to be studied. Peptides without N-terminal cysteines, but containing lysines, are cross-linked to BSA with glutaraldehyde.
2. Dissolve 1 μmol of peptide in PBS, and add to this 0.15 μmol of BSA (Sigma, Fraction V) dissolved in PBS.
3. Add 0.4 μmol fresh EM-grade glutaraldehyde to the mixture drop-wise with stirring. Continue to stir mixture 1 h at room temperature.
4. Dialyse three times against 500 ml of phosphate-buffered saline (PBS), pH 7.2, at 4°C. Store in aliquots at −70°C.

Protocol 3. Immunization of rabbits

Screen by immunofluorescence and immunotransfer blotting (see *Protocols 4, 7,* and *8*) the pre-immune sera of individual rabbits for the presence of antibodies that react with the cells to be used. The animal vendors will usually supply serum samples for such testing.

1. Immunize three female, non-pregnant rabbits by multiple-site intradermal injection (3, 4) with either KLH- or BSA-conjugated peptides (see

Protocol 3. *Continued*

Protocols 1 and *2*) or purified protein, such as mixture of G_i and G_o purified from bovine brain (5) or G_i purified from rabbit liver (6). Use conjugated antigens at amounts ranging from 0.25 to 2.0 mg per rabbit, in combination with Freund's complete adjuvant at a 1:1 (v/v) ratio.

2. Test the sera 3 to 4 weeks post-injection by reaction with purified G_i in immunotransfer blotting assays, as described below (see *Protocol 4*). Bleed those rabbits with positive titres at 2 week intervals.
3. If the titre drops, such that an antiserum diluted 1:100 cannot readily detect 100 ng purified G protein α subunits on immunoblots, boost the rabbit with antigen.
4. Perform the boost in identical fashion to the primary injection, except substitute incomplete Freund's adjuvant for the complete, and administer injections into the femural and subscapular muscles.

2.2 Test of antibody specificity by immunotransfer blotting

Immunotransfer blotting provides a means to assess the ability of antisera directed against peptides to react with intact G protein subunits. In addition, it is important to screen antisera for reaction with proteins other than G proteins subunits within lysates of the cells to be used in immunocytochemistry experiments.

Protocol 4. Immunotransfer blotting

1. Following SDS–PAGE, transfer the samples to 0.45 μm nitrocellulose (Schleicher and Schuell) by exposure to 60 V for 2 h, in 25 mM Tris–HCl and 192 mM glycine containing 20% methanol.
2. Wash the nitrocellulose in 20 mM Tris–HCl, pH 7.5, containing 500 mM NaCl ('TBS').
3. Block non-specific sites of antibody binding with 3% gelatin in TBS for 30 min at room temperature,[a] with vigorous agitation on a rotary shaker.
4. Incubate the nitrocellulose with 1% gelatin in TBS containing the rabbit antiserum at 1:100 dilutions for 45 min at room temperature, with shaking.
5. Wash the nitrocellulose membranes for 3 × 5 min with TBS containing 0.5% Tween-20 (Sigma).
6. Incubate the membranes with biotinylated goat anti-rabbit IgG (Vector Laboratories) diluted 1:2000 in TBS containing 1% gelatin; wash as in step 5.

7. Incubate the membranes with horseradish peroxidase-conjugated streptavidin (Vector Laboratories) diluted 1:2000 in TBS containing 1% gelatin; wash as in Step 5.
8. To initiate the colour reaction, incubate the membranes in 100 ml TBS containing 60 μl H_2O_2 plus 60 mg 4-chloro-1-naphthol dissolved in 20 ml methanol.

[a] Note: 'room temperature' should be warm enough to maintain gelatin in a liquified state.

2.3 Purification of antibodies

In addition to antibodies, rabbit antisera contain other glycoproteins which compete non-specifically with antibodies for binding to antigens. Serum also contains proteases, which can break down the antibody or antigen during incubations with sample. Affinity chromatography of IgG using protein A–Sepharose is the easiest way to eliminate these contaminants.

Serum contains contaminating IgG, as well. By eliminating both IgGs directed against epitopes within KLH or BSA and naturally occurring IgGs that are directed against components of the cells to be used, peptide-based affinity purification of antibodies provides a means to assess the specificity of the antibodies.

Protocol 5. Purification of IgG from antiserum

1. Equilibrate 2 ml of protein A–Sepharose (Pierce) with 10 mM Tris–HCl, pH 7.4, at room temperature.
2. Dilute 1 ml of rabbit serum to 10 ml in the equilibration buffer. Apply to the column and collect 2 ml fractions. Wash the column with 10–20 ml of the equilibration buffer.
3. Elute the IgG with citric acid, pH 3.6. Collect 666 μl fractions in microfuge tubes containing 333 μl of 1 M Tris–HCl, pH 8.0 to neutralize the samples immediately.
4. Where further purification is not needed, concentrate the pooled fractions of IgG five to 10 times in Centricon 10 microconcentrators (Amicon); dialyse the salts out by diluting once again with PBS and then reconcentrate the sample. Store either at 4°C or in aliquots at −20°C to avoid repeated freezing and thawing.

Protocol 6. Affinity purification of antibodies

1. Dissolve 2 mg of the peptide used as antigen in 1.0 ml of 100 mM sodium bicarbonate, pH 9.6.

Protocol 6. *Continued*

2. Add this to 1.5 ml Reacti-gel 6X (Pierce), activated according to the manufacturer's directions, in a 2 ml microfuge tube.
3. Incubate the mixture for 16 h at 4°C on an inversion mixer.
4. Apply fractions containing IgG (see *Protocol 5*, step 3) directly to a column containing 1–2 ml of Reacti-gel 6X to which the peptide antigen has been coupled.
5. Recycle the mixture of IgGs through the column for 1–2 h at room temperature.
6. Collect the unbound material, and elute the specifically bound antibodies with either 0.5 ml of 100 mM glycine, pH 2.8, or 0.5 ml of 8 M urea.
7. Neutralize the purified IgG solution with HCl or NaOH, added drop-wise while vortexing. Test the pH by applying 10 μl samples to pH paper. Dialyse against Tris-buffered saline, pH 8.1.
8. Store the affinity-purified antibodies either at 4°C, or in 50% glycerol at −20°C.

---

3. Light microscopy

Indirect immunofluorescence experiments permit one to assess rapidly the localization of different G protein subunits within different types of cells; the results of each experiment can be obtained within one day. An example of G protein distribution within Swiss 3T3 fibroblasts is shown in *Figure 1*.

3.1 Fixation and staining of cells

Because cell material that has dried results in non-specific reactions, care must be taken to ensure that the cell samples are not permitted to dry out at any step; transfer the coverslips rapidly from one solution to the next.

3.1.1 Controls for immunofluorescence

The native antibody complement of rabbit serum varies with time; therefore, comparison of immune serum with pre-immune cannot by itself establish specificity. The following control experiments are essential for correct interpretation of results (see also *Table 1* for a troubleshooting guide):

- pre-immune sera corresponding to each antiserum
- antibodies directed against purified protein, such that epitopes formed by secondary and tertiary structure are detected, as well as those within primary structure
- antibodies produced by different animals
- specific peptide or antigen pre-adsorption of the antibodies

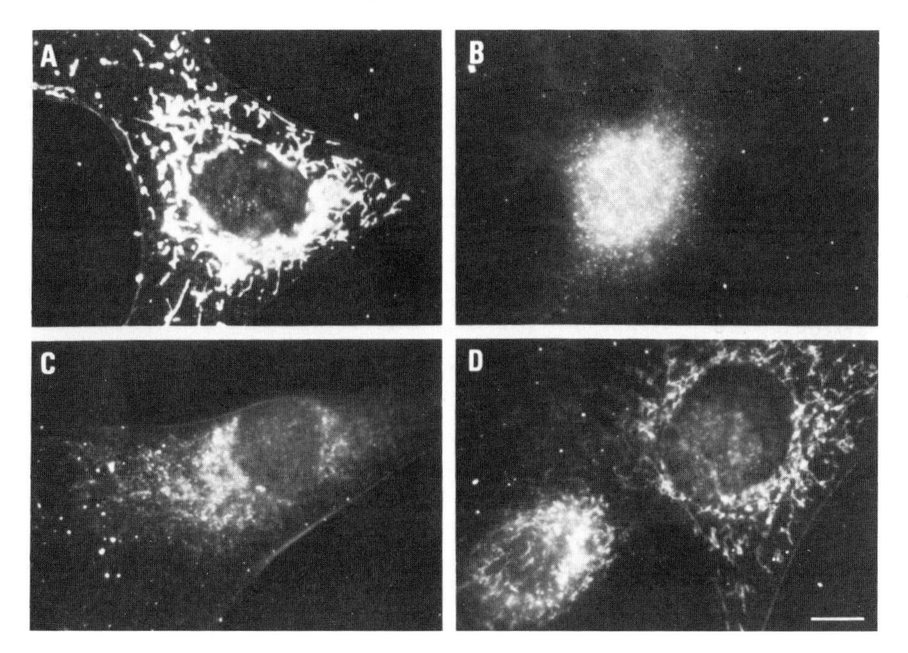

Figure 1. Effects of thrombin on the distribution of $G\alpha_i$ within mouse 3T3 fibroblasts. Prior to fixation and immunolabelling, cells were either untreated (*panel* ***A***) or treated with 10 units/ml thrombin at 37°C for 5 sec (*panel* **B**), 10 sec (*panel* **C**), or 20 sec (*panel* **D**). Bar, 10 μm.

- alternative means of visualizing the antibodies, such as peroxidase products in place of the fluorophore
- testing of each secondary antibody with no primary antibody present
- different methods of fixation (fixation can create epitopes, obscure epitopes through cross-linking of adjacent proteins, or fail to fix the protein of interest in its native position)
- use of antibodies directed against proteins of known localization, such as tubulin, to assess the integrity of the cell and organelles after fixation and permeabilization

3.1.2 Fixation and permeabilization of cells

The following fixative, prepared freshly before use, results in good preservation of intracellular fine structure such that the final immunofluorescent images have optimum resolution.

(a) Add 2 g paraformaldehyde (Sigma) to 50 ml 0.1 M Pipes buffer, pH 6.8, containing 2 mM EGTA and 2 mM $MgCl_2$, in a 250 ml Erlenmeyer flask.

(b) Heat for about 20 min to dissolve, in a gently boiling water bath in a fume hood; swirl occasionally to mix; cool to room temperature before use.

Alternatively, use commercially prepared methanol-free formaldehyde at 10% (Polysciences), diluted to 3.7% with PBS. To control for fixation by aldehydes, cells can be fixed and permeabilized by treatment with a solution of 95% methanol and 5% acetic acid for 5 min at room temperature.

Protocol 7. Fixation of cells

1. Sterilize glass coverslips by autoclaving for 20 min, which leaves them less fragile than after flaming with 95% ethanol. Separate individual coverslips with pieces of metal foil to prevent sticking.
2. 24–48 h prior to processing, trypsinize cells and plate at low density (e.g. 1.5×10^5 cells/10 mm^2 for fibroblasts) such that cells are examined prior to confluence.
3. Plate fibroblast and glioma cell lines directly on the coverslips in 35 mm tissue culture dishes. Plate other types of cells, such as early passage endothelial cells or neuronal explant cultures, on to glass coverslips coated with laminin or fibronectin; substrates will adhere best to glass coverslips treated first by washing for 1 h in gently heated 5% H_2SO_4.
4. Wash the cells grown on coverslips once with phosphate-buffered saline (PBS) at room temperature, and using fine forceps, place into Coplin staining jars containing fixative for 20 min at room temperature. Keep cell sides of coverslips facing the front of the jar.
5. Pour the fixative out of the jars; wash coverslips in the Coplin jars three times with PBS at room temperature, without rinsing the cell sides of the coverslips. Samples may be stored at this step in sterile PBS at 4°C for 1–2 days, although best results are obtained by proceeding to the mounting step on the same day (see *Protocol 8*).

3.1.3 Staining of fixed cells

i. Optimizing antibodies

For each primary antiserum and the corresponding pre-immune serum, test various dilutions ranging from neat to 1:1000. To minimize non-specific background, choose the highest dilution at which the signal is still bright but at which staining of the corresponding pre-immune serum is negligible. Optimize and refine the dilution.

Using affinity-purified antibodies such as goat anti-rabbit IgG conjugated to either fluorescein or rhodamine (Cappel Research Reagents), optimize the dilution of the secondary antibodies. With the directions of the manufacturer as a guide (and using primary antibodies of known localization and optimum dilution), select the highest dilution at which the signal is still bright but at which there is no fluorescence observed for the secondary antibody alone.

Protocol 8. Staining of fixed cells

1. Permeabilize fixed cells on coverslips (see *Protocol 7*) with 0.2% Triton X-100 (Sigma) in PBS, 10 min at room temperature in Coplin staining jars.
2. Block non-specific sites of antibody binding with 10% sterile normal goat serum (Gibco) in PBS, 30 min at room temperature.
3. Dilute primary antibodies with the blocking solution.
4. Place each coverslip face-down on to 50 μl of diluted primary antibody solution on a piece of Parafilm.
5. Incubate the samples for 45 min at 37 °C in a humidified chamber, such as a cell incubator.
6. Return the coverslips to the Coplin jars and wash for 5 × 5 min in PBS.
7. Dilute secondary antibodies with blocking solution.
8. Place each coverslip face-down onto 50 μl of diluted secondary antibody solution on a piece of Parafilm; repeat steps 5 and 6.
9. To mount the coverslips, remove each one, touch one edge to tissue paper to drain excess buffer, and briefly and gently blot the bare side of the coverslip with tissue paper.
10. Place the coverslip cell-side down on to a microscope slide, on a 14 μl drop of Fluoromount (Fisher) to which 2.5% (w/v) 1,4-diazabicyclo-(2,2,2) octane ('DABCO'; Polysciences) has been added to stabilize fluorescence.
11. Seal the coverslip to the slide with nail varnish, and store the slides at 4°C.
12. Observe the samples by phase-contrast and fluorescence microscopy, on a microscope equipped with a mercury arc lamp and appropriate filters; in order to examine the localization of proteins at the level of subcellular organelles, oil immersion objectives of at least 60× magnification are recommended.
13. Photograph fluorescent images with Ektar 1000 ASA film (Kodak) with the camera set either to automatic exposure time for fluorescein (generally resulting in exposures of 1–2 sec), or set to 30 sec or 1 min for rhodamine. Film can be printed commercially with colour corrections such that the backgrounds are black.

ii. Alternative method for detection of labelled antigens: peroxidase products

To eliminate the possibility of artefacts produced by fluorescent compounds, substitute biotinylated goat anti-rabbit IgG (Vector Laboratories) for the

Table 1. Troubleshooting guide

Problem	Cause	Solution
No cells	under-fixation	fix for longer
	detergent too concentrated	reduce the detergent concentration
	washing too forceful	wash more gently
No fluorescence	over-fixation	fix for less time
	epitope cross-linked by fixative	use a different primary antibody or fixation method
	cells not permeabilized	permeabilize for longer
	primary or secondary antibody decomposed or bleached	use fresh antibody; add fluorescence stabilizer (DABCO)
	secondary antibody directed against wrong animal or Ig subtype	use correct antibody
	antigen of interest not in cells	use different cells
	mounting medium below pH 9.0	use different mounting medium
Fluorescence with pre-immune serum	antibody concentration too high	optimize dilutions
	cells dried out	minimize exposure to air
	incomplete washing after secondary antibody	wash for longer
	serum contains antibodies against cell structures	affinity-purify corresponding immune serum
	autofluorescence of cells	usually very low level
High background (nuclear) fluorescence with immune serum	primary or secondary antibody too concentrated	optimize antibody dilutions
	primary or secondary antibody contain contaminating antibodies	affinity-purify antibodies
	incomplete washing after primary and secondary antibody incubations	wash for longer
High background fluorescence on cells and substrate	primary antibody reacts with extracellular matrix components	affinity-purify antibody or use alternative antibody
Fluorescence with secondary antibody alone	secondary antibody too concentrated	optimize dilution
	secondary antibody not purfied	use affinity-purified antibody
Fluorescence in absence of primary or secondary antibody	autofluorescence of cells	usually insignificant; if not, use different cell type
Constellations of fluorescence	aggregates of secondary antibody	briefly centrifuge secondary antibody

fluorophore-conjugated secondary antibody (see *Protocol 8*, step 5). Following the washing step, incubate with horseradish peroxidase-conjugated avidin (Vector Laboratories) and wash again. Initiate the colour reaction by adding a solution of 0.5 mg of diaminobenzidine in 50 mM Tris–HCl, pH 8.0, to which 0.01% H_2O_2 has been added just before use. Wash again and view as in *Protocol 8*.

iii. Double-labelling of cells

It is often of interest to observe the co-distribution of two proteins. When primary antibodies are from different sources, such as from rabbit and mouse, combine the antibodies at the appropriate dilutions and incubate simultaneously. Choose secondary antibodies also from two different animals, such as rhodamine-conjugated sheep anti-mouse IgG and fluorescein-conjugated goat anti-rabbit IgG; incubate these also simultaneously. Controls should include labelling cells with each pair of primary and secondary antibodies alone, and reversing the secondary antibodies to ensure that no cross-reaction occurs. If the primary antibodies are from the same animal, then the incubations are performed consecutively.

3.1.4 Specific pre-adsorption of antibodies with peptides

This type of control experiment is especially important to perform when using either unpurified antisera or IgG fractions, which can contain either auto-antibodies directed against components of the cell, or antibodies directed against KLH which bind to epitopes within the cell.

Protocol 9. Peptide pre-adsorption of antibodies

1. In PBS, dissolve either the peptide used as antigen or, as a control, a non-related peptide, at 4 mg/ml.
2. Dilute peptide stock solution with PBS to concentrations ranging from 1 to 100 μg/ml. Add 10 μl of peptide solution to 1 μl of antiserum in a 1.5 ml microfuge tube.
3. Incubate the mixture either for 5 min at room temperature, or with moderately vigorous shaking for 1–16 h at 4°C. The time of incubation will vary with each peptide–antibody pair.
4. Bring the volume up to 100 μl with PBS, and use the sample directly in the immunocytochemistry experiments.
5. To test the efficiency of the pre-adsorption on immunotransfer blots, increase the reaction volumes appropriately and use in place of the diluted primary antibody at step 4 of *Protocol 4*.

3.1.5. Treatment of cells with agonists or other agents

Once the localization of a protein within the cell has been established, it may be of interest to observe changes in localization upon treatment of the cell with agents that activate signal transduction pathways. As an example, note the dramatic changes in localization of $G\alpha_i$ occurring in Swiss 3T3 fibroblasts treated with thrombin for as little as 5 sec, as shown in *Figure 1*.

Protocol 10. Treatment of cells with activating agents

1. Expose cells grown on coverslips to the activating agents immediately prior to fixation.
2. Add agents, such as the calcium ionophore 4-bromo-A23187 at 5 μM (Molecular Probes), directly to the cell medium and incubate at 37°C for the appropriate time, such as 30 or 60 min.
3. Stop the reaction by immersion in fixative, and process the samples for immunofluorescence (see *Protocols 7* and *8*).
4. For very short times of incubation, such as 5–20 sec for thrombin or platelet-derived growth factor, gently agitate the tissue culture dish on a rotary shaker. Grasp the coverslip with forceps, and then add the reagent. At the appropriate times, immerse the coverslip into fixative contained in a Coplin jar and process for immunofluorescence (see *Protocols 7* and *8*).

---

4. Electron microscopy

Immunogold labelling using embedded thin sections provides an alternative means of visualizing G proteins, one in which the access of antibodies is not dependent upon permeabilization with detergents, so that the plasma membrane remains intact. The higher resolution of electron microscopy permits study of localization of proteins within organelles. The limitations are that each experiment can take several days to complete, so the screening of many different antibodies and types of cell is not practical and should first be done at the light microscopy level (see *Protocols 7* and *8*). It is also important to note that during fixation with glutaraldehyde, which takes longer than formaldehyde, the plasma membrane becomes transiently permeable to calcium, so the localization of proteins involved in signal transduction could change.

A comprehensive description of techniques of electron microscopy is beyond the scope of this chapter (see reference 7 for more information), but I have included a partial list of protocols required to prepare samples for analysis of G protein localization which can then be observed by standard techniques of electron microscopy.

While the reactions are essentially the same as for light microscopy, the necessary sectioning of the cells requires more complete fixation and thus harsh treatment of cellular proteins, but eliminates the need for permeabilization so that the plasma membrane remains intact.

Protocol 11. Preparation of gold-conjugated IgG

1. Adjust the pH of 10 ml of a colloidal gold solution, with 6 or 15 nm diameter gold particles (Polysciences), to 9.5 with fresh 0.1 M K_2CO_3; filter through a 0.22 μm Nalgene filter.
2. Mix 150 μg of protein A-purified IgG (see *Protocol 5*) with 10 ml of the colloidal suspension of gold particles.
3. With vigorous stirring, add 15 μl of the IgG diluted into 85 μl of 2 mM boric acid. Incubate the mixture for 10 min at room temperature.
4. Stabilize the colloid by addition of 100 μl of 10% BSA, pH 7.4, with vigorous stirring.
5. Concentrate the mixture to 1 ml in a Centricon 10 (Amicon).
6. Apply the conjugate to a column containing Ultragel AcA-44 (LKB), previously equilibrated with 1% BSA in PBS. The colloid will elute in two peaks; discard the first peak which consists of gold aggregates.
7. Concentrate the second peak in a Centricon (Amicon), resuspend in 0.1% BSA in PBS, and concentrate to 1.5 ml.
8. Sterilize the gold–IgG conjugate by filtration and store at 4°C.

Protocol 12. Preparation of samples

1. Plate cells on Thermanox plastic coverslips (Miles Laboratories, Lab-Tek Division) for 1–2 days before use.
2. Wash two or three times with PBS.
3. Fix for 1 h at room temperature with 1% glutaraldehyde (EM grade), 0.2% picric acid, and 6 mM eserine in PBS.
4. Rinse coverslips twice quickly with ice-cold deionized water.
5. Dehydrate cells with 50% ethanol for 15 min, followed by three 60 min incubations in 75% ethanol.
6. Infiltrate cells with 1:1 mixture of LR white (Polysciences) and 75% ethanol (made by adding 75% ethanol to LR white slowly, while stirring), for 1 h.
7. Incubate samples overnight at 4°C in 100% LR white resin, followed by 2 h at room temperature in fresh 100% LR white.

Protocol 12. *Continued*

8. Place coverslips over gelatin capsules filled with resin and cure at 50°C for 2 days.
9. Cut sections parallel to coverslips (60 nm thickness) and mount on nickel grids.

Protocol 13. Staining of sections for electron microscopy

1. Incubate mounted sections at room temperature for 1–2 h with either 1% ovalbumin or BSA in PBS, to block non-specific sites of antibody binding.
2. Incubate mounted sections with gold-labelled antibodies (see *Protocol 11*) for 36 h at 4°C.
3. Wash in 10 mM Tris–HCl in PBS.
4. Stain labelled sections with neutralized 2% aqueous uranyl acetate for 3 min followed by 1 min with bismuth subnitrate.

Acknowledgements

I would like to thank the following people for their contributions to this work: Marilyn Woolkalis, George Gerton, Robert Smith, Leonard Jarett, and in particular, David Manning. This work was supported in part by National Institutes of Health grant GM 34781 to D.R.M.

References

1. Lewis, J. M., Woolkalis, M. J., Gerton, G. L., Smith, R. M., Jarett, L., and Manning D. M. (1991). *Cell Regulation*, **2,** 1097.
2. Ellman, G. L. (1959). *Arch. Biochem. Biophys.*, **82,** 7077.
3. Vaitukaitis, J. L. (1981). *Methods Enzymol.*, **73,** 46.
4. Mumby, S. M., Kahn, R. A., Manning, D. R., and Gilman, A. G. (1986). *Proc. Natl Acad. Sci. USA*, **87,** 265.
5. Sternweis, P. C. and Robishaw, J. D. (1984). *J. Biol. Chem.*, **259,** 13806.
6. Bokoch, G. M., Katada, T., Northrup, J. K., Ui, M., and Gilman, A. G. (1984). *J. Biol. Chem.*, **259,** 3560.
7. Hayat, M. A. (ed.) (1981). *Techniques of Electron Microscopy: Biological Approach*. University Park Press, Baltimore, MD.

7

Technology for real time fluorescent ratio imaging in living cells using fluorescent probes for ions

W. T. MASON, J. HOYLAND, I. DAVISON, M. A. CAREW, J. JONASSEN, R. ZOREC, P. M. LLEDO, G. SHANKAR, and M. HORTON

1. Introduction

In this article, we shall discuss the means for applying fluorescent probes and computer-controlled image acquisition and analysis to previously intractable problems. Image processing is at the heart of virtually all of these new approaches. This article will focus mainly on how image acquisition and analysis can be used to study ionic gradients in living cells with optical probes. Some specific applications from our laboratory using this technology will be used to illustrate the potential for such techniques.

The development of chemical probes to image ions has made it possible to study specifically biological activity in single living cells. Computer-controlled instrumentation for rapidly acquiring data from living tissue has in turn made it possible to acquire ultra-low light level data from these probes and to analyse either images or temporal changes in light emission, or both.

The ability to interface fast video cameras and computer technology to the conventional microscope has made it possible not only to make qualitative observations, but to derive quantitative image data from single cells, at speeds of up to 30 video frames per second with video cameras or confocal laser scanning technology, or many hundreds of samples per second if photon counting technology with photomultipliers is used.

2. Multidisciplinary advances

A number of important scientific advances in chemistry, biology, physics, electronics, microscopy, and computing have produced an exciting range of new technologies for single cell study of living biological systems.

These developments have included:

- fast microelectronic circuitry for capturing video images in digital format in real time for later processing
- development of low-light level detectors and ultra-high sensitivity photomultiplier tubes which can detect and image faint fluorescence at high speeds
- synthesis of chemical probes for cellular function, ranging from chemiluminescent and fluorescent dyes sensitive to ions such as calcium, sodium, chloride, and protons, and to intracellular proteins
- image analysis software to enable quantitative measurements on digital image or photon data.

These approaches allow measurement in real time of the fast changes in intracellular ionic concentration in living cells, with limited disruption of normal cell function.

3. Chemical probes for imaging functions of living cells

3.1 Intracellular ions

The first measurements of ionic activity used photoproteins such as the calcium sensitive molecule aequorin, which emitted light when combined with calcium ions. Then calcium-sensitive dyes included arsenazo and murexide, but these dyes had major disadvantages in that they required microinjection. The development in the 1980s of new fluorescent dyes by Tsien and colleagues provided the ability to incorporate dyes into single cells and to investigate ionic activity (1). The dyes are sensitive to nanomolar concentrations of calcium such as occur in single cells, and other dyes for different ions provide sensitivity appropriate to physiological requirements. The acetyoxymethylester form of the dye can be loaded into single cells, and most cells contain endogenous esterases which rapidly (~5–30 min) hydrolyse the dye to form the free acid, which is trapped in the cell and is ion-sensitive.

A number of such dyes are shown in *Table 1*. These dyes have differing wavelengths of light output, but they all emit photons of light in the visible spectrum. Their light output is well matched to available detectors, including both photomultipliers and intensified video cameras.

The two most commonly used dyes for measuring intracellular calcium are Fura-2 (*Figure 1*) and Indo-1. The dyes have high quantum efficiency and are sensitive to calcium at concentrations from 30 nM to 5μM or so. Fura-2 displays a single emission peak at 510 nm, but two calcium dependent absorption maxima, one at 340 nm which increases with increasing ionized calcium and a second at 380 nm which similarly decreases with a rise in ionized calcium. Indo-1 is generally excited by only a single wavelength of light (340–

Table 1. Table showing the different optical probes currently in common use

Ion	Dye	Excitation wavelength (nm)		Emission wavelength (nm)	
		1	2	1	2
Dual excitation dyes					
Ca^{2+}	Fura-2	340	380	510	
	Fura-5	340	380	510	
	Fura red	480–500	425–450	660	
H^+	BCECF	440	490	530	
	SNARF-6	500	560	610	
	SNAFL-1	500	560	600	
Na^+	SBF1	340	380	510	
K^+	PBFI	340	380	510	
Mg^{2+}	Mag-Fura-2	340	380	510	
Dual emission dyes					
Ca^{2+}	Indo-1	350		405	480
H^+	SNAFL-1	500		540	635
	SNARF-2	500		550	640
	DCH	405		435	520
Na^+	FCRYP-2	350		405	480
Single wavelength dyes					
Ca^{2+}	Fluo-3	505		530	
	Ca Green	505		530	
	Ca Orange	550		575	
	Ca Crimson	590		610	
Cl^-	SPQ	350		440	
	Fluorescein	495		535	
	Rhodamine	550		595	

Dyes fall into three categories, namely those with ion-dependent dual excitation spectra, those which are dual emission, and those which show ionic dependence only at a single wavelength.

360 nm), but emits light at two different calcium-sensitive wavelengths (405 and 490 nm). With all such probes, choice of excitation wavelength will influence the wavelengths of light emitted and the specific dynamic range of the dye response.

The field of calcium research is attracting considerable attention and new probes for calcium ions are being developed including Calcium Green, Calcium Crimson, and Calcium Orange, and Fura Red. Fura Red for instance exhibits fluorescence decreases as Ca^{2+} increases and it has a large Stokes shift of ~175 nm. It may be combined with probes such as Fluo-3, a single wavelength Ca^{2+} probe whose fluorescence increases with Ca^{2+} activity. Because these dyes are excited near convenient visible laser lines, they also have the potential to be simultaneously loaded into a single cell for ratio-

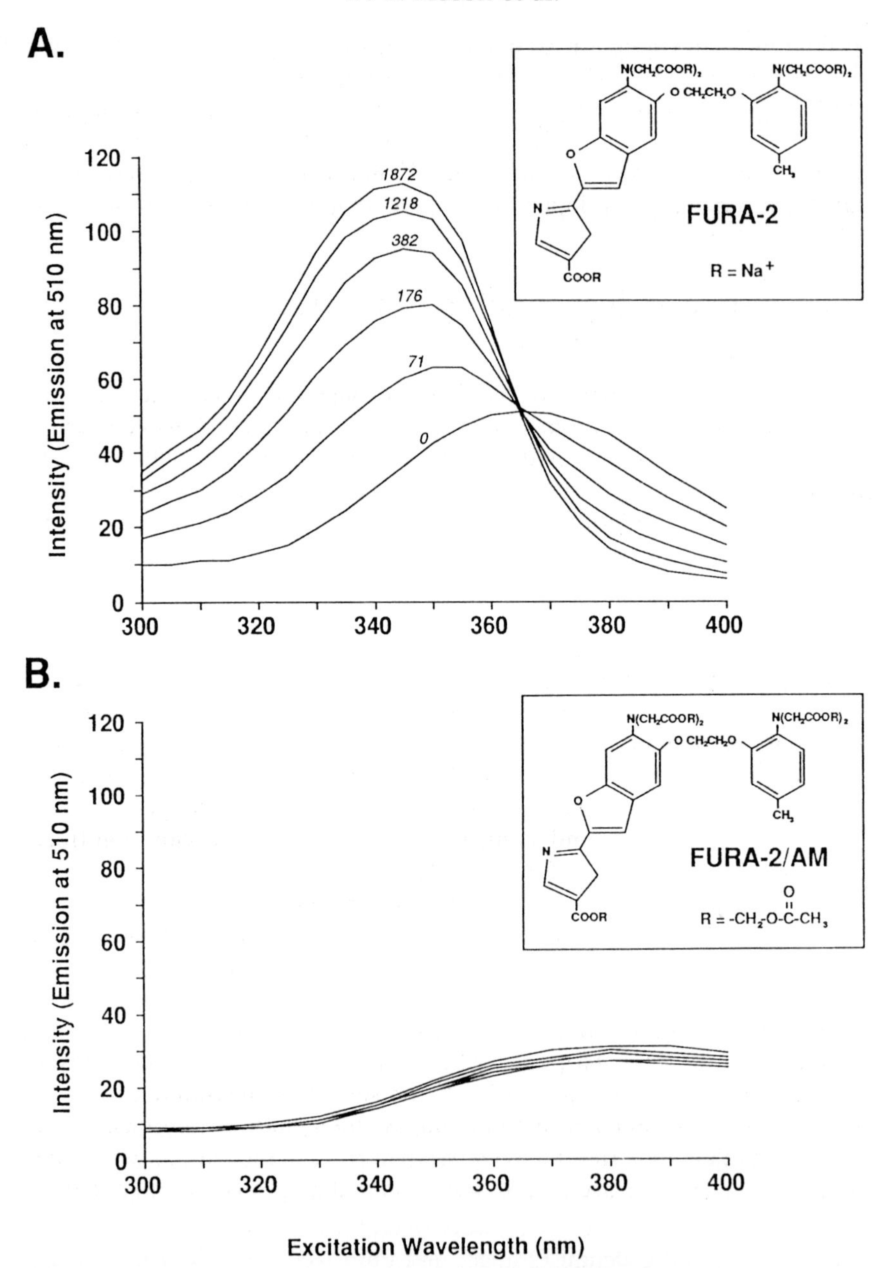

Figure 1. A. Spectra of Fura-2 free acid, the calcium-sensitive optical probe. Note that as ionized calcium concentration increases, 340 nm fluorescence (measured at 510 nm) increases, but 380 nm fluorescence decreases. This permits ratio measurements to be made. **B.** Similar spectra of Fura-2 acetoxymethylester, used to load the cells, showing relative calcium-independence of fluorescence. Esterases inside the single cell hydrolyse this version of the dye to form the free acid shown in (A) (**inset**).

metric confocal imaging, or for dynamic video imaging with CCD cameras in systems not configured with UV-transmitting optics.

Other work with new calcium probes is focusing on indicators with higher and lower affinity for Ca^{2+}. Fura-5, for example, has an affinity for Ca^{2-} of 40 nm while Fura-2 has an affinity of 135 nm at room temperature. Other probes are also being developed for magnesium ions, with excellent sensitivity and specificity.

4. Real time video imaging of ion-sensitive fluorescent dyes

Fluorescence ratio imaging is the key to dynamic, video-enhanced light microscopy for optical probes detecting ions and second messengers. If fluorescent images are obtained as a pair at 340 and 380 nm (with Fura-2 for instance), and the images are ratioed on a pixel-by-pixel basis (a pixel is the resolving unit of a video camera, many thousands of which are combined together to give an overall image). The resulting 'ratio image' is proportional to ionized calcium concentration and reduces the chance of possible artefacts due to uneven loading or partitioning of dye within the cell, or varying cell thickness and dye concentration.

Indo-1 is well suited for photometric measurements—it has a faster temporal response to calcium changes (5–16 ms), and can be used with static optical beam splitters to separate the emitted light and focus it on to two photomultiplier tubes as a continuous signal. Ratio measurements can also be employed. This approach has the advantage that no movement need take place in order to change filter position, and so measurements are fast and vibration free.

Another dye (BCECF), for example, measures intracellular pH as an optical signal, using similar dual wavelength imaging technology. This dye is excited at 440 and 490 nm and measured at ~510–520 nm. An increase in pH increases the fluorescence of the dye excited at 490, but this has little effect on 440 nm fluorescence. BCECF can be combined with Fura-2 to provide simultaneous imaging of Ca^{2+} and pH. A four-position filter wheel is used to excite the probes at 340, 380, 440, and 490 nm, after both dyes have been loaded into the single cell simultaneously. Fluorescence is measured with a 515 dichroic long pass filter and a 535 bandpass filter, and this provides strong signals for both dyes with minimal overlap or interference of the two probes.

Other optical probes are also available for sodium, chloride and potassium ions.

5. Comparison of photometric detection, laser scanning confocal, and dynamic video imaging

The technologies for studying optical probes in biological systems fall into three categories. **Photometric technology** permits temporal measurements of photons on fast time scales.

Table 2. The pros and cons of imaging and photometry measurement of optical probes in living cells

	Imaging	**Photometry**
Spatial information	Good	Limited
Speed	Slow (1–30 samples/sec)	Fast (5000 samples/sec)
Sensitivity	Lower	Higher
Data content	High	Low
Multiple parameter acquisition	Difficult	Easy
Results display	Usually off-line	On-line
Cost	Higher	Lower

Video cameras and laser scanning confocal technology permit a second dimension of observation. **Video-enhanced light microscopy**, or **dynamic video imaging** permit both *temporal* and *spatial* measurements of biological activity. The comparative characteristics of imaging and photometry are discussed in *Table 2*. For most work with ion-sensitive optical probes, intensified cameras are used since the working levels of light emitted by these dyes are not detectable by normal video cameras alone. A photosensitive array combined with a front end intensifier to obtain light amplification is used to image the cells under study. Typically these arrays provide up to 512 × 512 pixel resolution at rates up to 30 images per second.

5.1 Confocal laser scanning microscopy (CLSM)

This also provides spatial and temporal resolution. The imaging method may vary from manufacturer to manufacturer. Images of 512 × 512 pixels may be obtained, constructed by sequentially scanning a small point of laser light across the sample, and detecting emitted photons with a photomultiplier tube. CLSM technology yields a very small depth of field, reduces out-of-focus photons and produces very fine detail as optical sections through the cell, which can be used to build up a 3D image. One disadvantage is that frame averaging may be required to reduce noise. Although some systems acquire confocal images at video frame rate and under excitation conditions where minimal bleaching occurs, it is usually necessary to average 20–100 images to obtain high quality data. New CLSM technology from some manufacturers also permits work with UV-excited probes.

CLSM technology offers several specific advantages:

- resolution is enhanced
- inaccuracy due to *z*-axis localization of ions or indicator dye is improved
- background due to stray light is eliminated
- background due to medium is reduced
- haloing is reduced

Thus both CLSM and conventional video microscopy potentially provide a quantitative approach to imaging optical probes. Both techniques have the advantage that digital image analysis techniques permit a wide range of image information to be obtained and can yield both quantitative and qualitative data.

Most importantly, ratio imaging eliminates artefacts due to probe localization and cell geometry. Many of the best ion-sensitive and the new nucleotide-sensitive probes change spectral properties at two wavelengths. Ratio analysis of the two images produces accurate quantification and reduces many artefacts associated with dye localisation and cell thickness.

5.2 Photometric technology

One method of measuring fluorescent probes for ions is by use of photomultiplier-based technology. The system used in our laboratory is called PhoCal (*Figure 2*, Applied Imaging, Hylton Park, Wessington Way, Sunderland, UK). Photometric detection can also be combined with electrophysiology making it possible to accumulate fast electrophysiological signals (up to 35 kHz) while at the same time recording the somewhat slower responses of calcium ions (*Figures 2–5*).

5.3 Dynamic video ratio imaging of ions in cells

Dynamic video imaging can resolve optical probes within cells in terms of both time and space. The MagiCal system used in our laboratory facilitates real time imaging experiments. A typical configuration is shown in *Figure 6*.

Living cells are loaded with fluorescent probe and mounted on a microscope, illuminated by a stable wide-spectrum light source such as a xenon lamp. An image is projected from the microscope on to the face-plate of an intensified video camera, which produces a standard analogue video signal. An intensifying camera is used because the fluorescent image is very faint, typically two or three orders of magnitude fainter than can be detected by the human eye. The faintness is because the dye concentration has to be low enough to avoid toxic effects to the cells under study, and the light source must be reduced so that it does not bleach the dye.

6. Low light level cameras for fluorescence ratio imaging

Cameras employed for real time fluorescence ratio imaging are similar to those used for astrophysics. Several different types of detector are available, consisting of either video frame rate detectors or slow scan charge coupled devise detector technology. *Table 3* summarizes the pros and cons of two of the most widely used cameras.

Generally, either intensified video cameras or cooled slow scan read-out

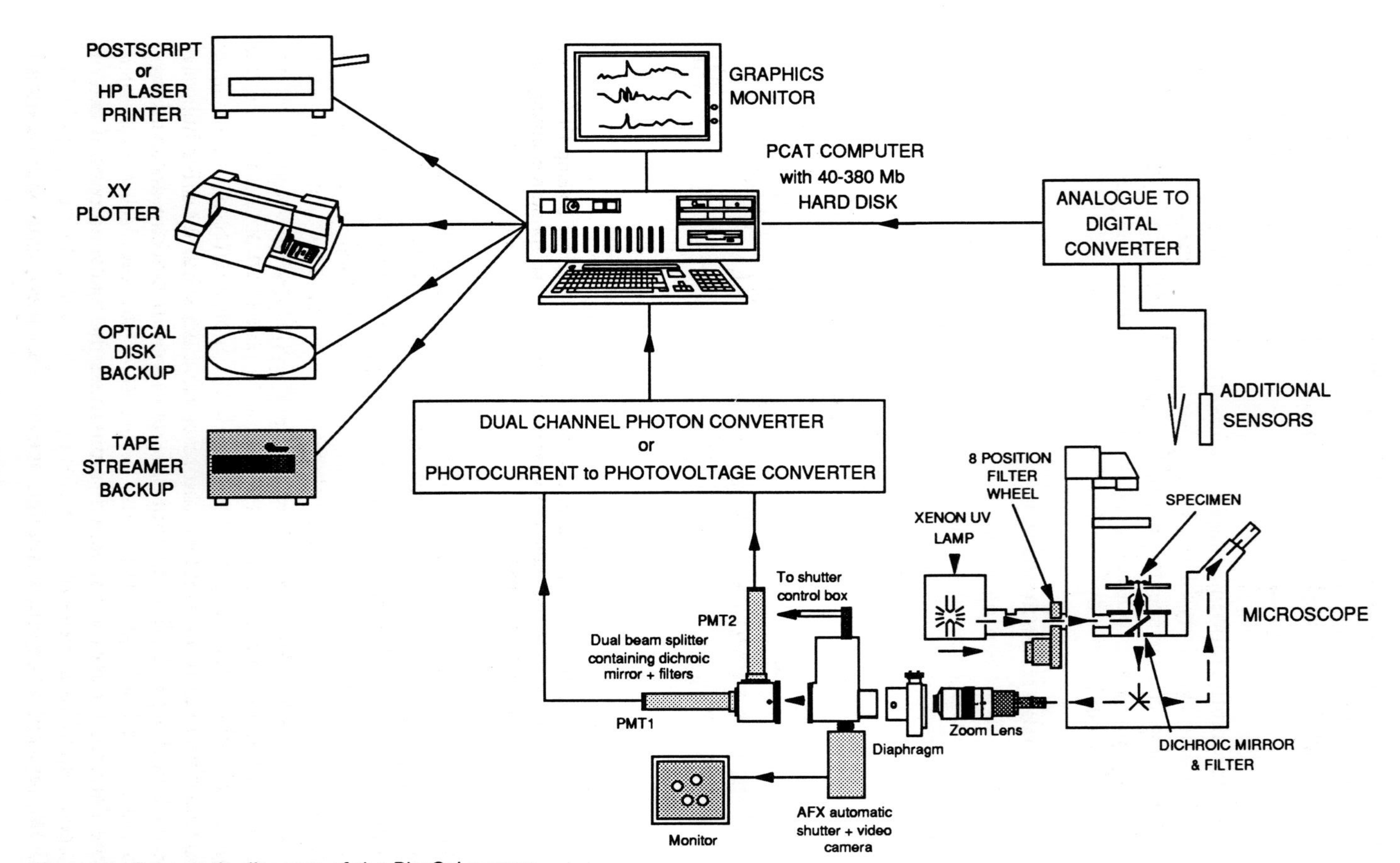

Figure 2. Schematic diagram of the PhoCal system.

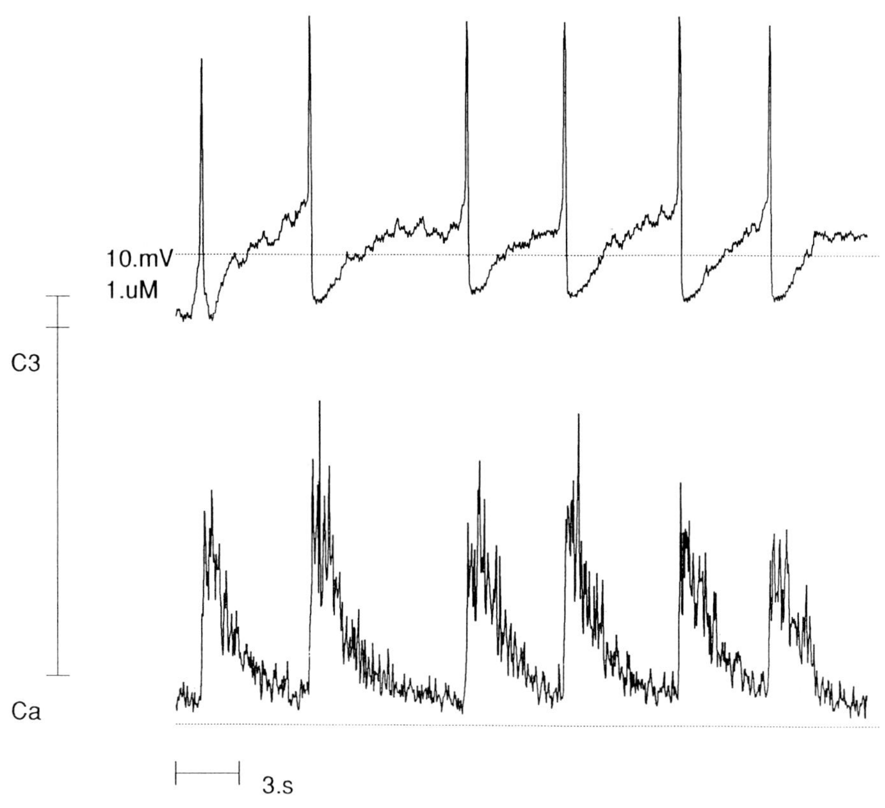

Figure 3. Photometric experiment using Indo-1 and the PhoCal system to measure simultaneously action potentials (top record) and calcium transients in a spiking GH3 clonal pituitary cell. Two emission wavelengths were recorded using fast photon counting, with about five sample points per msec.

cameras are employed. Both employ charge-coupled device detectors, but with different electronic outputs and different noise levels.

Intensified video cameras are used for fast applications where video signals are required. They are generally two-stage, with an optically coated front end intensifier which governs the spectral sensitivity of the camera and this in turn is coupled optically with a lens or with a fibre-optic taper to the video camera stage. Coupling with a fibre-optic taper is preferable to an optical relay lens as light loss is minimized. Typically, a relay lens coupled detector is 5–10 times less sensitive than a fibre-optic coupled system. Most cameras for this work are custom-designed. The first stage of the camera provides intensified input via a micro-channel plate. Typically 10^{-5} or 10^{-6} lux is the light level required to be detected. A fibre-optic taper then reduces the image area on to a CCD image sensor. These devices generally put out a standard video signal which can be captured using video frame grabbers. The video signal is composed of

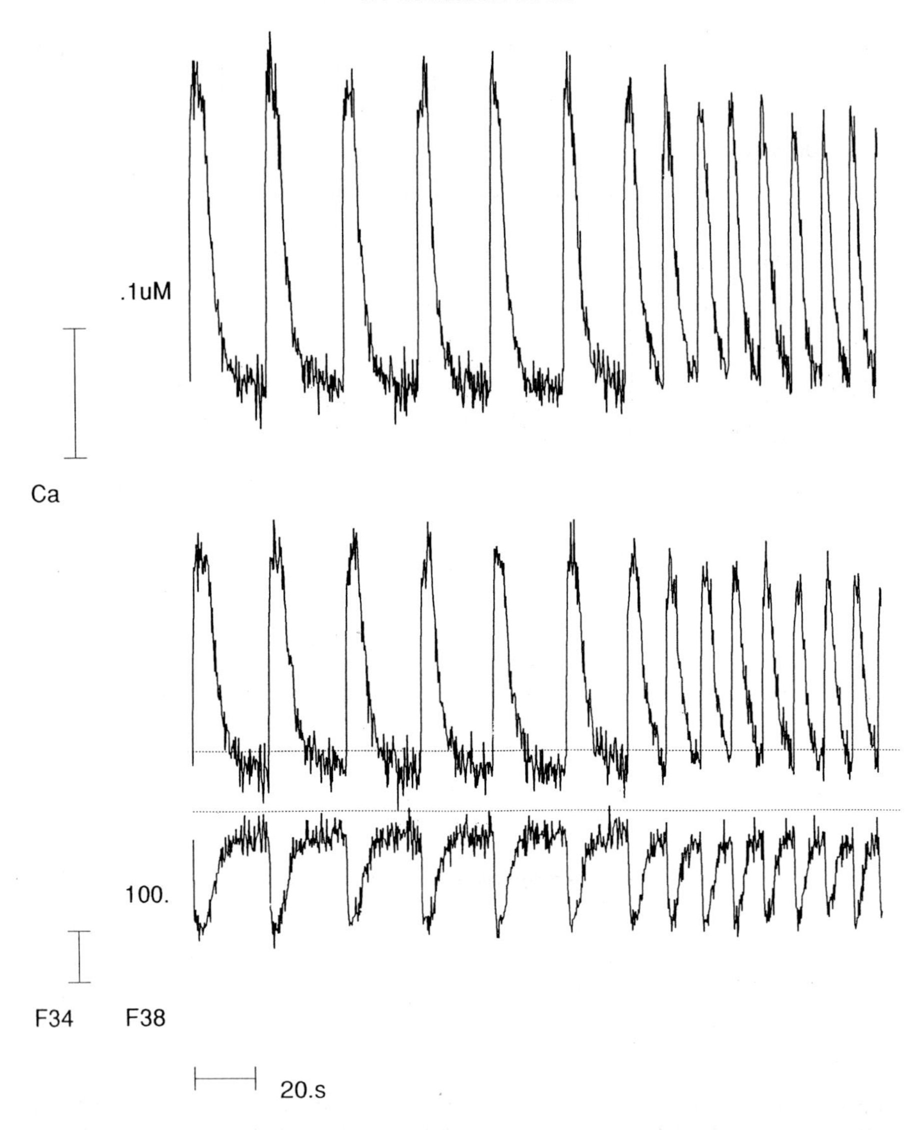

Figure 4. Dual wavelength photo current measurements of intracellular calcium in an isolated rat muscle cell, loaded with Indo-1. The top trace is the calculated intracellular calcium concentration; the bottom two traces show the 405 and 490 nm traces of light emission from Indo-1 respectively. Note that one wavelength increases in intensity and the other decreases as calcium rises. Muscle cell was stimulated with electrical field stimulation.

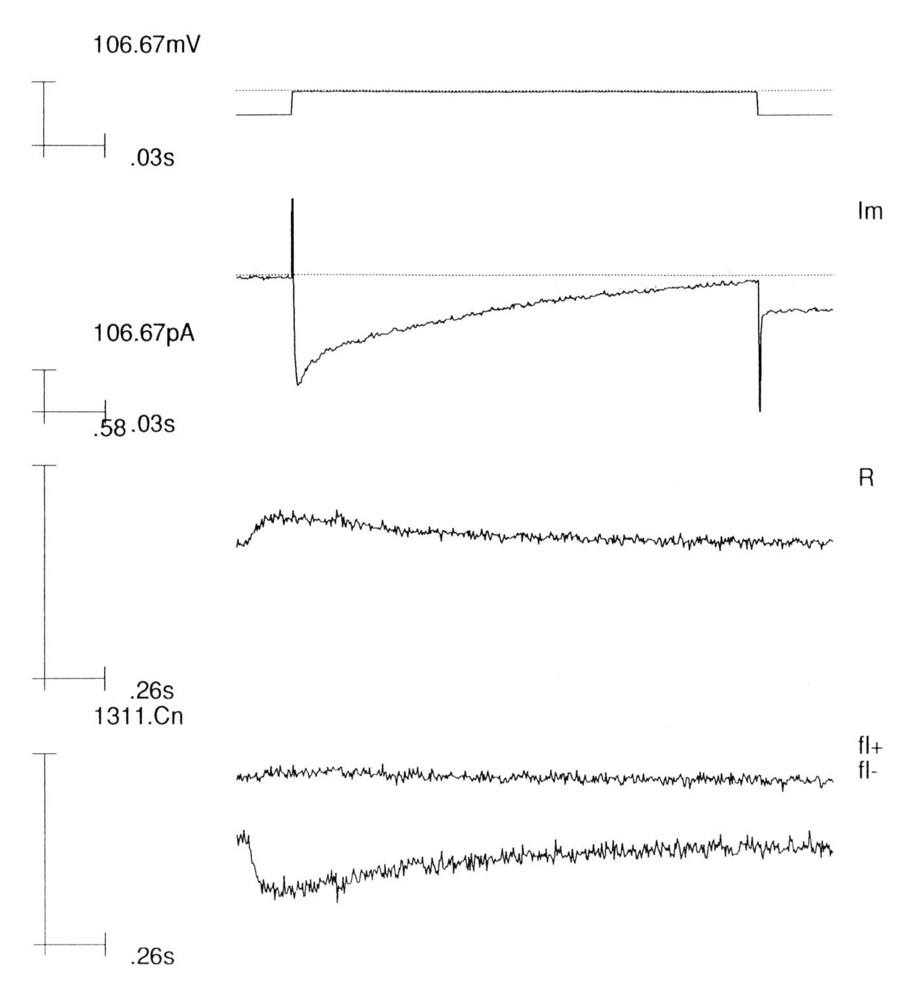

Figure 5. Measurement of simultaneous intracellular calcium, together with membrane current and voltage using PhoClamp, which permits voltage or current clamp measurements together with two wavelengths of photon information using photon counting. Here a voltage clamp step is imposed which elicits a long-lasting rise in intracellular calcium, the latter persisting for about ten times longer than the current transient. Note the intracellular calcium trace is at a different time scale than the voltage and current traces.

Table 3. Comparison of the properties of intensified CCD cameras and slow scan CCD cameras

	Intensified CCD	**Slow scan CCD**
Operating temperature	ambient room temperature	cooled to −50 to −150°C
Readout speed	fast (25 frames/sec)	slow and variable (1 frame/sec)
Dynamic range	limited ($\sim 10^3$)	high ($\sim 10^6$)
Easy to saturate?	yes	no
Noise level	noisy due to intensifier	ultra quiet
Averaging	fast averaging circuitry required	averaging on CCD face
Integration	none except in hardware	integration by CCD
Resolution	generally low 8-bit (1 part in 256)	very high 12- to 16-bit (1 part in 4096 to 1 part in 65 536)
Spectral response	blue-green or red	red
Readout area	must readout complete frame	programmable pixel readout
Line scanning	none	fast
Usable for bright field?	no	yes

odd and even lines (interlaced), producing 625 lines per frame and 25 frames per second in Europe or 30 frames per second in North America. Many systems use only the odd or even lines, permitting a filter to be changed in between for very fast applications. For slower applications, both odd and even lines may be acquired. New generation CCD technology may allow single frame images to be obtained without signal averaging to reduce noise and thus provide very usable data. Averaging or integration is performed after an analogue image has been acquired, and following averaging the image is read out into computer memory through a high speed analogue to digital converter, typically at 8-bit accuracy. Intensified cameras have limited dynamic range (about 10^3), but this is well suited to most available optical probes such as Fura-2 which have dynamic ranges of about 30.

'Slow-scan' CCD cameras typically consist of a surface mounted chip which is subjected to cooling to −20 to −45°C. This reduces dark current and provides the capability to accumulate photons on the chip face for long times without elevating the background signal. These cameras are valuable for studying optical signals which do not vary greatly with time. The cameras produce lower noise images and possess higher dynamic range but can take many seconds to integrate and readout the image. Dynamic ranges of 10^5–10^6 are achievable by using 8-, 12-, or 16-bit conversion, providing up to 65 536 grey levels.

7. Computer hardware for fluorescence ratio imaging

Our laboratory uses two different systems for fluorescence ratio imaging. Called MagiCal and MiraCal (Applied Imaging, UK), their respective features are compared in *Table 4*. Whereas MagiCal is capable of grabbing up to 25 frames per second from a CCD camera, MiraCal uses a cooled slow scan readout camera producing up to five images per second although typically operating at one to two images per second. There is a cost differential of more than 2:1, with the complexity of faster real time imaging costing substantially more.

Table 4. Comparison of MagiCal and MiraCal, two typical high and low cost imaging systems

Characteristic	**MagiCal**	**MiraCal**
Description	Top of the range product	Entry level—upgradable to MagiCal
Cost	Above £70 000 for basic system	Below £35 000 for basic system
Performance	High performance; fast	Mid performance; slower
Capture basis	Video-based capture system; can access most video cameras	Line scan readout system; dedicated non-video or video camera system
Recommended detector	Intensified video frame rate CCD	Slow scan cooled CCD or intensified video frame rate CCD
Interface to video camera	Yes	Yes
Interface to cooled readout CCD	No	Yes
Image size	Up to 512 × 512 pixels; typically use 256 × 256 pixels	192 × 165 pixels up to 512 × 512 (video)
Sub-array scanning	No	Yes
Exposure time	Video-locked to 40 msec (PAL); 30 msec (NTSC)	Variable from software
Absolute usable capture speed for low light levels	25 images per second	5 images per second
Time between ratio images	80 msec	500–2000 msec
Image storage	Up to 32 Mb of DRAM for fast access and display; thence to hard disk	Storage to computer memory and hard disk
Playback/animation speed	Fast; up to 10 times real time if required and variable from keyboard	Slow; dependent on disk controller speed but typically 10 times slower than MagiCal
Display monitors	Split screen display; two colour monitors—one for text and one for images	Single colour display for text and images

Table 4. *contd.*

Characteristic	MagiCal	MiraCal
Software options	About 600; full morphometric analysis available in package	Limited to about 100; offers limited morphometry —only linear spatial measurements
Software relationships	Can network to ORACal and other software packages for Magiscan	Can network to ORACal, but is also Stand-alone
Single pixel access and user interface with images	Light pen	Mouse
Tape recorder mass storage	Available, for up to 60 min of continuous data	Not available
Software image averaging	Available as standard	Available as standard
Hardward image averaging	Available as option	Detector used for integration
Image digitization	8 bit, 256 grey levels	8 bit, 256 grey levels
Spectral sensitivity	Generally good blue sensitivity	Optimal for red and blue sensitivity
Frame integration capacity	256 frames	Unlimited
Tape streamer backup	Yes	Yes
Optical disk backup	Yes	Yes

The MagiCal system (*Figure 6*) is based on an 80486 control processor in a PC linked by a fast parallel link to an image processing unit (IPU). The IPU contains dedicated hardware for video input and output, and image memory which can contain up to 512 images at full resolution, or more at lower resolution. Image memory access is through a special pipe-lined processor. Purpose-designed hardware performs real time averaging, subtraction, ratioing, control of filter changer, and video tape recorder, etc. A montage image from pituitary cells imaged with this system during growth hormone releasing hormone stimulation is shown in *Figure 7*.

MiraCal (*Figure 8*) is also configured on an 80486 system processor, making use of advanced graphics display technology. In its standard form, it uses a custom-designed, cooled read out camera with 8-bit analogue to digital conversion, having over 32 000 pixels. Other higher dynamic resolution 12- and 16-bit detectors up to 512 × 512 pixels are also supported, although these are not required for most ratio imaging. Image integration and filter wheel control can be controlled from the icon-like interface, such that variable exposure times for each wavelength image can be requested to make full use of the camera dynamic range. Images are written either to hard disk or to system memory. The system can either capture up to five images per second, or log data from 20 irregular regions of interest to a file available for graphing. The resulting data provides ratiometric images suitable for most biological applications.

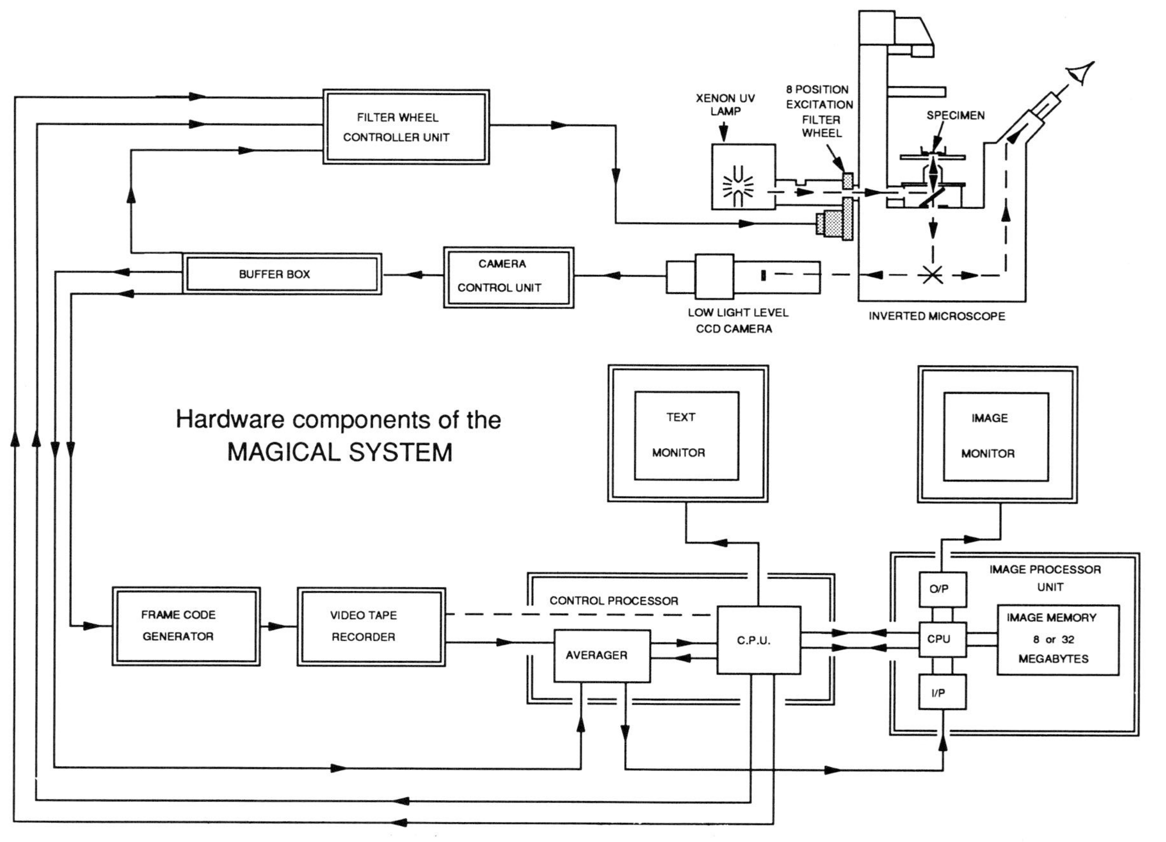

Figure 6. Schematic diagram of the MagiCal system used in our laboratory for dynamic video imaging. This shows interaction and control of various components. Images can be stored directly into dynamic random access memory in the image processor unit, or on magnetic video tape together with a unique frame code for off-line playback and analysis. Output of data and archiving of records is available on a number of common peripherals.

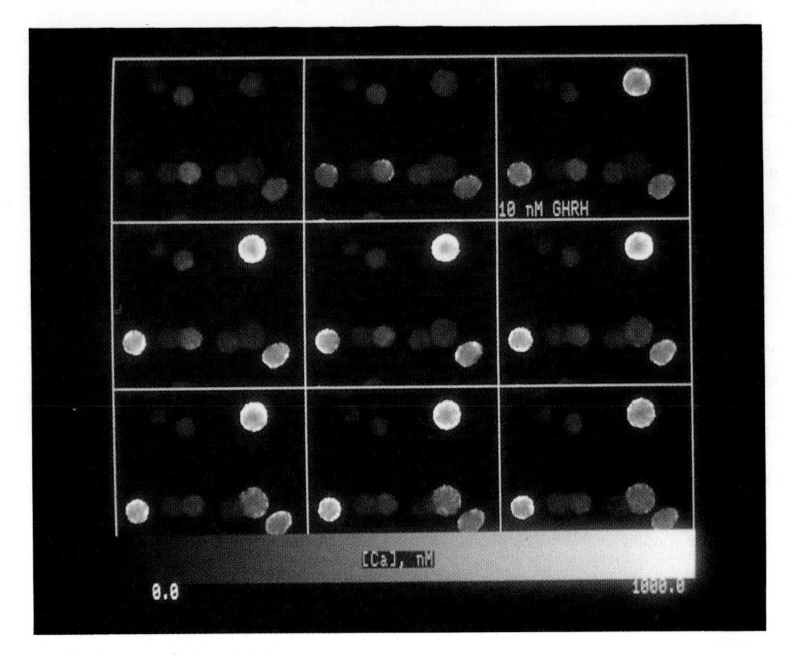

Figure 7. Imaging of rat anterior pituitary cells loaded with Fura-2. Note that the whitest regions are high calcium, localized close to the plasma membrane during stimulation with growth hormone releasing hormone.

8. Capturing and real time processing of video signals

In the digitization process of an analogue video signal, each video line is quantized into discrete time intervals and in each interval the signal intensity is read as one of 256 discrete grey levels (8-bit digitization). Digitization is performed in order to record the video signal as an array of digital values into the IPU for processing and analysis; this can be performed at the rate of 25 full video frames (with up to 512 × 512 pixels) per second.

Standard European video frame rates are 25 frames per second (one frame every 40 milliseconds), and at maximal resolution of 512 × 512 pixels, image sizes are about 262 000 pixels. Dedicated hardware has been developed to average and ratio images at this rate in real time.

The MagiCal hardware allows selection of the number of frames to be averaged, and by selecting the number bits to be shifted out, the signal can be averaged or integrated. MiraCal performs integration on the imaging chip and stores images either directly to a RAM drive in system memory or to hard disk as a single step.

Input and output look-up tables provide the ability to output the values at

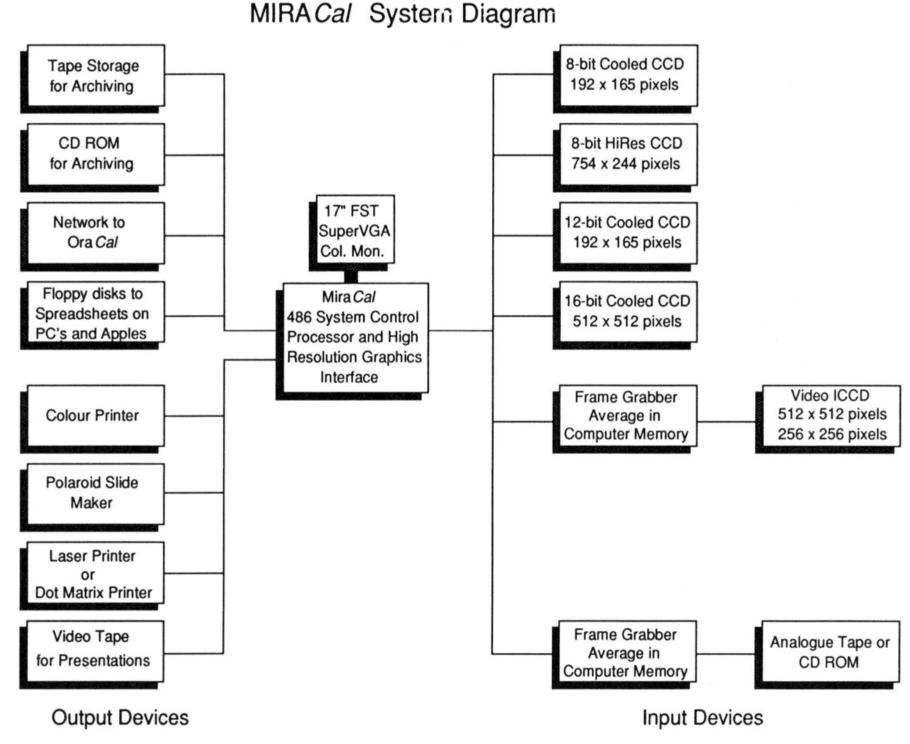

Figure 8. Schematic diagram of the MiraCal low light level imaging system.

each pixel as a colour. This is called 'pseudocolour' and 'false colour' because the user chooses the colour to be displayed to suit his/her own preferences. Hence, in the examples in this article, blue represents low calcium concentration and red is high calcium concentration.

8.1 Image processing: averaging, background correction, and ratioing

Once images have been captured, background fluorescence is removed by capturing images at each desired wavelength, without cells, but in conditions identical to the experiment. These background images are subtracted pixel-by-pixel from each of the cell images before further processing. 'Shade correction' allows for corrections in camera uniformity.

Ratioing to give a calibrated ion concentration involves applying a mathematical formula at every pixel, which takes into account an experimentally determined constant appropriate to the dye–ion interactions, dye quantum efficiency and system optics, intensity ratios for the pixel in each of the two images, and calibratable extremes of ratio intensity measured in the experimental arrangement.

Ratioed images contain whole numbers representing ratios or ion concentrations and measurements made on these images use the tables to look up the true values.

9. Cell culture and loading of fluorescent probes

The cells intended for study are plated on to thin glass coverslips, usually no. 1 or 1.5. This permits focusing from below using an inverted microscope and allows ultraviolet light to be used to excite the probe. Plastic media cannot be used as plastic absorbs ultraviolet excitation light.

Cells may be loaded with ion-sensitive fluorescent probe in one of two ways. The free acid form of the probe may be directly loaded at 10–100 μM through the cell membrane by micropipette or patch pipette. Many probes are available in the acetoxymethylester form. These non-polar ester derivatives will diffuse across the membrane to be hydrolysed by non-specific cytoplasmic esterases, to yield the membrane-impermeable free acid form (see Chapter 8, this volume). The ester form of the dyes is largely insensitive to changes in ion concentration. Unfortunately most plant cells have low esterase activity, and this has made experiments difficult unless micro-injection is used.

For experiments measuring intracellular free calcium using Fura-2 or measuring pH using BCECF or other probes for other ions, the probe can be initially made up as a stock solution of 2 mM in DMSO. Cells are normally incubated in a 4 μM solution of the acetoxymethylester form (Fura-2 AM, BCECF-AM, or other) made up in a standard extracellular medium for 30 min at 37°C. FURA-2 and BCECF can be simultaneously loaded into single cells, and imaged using a 4-wavelength protocol. This provides an estimate of intracellular calcium and hydrogen ion concentration (*Figure 9*).

Data collection from cells should start within ~2 h of loading as they have been seen to lose responsiveness when loaded for extended periods.

9.1 Calibration of ion-sensitive dyes in living cells and in solution

The following method uses as an example intracellular free calcium measurements using the dual excitation fluorescent probe Fura-2 which is available in both the free acid and acetoxymethylester forms. Many other probes for different ions are now available but the general methods are valid for any dual excitation probe. For pH sensitive dyes such as BCECF, permeabilization with nigericin could be used in place of ionomycin as in the following example.

Data required for calibration may be entered into the TARDIS software by entering the maximum (R_{max}) and minimum (R_{min}) ratios obtained into the following equation which is pre-programmed into the software.

$$\text{Ion concentration} = K_d \,.\, \beta \,.\, [(R - R_{min})/(R_{max} - R)]$$

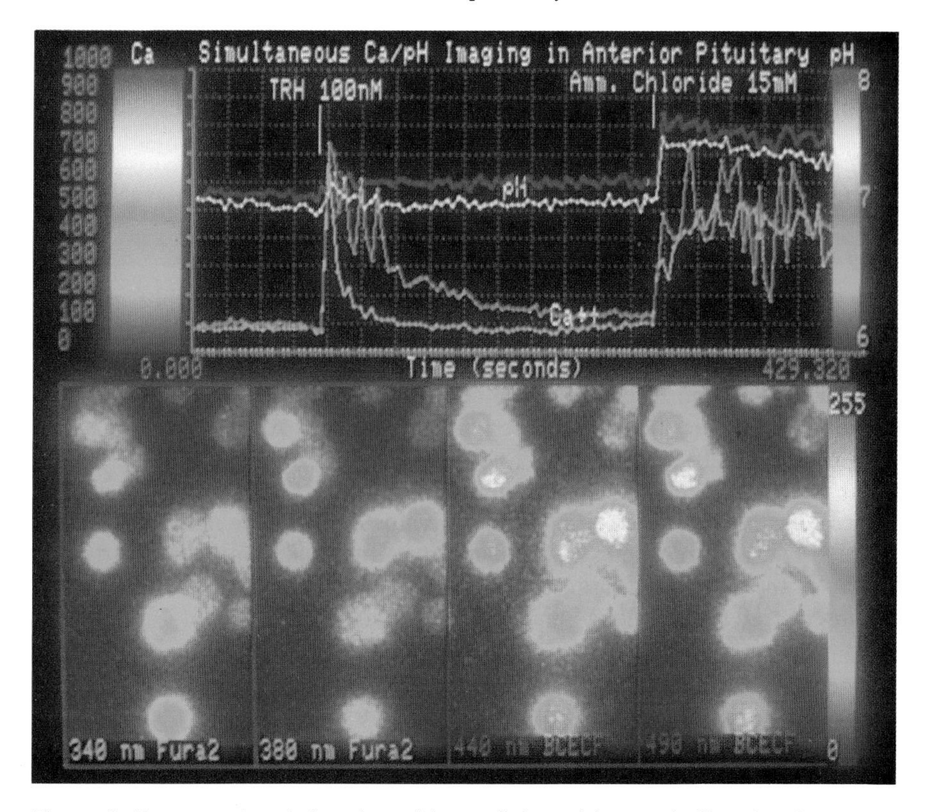

Figure 9. Four wavelength imaging of intracellular calcium and pH, using Fura-2 and BCECF in the same cell. Upper two traces (red and yellow) are pH; lower two traces (blue and green) are calcium. This shows that thyrotropin-releasing hormone (TRH) causes a typical calcium transient in a bovine prolactin cell, but little significant pH change. Resting pH in these cells varied between 6.9 and 7.2. In the continued presence of TRH, no further discharge of the IP_3-sensitive calcium store could be measured, even when TRH was increased to submaximal doses in excess of 1 μM. This indicated that the IP_3-sensitive pool is fully discharged. We use the application of a weak base, namely 5–15 mM NH_4Cl, to promote alkalinization of the cell interior. 15 mM NH_4Cl caused a transient pH change to 7.5–7.8 from the resting value. Of most significance, however, was the observation that this treatment also appeared to trigger a simultaneous release of intracellular calcium in the presence or absence of TRH. A similar observation was made when extracellular calcium was removed, suggesting that the source of calcium rises from inside the cells.

where K_d is the dissociation constant of the probe, β is (intensity at the upper wavelength at R_{min})/(intensity of the upper wavelength at R_{max}), R is the measured ratio, R_{max} is the ratio when the probe is saturated with calcium ions, and R_{min} is the ratio with no free calcium ions present.

9.1.1. Free acid solution method

The simplest but arguably least accurate method for initial calibration is to prepare two solutions of medium, one containing 5 mM $CaCl_2$ which will saturate the probe with free calcium ions and result in the maximum ratio obtainable, and the other having the $CaCl_2$ substituted by 1–10 mM EGTA which will bind all free calcium ions to result in the minimum ratio obtainable. Both must contain the free acid form of the probe at a concentration of ~50–100 μM which is the approximate concentration found in cells loaded with the ester derivative.

9.1.2 Intracellular method

Cells are loaded with the acetoxymethylester derivative of the probe and the cell membrane is permeabilized. Ionomycin at a concentration of ~2 μM is suitable for calcium. Some laboratories have also used low concentrations of digitonin or saponin to permeabilize cells for calibration.

The R_{max} measurement should be made in elevated extracellular $CaCl_2$. A concentration of 10 mM has been found to be sufficient to saturate the probe within the cell. Addition of the ionomycin causes a rapid, sustained rise in intracellular calcium. Measuring the R_{min} value on the same field by washing the cells two or three times with calcium-free medium containing 1–10 mM EGTA is the next step. Transport across the cell membrane—resulting in binding of free calcium ions—may require at least 15 min.

10. Imaging biological activity

Image processing software in the MagiCal and MiraCal image analyser can present results in a wide variety of ways, which allow analysis of biological activity. Examples include:

- montaging the same cell at different times during stimulation (*Figure 10*)
- comparing large numbers of cells to assess anisotropy
- superimposing graphs of different regions, either of the same or of different cells, to compare their behaviour (*Figure 11*)
- histograms of the frequency of occurrence of ion concentrations at all of the pixels in a region
- profiling pixel intensity along lines defined through cells and comparing these with profiles of the same lines from other images in the sequence, or plotting pixel profiles as a function of time
- 3D views of ion concentration profiles across a region (*Figure 12*)
- examining ratio images to reveal subcellular changes in ion concentration in the cytoplasm or nucleus
- animating sequences to compare the changing ion profiles as a function of time
- plotting images in a stack, using time as a third dimension, to follow where changes occur

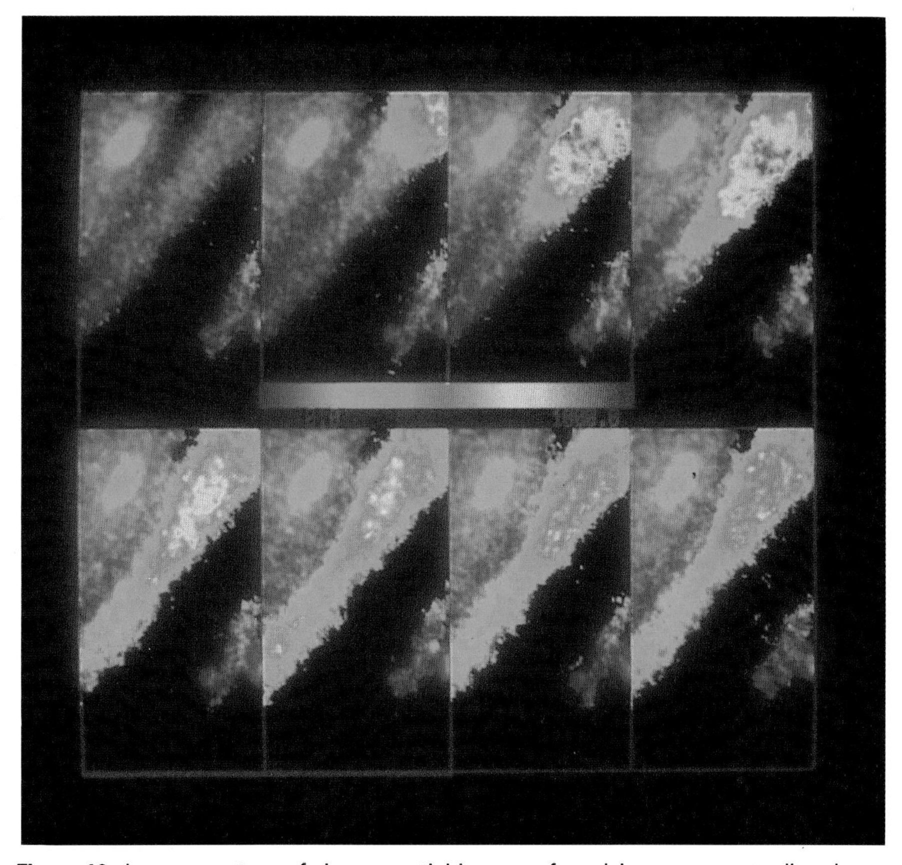

Figure 10. Image montage of six sequential images of a calcium wave spreading down the length of a human smooth muscle cell stimulated by thrombin, measured using MagiCal. The images shown here were measured with Fura-2.

- allowing tables of results to be analysed statistically, studied in spreadsheets, graphed in different ways, etc.

10.1 Imaging intracellular calcium store refilling

The nature and control of the plasma membrane calcium channels that mediate entry of extracellular calcium in non-excitable cells is only partially understood, and the method of isobestic point sampling described above is one possible approach which can be used with optical probes. Normally calcium influx follows the agonist-induced discharge of intracellular calcium stores. The stores then refill as a consequence of the increase in cytosolic calcium concentration. Models for refilling of stores include control by the fullness of the store *per se* (e.g. the capacitance model (2)) as well as control by various second messengers produced by the agonist (3).

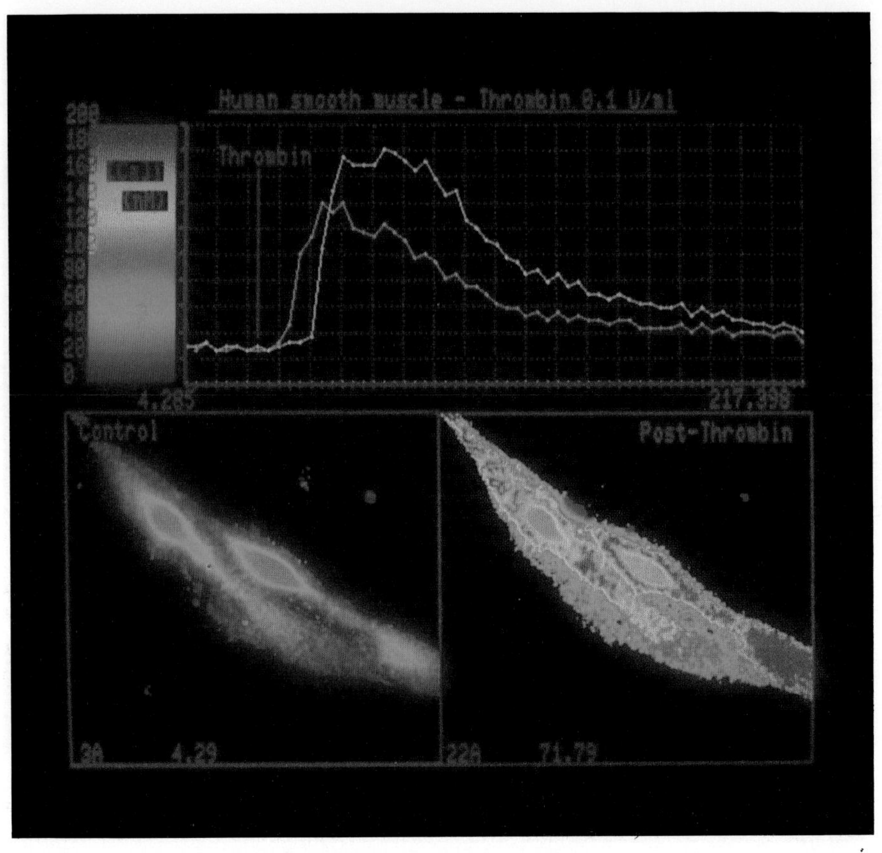

Figure 11. Measurement of intracellular calcium in different areas of a human smooth muscle cell loaded with Fura-2, during stimulation with 1 μM thrombin which causes the generation of IP_3. A light pen is used to identify irregular regions of interest which can then be individually graphed simultaneously.

Activation of calcium influx by store depletion can be studied by allowing the intracellular calcium stores to empty (if only partially) by incubation of the cells in nominally calcium-free medium. Addition of calcium to the medium causes an increase in $[Ca^{2+}]_i$ which may enter the cell through the channels involved in a normal agonist-mediated response.

An example of this approach is shown in *Figure 13*. Single rat adenohypophyseal cell were loaded with Fura-2 (4 μM, 30 min at 37°C) and $[Ca^{2+}]_i$ measured by ratiometric video imaging. Cells were incubated for 50 min in medium to which no $CaCl_2$ had been added, in order to deplete calcium stores. The concentration of calcium in the low-calcium medium was less than 1 μM, measured with Fura-2 free acid. In these experiments, the ratio of 340/380 nm fluorescence excitation is displayed, a measure of $[Ca^{2+}]_i$.

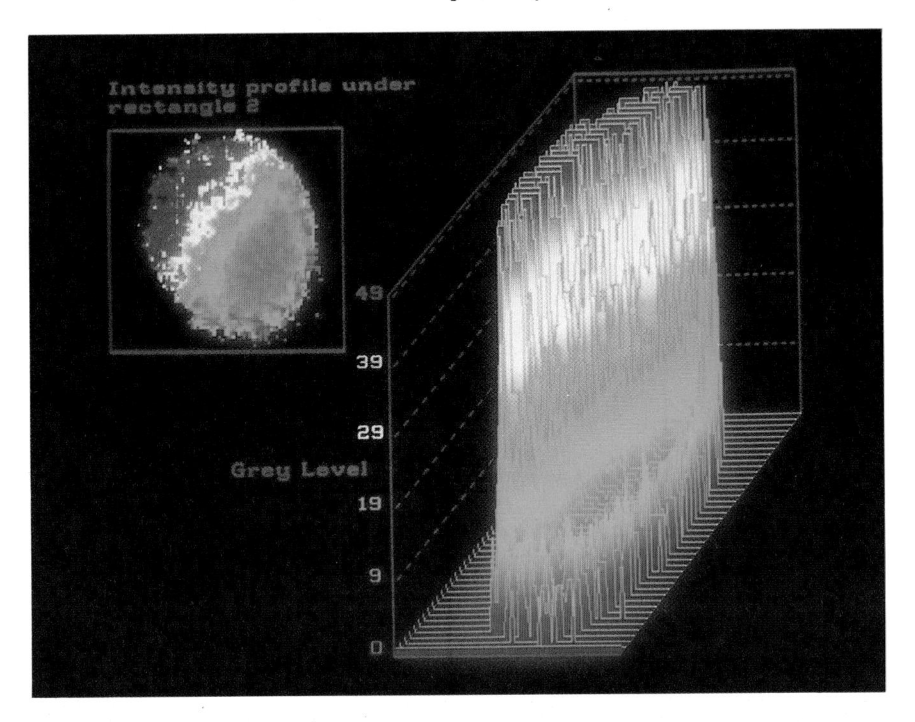

Figure 12. 3D profile of calcium concentration in a bovine anterior pituitary cell, with *x* and *y* dimensions as micrometers and $[Ca^{2+}]$ as the *z* axis.

Recordings of $[Ca^{2+}]_i$ in two cells in separate experiments (*Figure 13A* and *B*) are shown. Cell A was challenged with TRH after incubation in low Ca medium in order to determine the time required for depletion of stores. Cell B was incubated for the same time, but the first TRH addition was omitted, and thapsigargin (THG, a blocker of the endoplasmic Ca-ATPase) was added instead of the second TRH challenge. In both cases, a large increase in $[Ca^{2+}]_i$ was seen only when calcium was returned to the cell. TRH (Cell A) and THG (Cell B) both elicited calcium responses, presumably from refilled stores. The refilling phase is large and prolonged and can be characterized in terms of modulators of calcium channels and signalling pathways.

In summary, ratio imaging can be used to study the depletion and refilling time courses and characteristics of intracellular calcium stores, important sites for cellular calcium homeostasis. The main advantage of this type of approach is that store-depletion-activated calcium influx can be separated from that activated by an agonist.

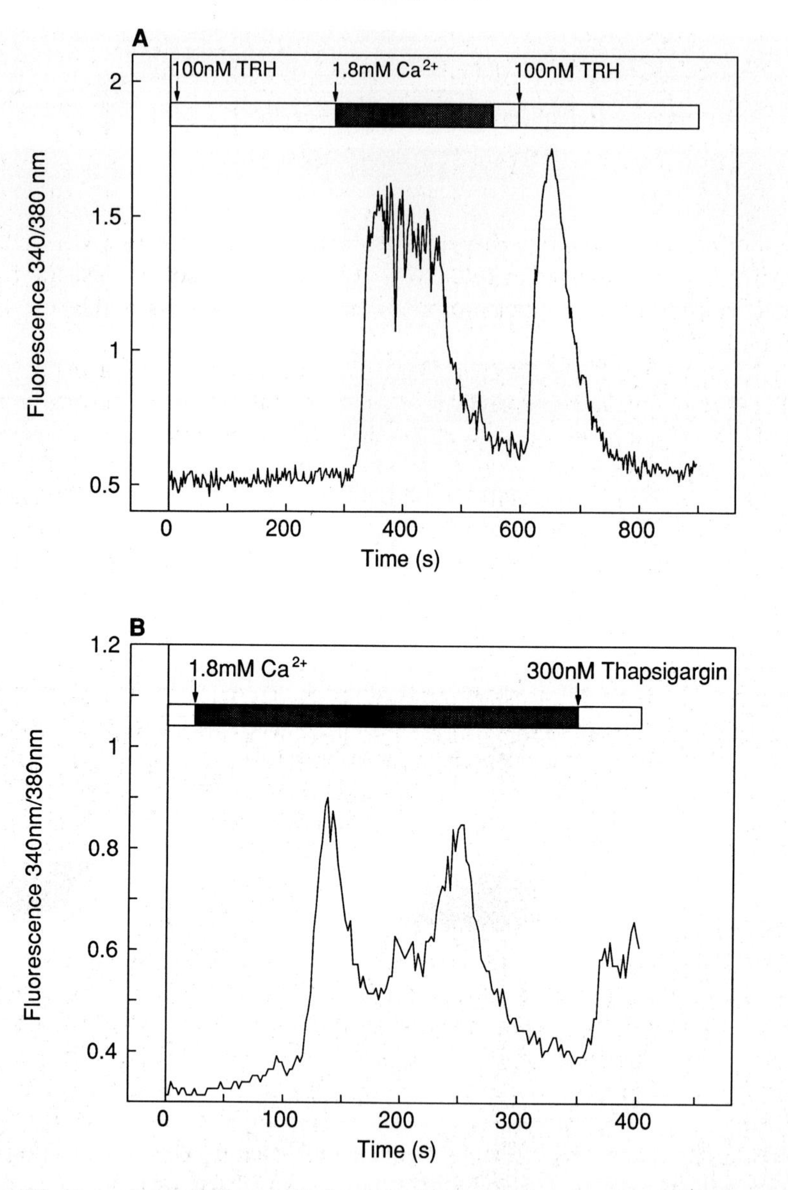

Figure 13. Normal rat pituitary cells loaded with Fura-2 using protocols to examine calcium store re-loading. Both cells were initially depleted of calcium using low Ca medium for 50 min, having been loaded with Fura-2. In (**A**), TRH treatment has no effect on [Ca^{2+}] after this treatment, but calcium restoration allows the stores to re-charge in the absence of TRH (during hatched bar). Subsequent TRH application causes a large rise in [Ca^{2+}] once stores have been re-charged. (**B**) Similar experiment showing oscillations in [Ca^{2+}] during re-loading, perhaps due to uptake into more than one compartment. Application of thapsigargin—a blocker of endoplasmic reticulum Ca-ATPase—also discharges this calcium store.

11. Manipulation of intracellular calcium responsive compartments—ratiometric imaging to detect nuclear calcium changes

Use of imaging technology can play a valuable role in detecting intracellular and intranuclear changes in calcium concentration. In some cells, the ratiometric calcium probes appear to be taken up by nuclear as well as cytosolic compartments.

In human smooth muscle cells (*Figure 14*), the nuclear compartment loads with calcium indicator and calibration of the signal revealed a resting ionized nuclear calcium level of about 300–500 nM, although large changes in cytosolic calcium do not perturb the level of calcium to any extent. Application of ionomycin to the cells showed convincingly that Fura in the nuclear compartment was accurately reporting calcium concentration changes, as a large rise in calcium could be induced. Thus, the nucleus was concluded

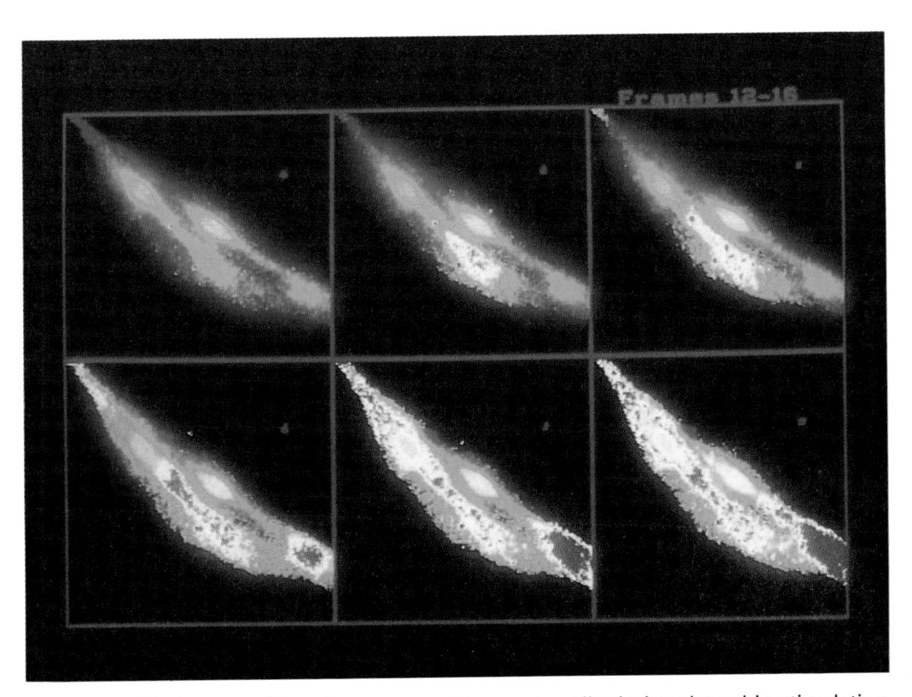

Figure 14. Six images of two human smooth muscle cells during thrombin stimulation. Note higher resting levels of calcium in nuclear region, which is unchanged as the calcium wave spreads along the wall.

to be relatively impermeant to small ions and presumably to other small molecules.

Recent work on multinucleated osteoclast cells has provided a different picture. In these cells, resting nuclear ionized calcium is low, probably about 50–100 nM. Agonists for the vitronectin receptor promote osteoclast retraction, and addition of peptides containing RGD-sequence caused a large rise in ionized calcium localized to the nucleus and mainly in the absence of any cytosolic changes.

This effect can be further dissected by the use of different agonists. Calcitonin, which causes plasma membrane entry of calcium, elicits a generalized cytosolic rise that is not localized to any specific compartment of the cell. On the other hand, tBuBHQ, an inhibitor of the Ca^{2+} ATPase in endoplasmic reticulum caused in the same cell a markedly different rise in cytosolic calcium presumably due to localized release of intracellular calcium from the endoplasmic reticulum. This study raises the possibility that specific intracellular signalling molecules exist which can evoke intranuclear calcium modulation in the absence of cytosolic changes *per se*, and that the nucleus may possess intranuclear stores of calcium ions.

The possibility that these changes are artefactual has been further eliminated by preliminary use of a real-time confocal microscope (Odyssey, Noran Instruments, Madison, Wisconsin). Using Fluo-3 as a single wavelength calcium probe under conditions where a focal place of 0.7–1.0 μm was achieved, discrete nuclear calcium changes evidenced by increases in Fluo-3 emission were observed after addition of several RGD-sequence containing peptides (*Figure 15*).

12. Potential artefacts in optical imaging with fluorescent probes

In discussing the potential artefacts and pitfalls which may be encountered in optical imaging of fluorescent probes, it is important to look at the general applications for which fluorescent probes may be used. For example, the calcium sensitive probe Fura-2 may be used either as a single wavelength qualitative probe to image the spatial distribution of calcium. Alternatively it may be used in dual wavelength ratio mode in a calibrated system to quantify additionally calcium concentrations and fluxes. The potential artefacts will be very different in the two systems; indeed, simply utilizing Fura-2 in its ratio mode corrects most of the problems associated with single wavelength measurements.

The following comments refer to the potential artefacts which may affect the *measurement* of ion concentrations. It therefore generally relates to imaging and photometry for *ratio* measurements rather than simple fluorescence. The

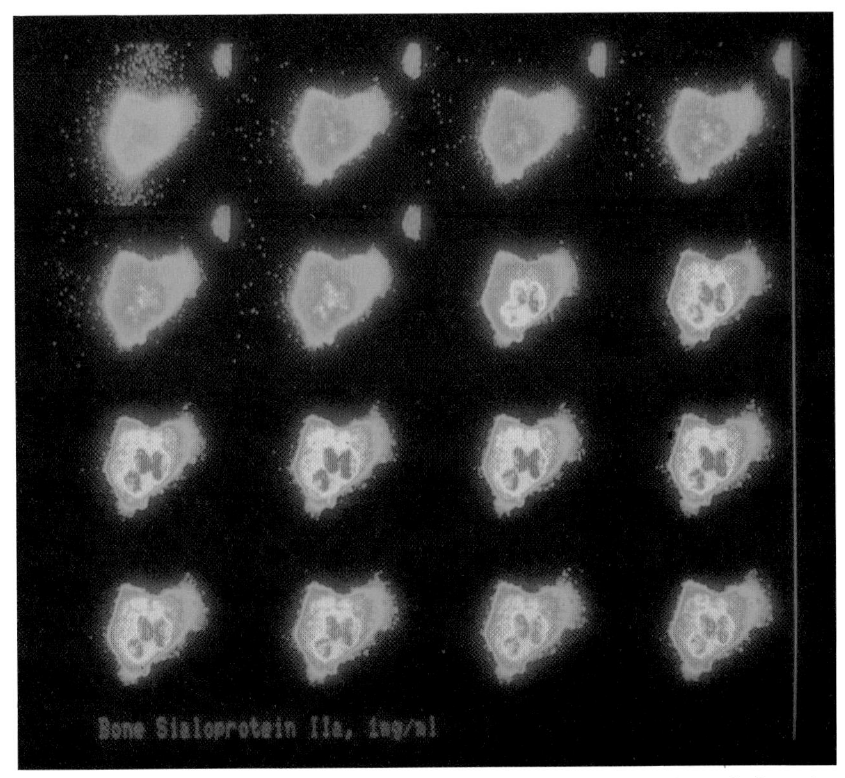

Figure 15. Confocal laser-scanning images of Fluo-3-loaded rat osteoclasts during stimulation with bone sialoprotein. Shown at ~2 second intervals, using an image plane of ~1.0μm resolution.

variety of difficulties experienced in dynamic video imaging are as diverse as the measurements we make. Potential artefacts discussed below can include:

- photobleaching of probe
- dynamic range: minimum and maximum illumination for effective images
- probe loading, de-esterification, compartmentalization, and buffering
- ion calibration, solutions or cells, choice of ionophore and pH changes
- cell movement and fast ion fluxes
- autofluorescence
- interactions between multiple probes
- averaging and intensifier noise
- probe leakage and exocytosis
- probe kinetics

12.1 Photobleaching

All fluorescent probes will photobleach to a greater or lesser extent when excited with a suitable wavelength, at a rate proportional to the intensity of the incident light. While this may not present a problem for some applications such as the simple spatial mapping of a fluorescent probe, it can seriously affect attempts to quantify ion concentrations using single wavelength probes.

The most obvious practical way to reduce photobleaching is to minimize the light reaching the probe. Unfortunately, reducing the incident light will also reduce the emitted light intensity so optimum conditions for image analysis must include:

12.1.1 Determination of optimal probe loading

Optimal loading of a probe in a particular cell type should be assessed by experiment. A maximal signal will only result from maximal loading but care must be taken to ensure that the probe does not significantly buffer the ion of interest. A compromise must therefore be sought between maximizing the signal and minimizing probe concentration. In practise, intracellular probe concentrations of 30–100 μM are usually suitable. Loading may be achieved either by direct microinjection or by acetoxymethylester loading across the cell membrane (33), as discussed above in this chapter.

12.1.2 Maximum sensitivity of data collection

The availability of suitably sensitive detectors has played a crucial role in the development of technology to measure dynamic ion changes in living cells. While adequately sensitive photomultiplier tubes have been available for some years, more recent developments in intensified CCD video cameras have allowed direct imaging of probes for measurement of ion concentrations and fluxes. In addition, fast computer access and processing now allows ratiometric techniques to be applied to video imaging, eliminating many of the possible artefacts and errors associated with single wavelength techniques. Other factors which will improve sensitivity are the use of excitation and emission filters that are well matched to the probe. The emission filter should have a bandwidth covering at least 90% of the emission spectrum. The use of an objective of the highest numerical aperture available will also improve sensitivity.

12.1.3 Ratiometric measurements

Utilizing ratiometric measurements will correct for uneven probe loading between cells and across each cell. It will correct for differences in cell membrane thickness, where the membrane unevenly absorbs some of the emitted light and it will correct for part of the photobleaching problem. Correction of these artefacts is, however, limited to the dynamic range of the detection system and any combination which takes either of the images used for the ratio outside its dynamic range will cause problems as described in the section below. Furthermore, significant photobleaching has been shown in Fura-2 not only to reduce the light output of the probe, but also to shift the emission spectrum (4). At just 8% photobleaching, using a standard ratio calibration, this results in an underestimation of up to 20% of 200 nM $[Ca^{2+}]_i$.

12.1.4 Effect of time

Minimize the time actually illuminating the probe by exciting it only while capturing data. Extraneous room light must be kept to a minimum as it will contribute to photobleaching not only during an experiment also but while loading and storing loaded cells before an experiment.

12.1.5 Effect of oxygen concentration

Minimize the oxygen concentration as it has been shown to play a major role in photobleaching of the Fura-2/calcium system (4). The lowest concentration concomitant with good cell viability will greatly reduce the rate of bleaching.

12.2 Dynamic range

Whatever type of system is employed to detect the emission from fluorescent probes, it will be limited to a specific dynamic range. That is, there will be a lower level at which detection is not possible and there will be an upper level at which the system saturates. The value at the maximum divided by that at the minimum is the dynamic range of the system. In addition to the values for the system as a whole, due regard must be paid to saturation in any single part of the system.

For example, an imaging system will have an overall dynamic range which is governed by the range of the probe, the range of the camera (which may have two intensifier stages, both capable of saturation and a CCD detector which may saturate) and the range of the analogue to digital (A/D) converter used to digitize the image. There will also be a maximum ratio set by the ratioing range in the calibration table. In an ideal system the effective 'gain' of each component is set so they all saturate at the same level. This, however, is not always possible and great care must be taken to ensure no part of the system will either saturate or fall below its minimum detection threshold during an experiment.

The most usual saturation artefact in ratio imaging is perhaps concerned with the setting of the maximum ratio level when ratioing. This is necessary to optimize the display of changes in ion concentration. If, for example, a cell undergoes a change of 200 nM during an experiment, it would be unwise to display it on an axis with a span of 5000 nM. In this case the maximum level set when calculating the ratios would be set to about 500 nM to make the changes more obvious. If this level is set, and ion concentrations exceed this value, then the system will simply show the maximum value for all higher ratios resulting in a serious under-reading error. The better image analysis systems software do allow 'test ratioing' and inform the user when this condition is approached. Care must, however, be taken to 'test ratio' the *highest* ratio images of a sequence.

Another frequent situation resulting in under-reading of ion concentration occurs when a component of the image collection system saturates. This may be the A/D converter, the CCD chip, or the intensifier stages. The safest way to eliminate this type of artefact is to collect and inspect the original images before they are ratioed, as the changes in fluorescence during an experiment cannot always be predicted. Some systems that perform the ratioing operation 'on line', and do not allow access to the original images, leave themselves open to this artefact.

The problems that occur when images fall below the minimum detectable value are less serious and more easily overcome. With the best software, a minimum threshold level is set below which all values are treated as zero and therefore do not show up on the image. In addition any pixels that then have a zero value automatically result in a zero value ratio and are not displayed on the ratio image. This problem may develop during an experiment due to a combination of changing ion concentration, photobleaching, and probe leakage or exocytosis. It causes a deterioration in the ratio image quality as pixels disappear but does not indicate erroneous values.

12.3 Probe loading

Most probes in general use are now available in acetoxymethylester forms which pass readily across most cell membranes. Once inside, non-specific esterases hydrolyse it to its free acid form. As the cell membrane is not permeable to this form of the probe it is trapped and concentrated inside. For example, loading with a solution of only 4 μM Fura-2 AM outside the cell for 30 min at 37°C may result in intracellular concentrations of Fura-2 free acid of 50–100 μM.

As with all experimental systems, care must be taken to ensure that the probe does not significantly affect the measurement of the parameter of interest. In the case of ion-sensitive probes there is a risk of the probe buffering the ion of interest and experimental controls must be devised to ensure that this is not the case. A simple 'dose response' curve for probe loading should reveal the maximum safe loading concentration for a particular probe in a cell type. In addition it should also be noted that probes and/or the carrier (often DMSO) are generally toxic to many cell types. They should therefore be used within two or three hours of loading, preferably as soon as hydrolysis is judged to be complete, which may be 10–20 min after washing the extracellular probe away. A further wash at this point is advisable to remove any unhydrolysed probe which has passed out of the cells across the plasma membrane.

It should also be noted that the use of AM probes for loading can result in loading of organelles within the cells (5, 6). While these may give a good imaging signal, the probe inside may not be available to cytosolic free ions. Muted responses in some cell types may therefore be caused by the presence of organelles rather than small changes in ion concentration as indicated. Loading the cells with free acid probe by microinjection would not load the organelles and would reveal the real extent of the cytosolic free ion concentration. The use of micro-injected dextran conjugates may also minimize this type of effect.

12.4 Ion calibration

Calibration of a ratiometric imaging system (see above) may be performed either by measuring the ratios of solutions of known ion concentration or by loading the cells with probe and forcing maximum and minimum ratios. Firstly the loaded cells are permeabilized with a suitable ionophore allowing free ions from the extracellular medium to saturate the probe. This will achieve the maximum ratio possible. A chelator of the ion of interest is then added to bind preferentially all the free ions to result in the minimum ratio obtainable. Maximum ratios obtainable by saturating the probe in *solutions* can be up to double that of saturated probe inside *cells*. It is therefore most desirable to calibrate a system using the cells to be used in experiments rather than any other method. The reasons for this difference are not entirely clear but evidence suggests that intracellular viscosity (7) and compartmentalization of the probe are the main causes. The choice of ionophore will be governed by the ion of interest but it should be noted that some have been shown to be fluorescent at the wavelengths used to excite some probes.

Difficulties in calibration may be encountered if the cells of interest contain secretory granules. Permeabilizing the cells with ionophorc has been shown also to permeabilize secretory granules, especially to H^+ ions (8). Internal compartments of secretory granules may be quite acidic so this can result in a significant acidification in intracellular pH during calibration. A large variation in the values obtained for maximum (R_{max}) and minimum (R_{min}) ratios have been reported even in supposedly homogeneous cell populations. While the ion readily saturates the probe for the R_{max} value, chelating the ion to measure R_{min} may be more difficult and can take 15–30 min to reach a plateau. An alternative method to acquire the (R_{min}) value is to use $MnCl_2$ to quench the Fura-2 signal (9).

12.5 Cell movement and fast ion fluxes

Cells used for image analysis of fluorescent probes are usually plated on thin glass coverslips to allow transmission of ultraviolet wavelengths and enable viewing on an inverted stage microscope. Cells used in ratiometric image analysis must remain stationary during the experiment. Usually, experimental protocol requires changing the extracellular medium either for the addition of an agonist or perfusion of the cells. It is therefore imperative that they adhere firmly to the coverslip. Even a small movement of the cells while adding an agonist is likely to result in a crescent shaped artefact of apparently high ion concentration caused by shifting of the denominator image when ratioing. In the worst case cells are completely lost from the field of view. Some cell types in culture on the glass coverslip adhere quite well whereas others require it to be coated with a substrate such as poly-L-lysine to aid adhesion. The choice of substrate will depend on the cell type and application; for example, fibrinogen has been used to bind human platelets (10) but care must be taken to ensure that it does not interfere with the normal function of the cell.

In a dual emission ratiometric system, images are captured sequentially and then ratioed. The resultant images are therefore not true ratios as the individual images are not captured at the same time. The shorter the time between capturing image pairs, the better the approximation to a true ratio measurement. The better image analysis systems (such as MagiCal described above) have capture rates up to 25 per second or 40 msec between images used to calculate a ratio. While the time between pairs may be set very much longer thus enabling slower, longer experiments to be performed while ensuring the ratio image is a close approximation to a true ratio.

Very fast changes in ion concentration will also be subject to a similar error. This will result in under-read values during fast transient increases in ion concentration and over-read values as the concentration decreases. The only solution in this case is to increase the data capture rate by moving to a faster system. While they have other limitations, ratio photometric systems may be more appropriate with capture rates of up 200 Hz. However, dye–ion dissociation time constants may become a dominant factor in determining the time response of the system in these conditions.

12.6 Autofluorescence

Many of the probes employed to monitor intracellular ion concentrations require excitation wavelengths well into the ultraviolet region. Unfortunately, a number of natural peptides and other biogenic amines are also fluorescent in this region. This problem is especially prevalent in plant tissue and mammalian liver, kidney, and pancreatic cells. If the component from autofluorescence is small compared with the contribution from the probe it may be disregarded. Otherwise, levels of autofluorescence must be recorded before loading so they may be subtracted or thresholded from the probe-loaded images. In practice this is rather difficult as the excitation intensity must be set, to measure autofluorescence, before loading with probe. Preliminary experiments must therefore establish a loading protocol that gives a consistent level of fluorescence. Plated cells must also be either kept on the microscope stage while loading or accurately repositioned afterwards.

12.7 Interactions between multiple probes

Currently, the better imaging systems allow simultaneous monitoring of multiple probes. Intracellular pH and calcium, for example, are known to interact and play pivotal roles in cell signalling and secretion. BCECF and Fura-2 may be used to monitor pH and calcium; their excitation wavelengths are 440 and 490 nm, and 340 and 380 nm respectively. Emission spectra from both probes overlap at about 520 nm so they may both be monitored using a single emission filter set. There will, however, be significant optical crosstalk between the probes both in the combined emission spectrum and as emission from the lower wavelength probe excites the other probe. Fura-2 in this case emits a broad spectrum peaking at 510 nm so there is a significant proportion of light emission available to excite BCECF at 490 nm, the upper excitation wavelength. Experimental protocols must therefore be devised to measure both types of crosstalk in the cells of interest so corrections may be made. It is also advantageous to keep the lower wavelength probe concentration as low as possible consistent with adequate signal for good signal to noise. This may usually be achieved by weighting the loading concentration of each probe (11).

Care must also be taken when loading the ester form of multiple probes. It has been found essential to load probes simultaneously as the probes appear to compete for the intracellular esterase activity. Great difficulty has been experienced with sequential loading of multiple probes.

12.8 Averaging and intensifier noise

Some applications, such as monitoring fast ion fluxes, call for maximal capture rates. Others, such as mapping the spatial distribution of ion concentrations within cells, may call for slower rates with higher spatial resolution. For the reasons mentioned above, most video imaging systems currently use intensified CCD cameras to capture images. These do, however, have an inherent problem in that microchannel plate intensifiers introduce a degree of electronic noise into the final image. The most effective way to improve the signal to noise ratio is to average a number of video frames (12). This may be achieved either in software after storing the frames or on-line prior to storage and subsequent analysis. There is of course a trade-off between temporal resolution and signal-to-noise ratio so the experimental objectives must dictate the balance. Generally averaging four or eight video frames gives adequate signal-to-noise ratio for many purposes which gives a temporal resolution of between one-third and two-thirds of a second between ratio image pairs.

12.9 Probe leakage and exocytosis

The cell membrane is not completely impermeable to the free acid form of fluorescent probes. There is a small leakage of probe to the extracellular medium (5). In a constantly perfused system this causes no problems other than a very small drift of intensity which is easily accommodated by ratiometric image analysis. In static systems with no perfusion, long term leakage may increase the background fluorescence slightly but the effect is minimal.

A potentially serious effect has, however, been reported in mast cells (8). Acetoxymethylester loading here appears to load secretory granules with probe. On stimulation, the granules are released from the cell membrane to deposit their high concentrations of probe to the immediately surrounding area. This appears as a ring of high ion concentration around the edge of the cell and may be misinterpreted as an ingress of ions around the periphery. Examination of the images prior to ratioing would reveal the difference, a loss of probe would result in a sharp reduction of emission at both ratio wavelengths.

12.10 Probe kinetics

Most of the work on probe kinetics has been performed in solutions rather than cells. Measurements of association and dissociation constants have been performed for Fura-2 and Azo-1 (13) by the temperature jump relaxation method (14) and for Fura-2 and Indo-1 (15) by stopped flow measurements. The practical implications of this work are that Fura-2 requires 5–10 msec to reach equilibrium at 20°C in solution of ionic strength 140 mM. While the response will be faster at 37°C, other factors such as viscosity and spatial microheterogeneity may slow intracellular measurements. Some fast calcium fluxes may therefore be misinterpreted as the probe kinetics may be the limiting factor. However, the main rate limiting factor for a system will be the maximum capture rate of the system. For imaging systems this will generally be video frame rate or 40 ms per image. Photometric systems can run much more quickly and, in some cases, exceed the response time of the probe.

13. Summary

The development of optical probes for biological activity combined with powerful image acquisition and analysis technology is having a major effect on the study of physiological parameters of living cells. Most of these probes appear not to interfere with normal cellular processes, but require caution in interpretation of data as the probes may localize within cells, rendering them insensitive or modifying their properties. A number of possible artefacts also require caution in utilizing the optical probe.

New probes for ions, cyclic nucleotides, cellular enzymes, and genetic material are under development and will further extend this capability. Photometric and imaging technology allow the application of these probes to yield both temporal and spatial data. The scientist is now presented with a variety of means for imaging these low light level probes, including use of a variety of different detectors and imaging systems. Video frame rate systems offer high spatial and temporal resolution generally at a high cost; often the high acquisition speeds may not be warranted by the biological characteristics of the system. Lower cost systems like the MiraCal system used in our laboratory provide a reasonable speed of acquisition well-suited to most biological events, and at a low cost. Confocal imaging technology provides somewhat slower image acquisition, but with improved spatial resolution and less out of focus information. New techniques for image deconvolution will permit resolution enhancement for both confocal and video frame rate systems, but at a cost of both time and money, given the extensive computational power required.

New and even more powerful image processing systems will be developed, and cost/benefit will probably improve in coming years. Prototype imaging systems with industrial cameras and direct computer memory access can now produce up to about 200 images per second at high resolution, permitting fast cellular processes to be revealed. In addition, novel probes such as for genetic material and more rapid observational and analytical capabilities will probably emerge in the coming decade.

Technologies for fluorescence imaging will become more accessible to a wider number of laboratories. With this, the power of quantitative image processing and optical probe technology will be increasingly realized by scientists in academia, industry, and government, and the authorities who provide funding for these establishments.

Acknowledgements

We thank the Agricultural and Food Research Council, Medical Research Council, Applied Imaging Ltd., Kabi Pharmacia, Wellcome Trust, Nuffield Foundation, Guggenheim Foundation, and British Heart Foundation for valuable funding which has supported various aspects of the work discussed here.

References

Further reading

A comprehensive bibliography of the intracellular ion measurement literature would be too extensive to detail here, but the references below will provide worthwhile reading. The February/March 1990 issue of *Cell Calcium* (Volume 11) provides a very comprehensive recent treatment of additional methods and applications for work in this area.

Brakenhoff, G. J., van Spronsen, E. A., van der Voort, H. T. M., and Nanniga, N. (1989). *Methods in Cell Biology*, Academic Press, Orlando.

Cheek, T. R., Jackson, T. R., O'Sullivan, A. J., Moreton, R. B., Berridge, M. J., and Burgoyne, R. D. (1989). *J. Cell Biol.*, **109,** 1219.

Cubbold, P. H. and Rink, T. J. (1987). *Biochem. J.*, **248,** 313.
Fay, F. S., Carrington, W., and Fogerty, K. E. (1989). *J. Microscopy*, **153,** 133.
Inoue, S. (1986). *Video Microscopy*. Plenum Press, New York and London.
Lichtman, W., Sunderland, S. J., and Wilkinson, R. S. (1989). *New Biol.*, **1,** 75.
Mason, W. T. (1993). Fluorescent and Luminescent Probes for Biological Activity—A Practical Guide to Technology for Quantitative Real-Time Analysis. Academic Press, (in press).
Mason, W. T., Sikdar, S. K., Zorec, R., Akerman, S., Rawlings, S. R., Cheek, T., Moreton, R., and Berridge, M. (1989). Chapter 14 in *Secretion and Its Control* (ed. C. S. Oxford), pp. 225–38. Rockefeller University Press, New York.
Mason, W. T., Rawlings, S. R., Cobbett, P., Skidar, S. K., Zorec, R., Akerman, S. N., Benham, C. D., Berridge, M. J., Cheek, T., and Moreton, R. B. (1988). *J. Exp. Biology*, **139,** 287.
Monck, J. R., Reynolds, E. E., Thomas, A. P., and Williamson, J. R. (1988). *J. Biol. Chem.*, **263,** 4563.
Monck, J. R., Oberhauser, A. F., Keating, T. J., and Fernandez, J. M. (1992). *J. Cell Biol.*, **116,** 745.
Neylon, C. B., Hoyland, J., Mason, W. T., and Irvine, R. (1990). *Am. J. Physiol.*, **259,** C675.
O'Sullivan, A. J., Cheek, T. R., Moreton, R. B., Berridge, M. J., and Burgoyne, R. D. (1989). *EMBO J.*, **8,** 401.
Putney, J. (1990). *Cell Calcium*, **11,** 611.
Schlegel, W., Winiger, B. P., Mollard, P., Vacher, P., Wuarin, F., Zahnd, G. R., Wollheim, C. B., and Dufy, B. (1987). *Nature*, **329,** 719.
Shotton, D. M. (1989). *J. Cell Sci.*, **94,** 175.
Taylor, D. L. and Wang, Y. L. (1989). *Methods Cell Biol.*, volumes **29 & 30**.
Tsien, R. Y. (1981). *Nature*, **290,** 527.
Williams, D. A. and Fay, F. S. (1990). *Cell Calcium*, **11,** 75.
Winigcr, B. P., Wuarin, F., Zahnd, G. R., Wollheim, C. B., and Schlegel, W. (1987). *Endocrinology*, **121,** 2222.
Winiger, B. P. and Schlegel, W. (1988). *Biochemistry J.*, **255,** 161.
Wright, S. J., Walker, J. S., Schatten, H., Simerly, C., McCarthy, J. J., and Schatten, G. (1989). *J. Cell Sci.*, **94,** 617.

Literature cited

1. Grynkiewicz, G., Poenie, M., and Tsien, R. Y. (1985). *J. Biol. Chem.*, **260,** 3440.
2. Rizzuto, R., Simpson, A. W. M., Brini, M., and Pozzan, T. (1992). *Nature*, **358,** 325.
3. Irvine, R. F. (1990). *FEBS Lett.*, **263,** 5.
4. Becker, P. L. and Fay, F. S. (1987). *Am. Phys. Soc. Special Comm.*, C613.
5. DiVirgilio, F., Steinberg, T. H., and Silverstein, S. C. (1990). *Cell Calcium*, **11,** 57.
6. Steinberg, S. F., Bilezikian, J. P., and Al-Awquti, Q. (1987). *Am. Phys. Soc. Special Comm.*, C744.
7. Poenie, M. (1990). *Cell Calcium*, **11,** 85.
8. Almers, W. and Neher, E. (1985). *FEBS Lett.*, **192,** 13.

9. Hesketh, T. R., Smith, G. A., Moore, J. P., Taylor, M. V., and Metcalfe, J. C. (1983). *J. Biol. Chem.*, **258**, 4876.
10. Heemskerk, J. W. M., Hoyland, J., Mason, W. T., and Sage, S. O. (1991). *J. Physiol.*, **446**, 204.
11. Zorec, R., Hoyland, J., Akerman, S. N., and Mason, W. T. (1993). *Pfluger's Archiv*, **423,** 223.
12. Mason, W. T., Hoyland, J., Rawlings, S., and Relf, G. (1990). *Methods in Neurosci.*, **3,** 109.
13. Kao, J. P. Y. and Tsein, R. Y. (1988). *Biophys. J.*, **53,** 635.
14. Eigen, M. and DeMayer, L. (1963). In *Techniques of Organic Chemistry* (ed. S. L. Friess, E. S. Lewis, and A. Weissberger), Vol. III, part 2. Wiley, New York.
15. Jackson, A. P., Timmereman, M. P., Bagshaw, C. R., and Ashley, C. C. (1987). *FEBS Lett.*, **216,** 35.

8

Flow cytometric selection of responsive subclones and fluorimetric analysis of intracellular Ca^{2+} mobilization

JOHN T. RANSOM, JOHN F. DUNNE, and N. A. SHARIF

1. Introduction

Transmembrane signalling frequently involves fluctuations in intracellular free Ca^{2+} levels ($Ca^{2+}{}_i$). Such changes in $Ca^{2+}{}_i$ may be mediated via activation of ion channels in the plasma membrane or via the release of intracellular pools of Ca^{2+} into the cytosol. Ca^{2+} mobilization responses are often measured in bulk assay that measure the average response of the population over time. The most common examples are analyses in a spectrofluorimeter of responses in cells that have been loaded with fluorescent Ca^{2+} indicator dyes (Indo-1, Fura-2, Fluo-3 etc.). These are satisfactory when there is a uniform and robust response by most or all of the cells in the population. However, if only a small percentage of the cells respond, or if the cells exhibit very transient responses with widely different onset times, it can be difficult to detect such responses above the background signal of non-responding cells. In practical terms, it can be difficult to resolve a true response above the inherent noise of the assay. Or, it may be difficult to perform consistently reproducible dose-response assays, especially at points where low concentrations of agonist or high concentration of antagonist are used.

With commonly utilized cell lines of clonal origin, we have observed that a substantial percentage of the population often fails to respond to a particular stimulus while a smaller percentage responds with a clearly detectable response. This suggests that, with multiple culture passages over time, genetic variants have developed that exist in the population and which exhibit heterogeneous response phenotypes that may not be morphologically apparent. It is desirable to have a cell line with a homogeneous response phenotype to enable enhanced resolution of responses at low concentrations of ligand or to increase the frequency of responding cells in single cell assays such as patch clamp or imaging studies. Here we describe techniques that allow selection of subclones that yield populations with increased response homogeneity and synchronicity from a heterogeneous parental population. Both methods involve isolation of responding cells from the parental population by monitoring a 'real-time' Ca^{2+} response. The simplest method allows selection of cells during all time points of the response. A more precise method allows selection of those cells that respond during a narrow time window following encounter of the cell with the agonist ligand. This allows selection of cells that respond relatively quickly or, in theory, relatively slowly. The utility of the subclones in pharmacological studies and signal transduction studies has been demonstrated previously (1–3). Those results are also summarized in this chapter.

2. Preliminary analysis of the response

To perform the selections it is necessary first to analyse Ca^{2+} mobilization in cells using the fluorescent $Ca^{2+}{}_i$ indicator dyes such as Fura-2 or Indo-1 to determine response kinetics and magnitude (4). If the initial analysis is performed in a spectrofluorimeter cuvette it may be observed that a small or slow average response occurs that is difficult to detect. For example, compare the response of NG108–15 cells to angiotensin II (AII) with the larger and faster response to bradykinin (BK, *Figure 1*). To resolve whether the relatively small average response was due to a modest response magnitude in all cells or due to robust responses in a small percentage of the cells and no response or minimal responses in the remainder, it was necessary to analyse the responses

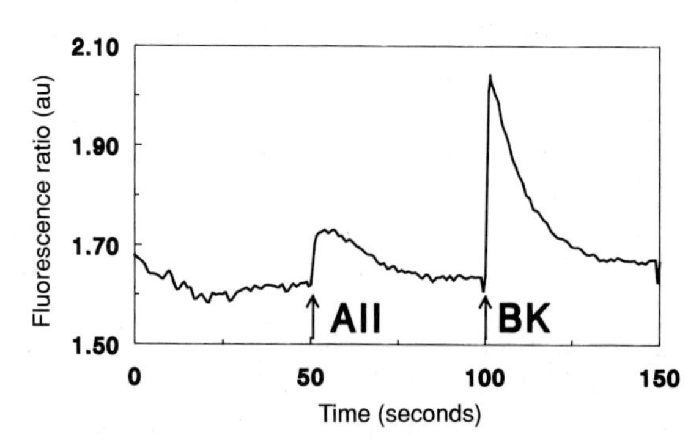

Figure 1. Spectrofluorimetric analysis of the fluorescence ratio of Indo-1-loaded NG108-15 cells during stimulation with AII or BK. NG108-15 neuroblastoma cells were loaded with Indo-1 and analysed on a dual emission spectrofluorimeter (see *Protocol 6*). Excitation was at 335 nm and emissions were simultaneously acquired at 400 and 500 nm. The ratio of the emission intensity at 400 divided by that at 500 nm was calculated and plotted in arbitrary units. AII (AII, 1 μM) and BK (BK, 1 μM) were added at the point indicated by the first and second arrows respectively.

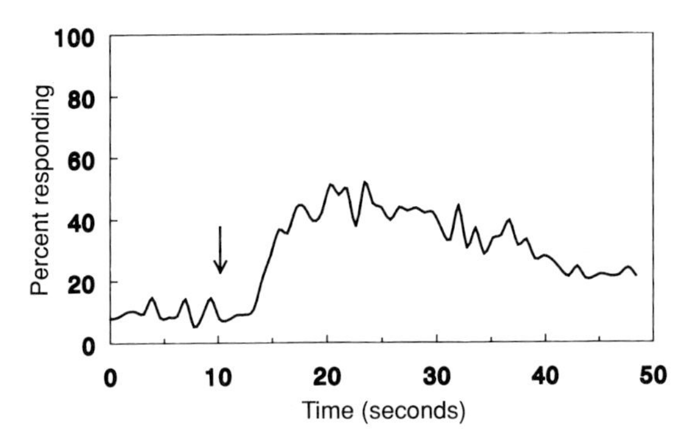

Figure 2. Flow cytometric analysis of the percentage of cells that respond to AII. NG108-15.C16 subclone cells were loaded with Indo-1 and analysed by flow cytometry (see *Protocol 3*). Using a time zero device, the stimulus (1 μM AII) was added at the 10 sec mark (arrow) without interruption of the sample flow. The percentage of cells with a detectable response was calculated using the arbitrary designation of 5–10% positive cells in the unstimulated portion of the analysis as background.

of all cells at the single cell level. This could be performed by flow cytometry or microscopic imaging (see Chapter 7, this volume). The advantage of a flow cytometric analysis is the acquisition of information about thousands of cells over several minutes. The disadvantage is that each cell is analysed for a fraction of a second at one point in time. The resulting data are a continuous plot of the percentage of cells in the population that exhibited a detectable response at different points in time after stimulation (*Figure 2*). Thus, a preliminary analysis of this type permits one to assess rapidly whether a small fraction of the cells responded to the stimulus. However, it may not be entirely clear if the cells responded simultaneously or asynchronously. If a microscope-based Ca^{2+} image analysis is performed it is possible to obtain information about the response phenotype of a single cell, or a field of several cells, for a prolonged period of time (see Chapter 7, this volume). It should be clear whether a fraction of the cells responded to the stimulus and whether the cells responded synchronously. However, much more time is required to analyse a significant number of cells. In either case, once it is clear that only a fraction of the cells exhibited a detectable response, it should be possible to 'sort clone' the responding cells by flow cytometry.

3. Overview of general hardware and software requirements

The instrumentation required for the efficient cloning of cells with particular calcium mobilization phenotypes deserves some explanation. A sorting flow cytometer with good optical and sorting performance is a necessity. A detailed description of such devices can be extensive and can be obtained elsewhere (5). However, special attention should be paid to the points that follow.

3.1 Dynamic range

The ratio of violet to blue emission should increase at least 4-fold when Indo-1 loaded cells are saturated with free calcium. Several factors influence the dynamic range of the assay, including dye loading. Too much or too little dye can dampen the response. Incomplete dye de-esterification during the loading procedure leads to poor signals since the ester form of the fluorophore is not calcium sensitive. Barrier and filters which discriminate violet and blue emission should be clean and scratch-free, with violet defined as 405 ± 20 nm and blue defined around 510 ± 20 nm. In some instruments the dichroic filter separating the two signals also steers the signal dramatically so that its alignment is critical.

3.2 Signal processing electronics

The nature of the signal processing can also affect the dynamic range. Some instruments create the ratio with analogue electronics while others calculate the ratio from digitized signals in a computer. In either case it is useful to verify the linearity of this process.

3.3 Cloning hardware and software

The cloning hardware and software should function well. If viable cells from a convenient cell-line that is known to be cloned easily (e.g. myelomas) are sort cloned into 96-well plates at 1 cell per well, at least 70 clones should be recovered. There should be less than 1 'double hit' per plate. Commercially available devices (Becton-Dickinson, Coulter) should routinely perform adequately.

4. Overview of accessory hardware and software

The aforementioned instrumentation would allow the selection of responding cells within a broad time span, and these cells may be acceptable to the investigator. However, calcium responses to many ligands occur within seconds, even at room temperature. Though the response may be sustained it is often the very earliest phase of the response which is most dramatic and the most useful from which to clone. Two recently developed devices greatly enhance the ability to resolve and sort kinetically defined phenotypes. They are available commercially (Cytek Development, Fremont, CA) or can be implemented from existing literature (6, 7).

4.1 Time-zero device

A 'time-zero' device adapts standard commercial flow cytometers so that a stimulus can be added to cells without interrupting their flow, and with minimum dead time between stimulus introduction and analysis. A minimum time-zero device consists of a sample chamber which can be mounted very near the flow cell and which allows stimulus injection through a septum or valve. Additional useful features include temperature regulation, mixing, and an electronic interface to time tag the data file at the time of stimulus onset (6).

4.2 Time-window device

A 'time-window' device adapts commercial flow cytometers so that stimulus is added to cells continuously at a fixed time before the laser intercept. This configuration allows a given 'time since stimulus' to persist, so that cells can be analysed and sorted from a pre-defined response time window. The device is described in detail elsewhere (7). Briefly, balanced air pressure pushes cells from one reservoir and stimulus from another toward a T-junction where they mix. The cells and stimulus then continue through tubing to the flow cell. The length of tubing and the summed air pressure control the delay between stimulation and laser interrogation/sorting.

5. Harvest and preparation of cells for loading with indicator dye

Analysis of Ca^{2+} mobilization by flow cytometry was first performed using lymphocytes as the target cells (8, 9). These cells are very compatible with flow cytometry since lymphocytes are small, non-adherent, and nearly round in shape. Application of flow cytometric techniques to adherent cells such as neuronal lines or epithelial cell lines could conceivably present several problems. Neuronal lines tend to have processes, which could be disrupted by flow cytometry. They also exhibit varying tendencies towards and dependencies upon adherence to each other or to the tissue culture plastic. This could make it difficult to remove the cells from the culture vessel and lead to clumping of the cells. Obviously if the goal is to isolate single cells to give rise to daughter populations, the presence of doublets, triplets, etc. invalidates the cloning effort in the initial stages.

Protocol 1. Harvest of adherent cells

1. Many neuronal lines (e.g. PC-12, NG108-15, and SH-SY5Y) can be dislodged from the culture vessel by gentle pipetting. Not all cells in a population will come off. Extremely adherent cells may be dislodged by gently scraping the vessel with a tissue culture grade scraper if necessary (e.g. Costar, Cambridge, MA).
2. Most clumps in the suspension can be removed by allowing the suspensions to settle in a narrow conical (15 ml) tube for 5–10 min. A primarily

Protocol 1. *Continued*

single cell suspension can be obtained by removing the cells in suspension above the sediment. Repeated pipetting of the suspension prior to analysis can disrupt remaining small clumps. Electronic gating based upon light scatter parameters during the analysis can also minimize the contribution of doublets to the analysis and sorting.

3. In the case of adherent cells that grow as confluent monolayers (e.g. MDCK cells), the use of dissociating agents such as a trypsin/EDTA solution (JRH Biosciences, Lenex, KS) to produce a single suspension has been useful. However, such treatment might alter the receptor of interest. It is very important to verify that the receptor remains structurally and functionally intact during the cell isolation procedure involving use of trypsin.

6. Loading of cells with Indo-1, a Ca^{2+}-sensitive fluorescent dye

Protocol 2. Dye loading

1. Pellet harvested cells in a centrifuge and resuspend in prewarmed (37°C) Indo-1 loading medium[a] at 1–5 × 10^6 cells/ml.

2. Dilute the stock of Indo-1 AM[b] to a final concentration of 2 μM and mix by gentle inversion to distribute the dye. Incubate cells in the dye at 37°C for 40 min and mix by inversion during the incubation to maintain cells in suspension.

3. After the incubation, pellet the cells in the centrifuge and wash twice in normal Ringer's[c] solution to remove excess unincorporated dye.

4. Resuspend the cells directly in normal Ringer's at 2 × 10^6 cells/ml for flow cytometric analysis.

[a] Indo-1 loading medium: 500 ml Hanks' balanced salts solution (HBSS), 10 mM Hepes buffer, 3 mM dextrose, pH 7.0. This solution should be filter sterilized for storage.

[b] Indo-1 AM stock: Indo-1 acetoxy methyl ester (Indo-1 AM, Molecular Probes, Eugene, OR) is prepared as a stock solution by dissolving it in dry DMSO to a concentration of 2 mM. The dissolved material can be aliquoted and stored at −80°C for months.

[c] 10× Normal Ringer's: 1.54 M NaCl, 55 mM KCl, 20 mM $CaCl_2$, 10 mM $MgCl_2$, 100 mM Hepes, 3 mM dextrose, pH to 7.3. This solution should be sterile filtered for storage. Prepare 1:10 dilution for normal mammalian Ringer's solution.

This procedure has provided adequate dye loading for PC-12 cells, NG108-15 cells, MDCK cells, and lymphocytes. We have never observed any benefit by extending the incubation with Indo-1 AM beyond 40 min. It is possible that other cells require variations in this simplified procedure for optimal loading. Such variations might include changes in the dye concentration, loading temperature, or duration of exposure to dye, or the inclusion of a post-loading incubation period to allow complete hydrolysis of residual intracellular Indo-1 AM. Non-hydrolysed Indo-1 AM will contribute Ca^{2+}-independent fluorescence signals during the analysis that will interfere with the Ca^{2+}-dependent signals generated by Indo-1.

7. Flow cytometric analysis of Ca^{2+} mobilization

Indo-1 stained cells are analysed at room temperature on a flow cytometer. If the software does not allow continuous data acquisition with time, it is possible simply to acquire and store data points as fast as possible using a timer. The files can be stacked in time later to recreate the kinetics of the response. However, it is best if continuous data acquisition software is available. The cells are illuminated with ~100 mW of UV light (355–361 nm) and analysed with respect to low and wide angle light scatter of the UV beam, violet and blue fluorescence emissions, and the ratio of violet to blue fluorescence. If point to point acquisition is required, meaningful information can be obtained from rapidly acquired serial files of ~1000 cells per time point. All files should be gated using incident beam scatter information and dye loading criteria to exclude debris and moribund cells so that each histogram includes only intact cells. The resulting files provide information on the percentage of detectable responding cells at points in time after stimulus addition and allow the investigator to decide from which time points in the response pattern the subclones should be selected.

Protocol 3. Flow cytometric analysis of the Ca^{2+} response

1. Aliquot the Indo-1-loaded cells into sample tubes prior to analysis (0.25–2.0 ml).
2. Analyse an aliquot of cells on the flow cytometer. Run the cells at 400–1000 cells/sec. This rate may be dictated by the compatibility of the cells with flow cytometric parameters such as the diameter of the flow nozzle orifice and the sensitivity of the cells to mechanical forces generated by the instrument fluidics.
3. If a time zero device is used, acquire a baseline reading of the cells for 10–15 sec. Then inject the stimulus into the sample chamber. Mark the time when stimulus is injected. (If such a device is not used, acquire a baseline file. Then remove the sample and add stimulus. Start the timer when the stimulus is mixed in and quickly return the sample to the instrument. A data point should be acquired as soon as the cells intercept the laser.)

Protocol 3. *Continued*

4. Continue analysis for several minutes after the maximal percentage of cells has responded.
5. If a response is not clear, then simulate one by adding 1 μM ionomycin to promote substantial Ca^{2+} influx (see *Figure 3E*).

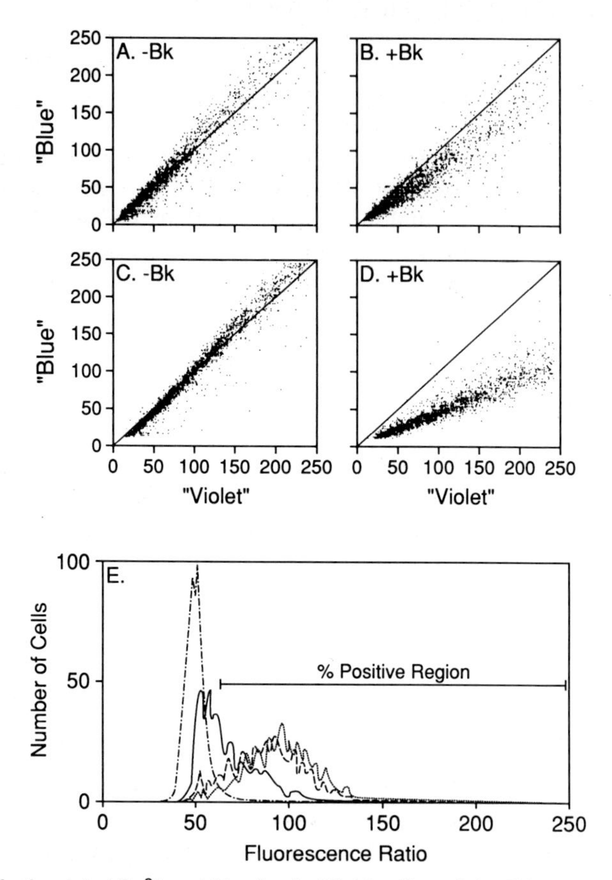

Figure 3. BK-stimulated Ca^{2+} mobilization in PC-12 cells and the BK1 subclone. Intracellular Ca^{2+} levels of Indo-1-loaded PC-12 cells (**A** and **B**) or BK1 cells (**C** and **D**) were analysed by flow cytometry. Panels A and C are dual fluorescence emissions cytograms of unstimulated cells. Panels B and D are cytograms of the cells 25 sec after stimulation with 1 μM BK. An increase in $Ca^{2+}{}_i$ causes an increase of violet intensity (abscissa) and a simultaneous decrease of the blue intensity (ordinate) for each responding cell, illustrated as a single point. The diagonal line was drawn as a visual reference. Panel E shows the violet/blue ratio histograms (cell number versus fluorescence ratio in arbitrary units) obtained with the BK1 subclone at rest (–·–) or 24 sec after stimulation with 10 nM (——) or 100 nM (– – –) BK or 2 μM ionomycin (. . . . .). The region indicated by the solid line is the domain taken as representing sufficient Ca^{2+} elevation to signify a positive response. (Reprinted from Ransom *et al.* (1) by permission.)

7.1 Data analysis and presentation

The results can be reported in terms of the percentage of the population which responded to a given stimulus at each time point (*Figure 2*). This is the only information necessary for sort-cloning. The sort decision rests solely on whether a responding cell exhibits a Ca^{2+} response of sufficient magnitude to be discriminated from the resting population. The question of response magnitude is relative. The dynamic range of the response will be clear and the selection of a region that includes cells with a relatively large response magnitude is straightforward. Thus, calibration of the instrument in terms of the absolute $[Ca^{2+}{}_i]$ is unnecessary. We have typically chosen an arbitrary boundary measurement and measured the change in frequency of cells above that boundary. Such a frequency is internally consistent and independent of experimental conditions. Our region that identifies a responding cell is defined as the region which includes ≤10% of the unstimulated cells. Thus there is always a background of 5–10% 'positive' cells even in the absence of stimulus.

8. Sort cloning

8.1 Isolation of subclones on the basis of Ca^{2+} mobilization responses at all times after stimulation

Protocol 4. Response dependent sort-cloning

1. In advance:
   (a) Configure the flow cytometer with an Automatic Cell Deposition Unit and associated software (Becton-Dickinson, San Jose, CA).
   (b) Sterilize the sorter with 70% ethanol.
   (c) Prepare several 96-well, U-bottom tissue culture plates by adding 100 μl of the appropriate tissue culture medium with 50 μg/ml of gentamycin sulfate. Antibiotic is added to inhibit growth of bacteria that might have contaminated the initial 96-well plate.
2. Harvest parent PC-12 cells, load with Indo-1, and prepare the cells for flow cytometric analysis of Ca^{2+} mobilization (see *Protocols 2* and *3*).
3. Define sort regions:
   (a) Define the first sort-region within the forward scatter and side scatter cytogram to include single cells.
   (b) Stimulate the sample with stimulus (BK, 100 nM) to determine where to define a second sort region within the Indo-1 fluorescence ratio, or the blue and violet emission cytogram, to include only cells which exhibited a maximal Ca^{2+} mobilization response. Typically this should include <20% of the responding cells.

Protocol 4. *Continued*

(c) Backflush the sample line to remove residual agonist.

4. Sort clone: using a new sterile aliquot of cells, add the stimulus and immediately return to the cytometer input port. Initiate sorting as soon as the cells intercept the laser beam.

5. Incubate the clones in the appropriate incubator. As colonies proliferate in the positive wells, transfer the subclone colonies to progressively larger tissue culture vessels.

6. When sufficient cells are available, rescreen the subclones by flow cytometry to determine whether they exhibit an increase in response homogeneity. Alternatively, rescreen the clones by conventional bulk spectrofluorimetry to determine if the response magnitude is increased compared to parental cells (see *Protocol 6*).

In the case of the BK-dependent cloning procedure that was used to isolate BK1 cells (1), preliminary experiments indicated that 25% of the cells could still be detected with elevated intracellular Ca^{2+} levels 90 sec after stimulation. Therefore, aliquots of PC-12 cells were stimulated with BK and analysed on the cytometer for up to 90 sec. During screening of the clones by flow cytometry, specific clones were selected for further study based on (a) the ability to exhibit a homogeneous and synchronous response to BK and, (b) morphological compatibility with flow cytometry. The BK1 (1) subclone was selected for study because nearly 100% of the cells responded to BK, because it grows as a non-adherent, single cell population, and because of its relatively small size and nearly round morphology which make it ideal for flow cytometry. The modest and heterogenous responses of the PC-12 cells are compared with the robust and homogenous responses of the BK1 subclone (*Figure 3, A–D*). When the ratio of the violet to blue fluorescence emissions are calculated a ratio histogram can be constructed (*Figure 3E*). The ratio is indicative of the relative intracellular ionized Ca^{2+} levels ($Ca^{2+}{}_i$). Notice that the ratio histogram of resting BK1 cells is very narrow. This indicates that $Ca^{2+}{}_i$ is uniformly regulated in BK1 cells. Resting $Ca^{2+}{}_i$ was more heterogeneous in the parental cells. The homogenous resting $Ca^{2+}{}_i$ in the BK1 cells made it very clear when the stimulated cells entered the positive response region.

Acetylcholine (ACh) responsive cells (2) were selected by stimulating PC-12 cells with ACh. The ACh2 subclone was selected for study because 70–80% of the cells responded to ACh and because it grows as a modestly adherent population. Cytograms from the resting and stimulated populations are compared with parental PC-12 cells in *Figure 4* (panels *A–D*) and ratio histograms of stimulated ACh2 cells are shown in *Figure 4E*.

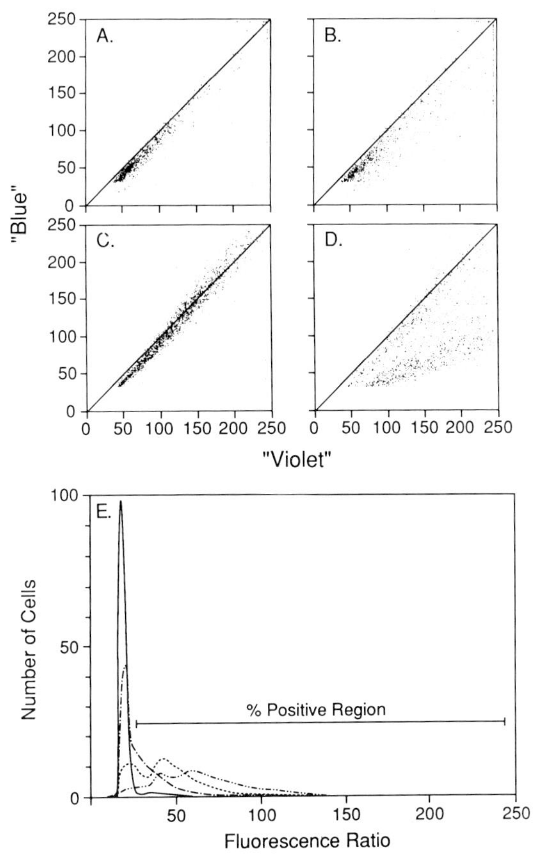

Figure 4. Dual fluorescence cytograms of PC-12 and ACh2 cells and the ratio histograms for ACh2 cells. Fluorescence emissions cytograms of Indo-1-loaded PC-12 cells (A and B) and ACh2 cells (C and D) were measured by flow cytometry before (A and C) and after (B and D) stimulation with 100 μM ACh. Panel E illustrates fluorescence ratio histograms calculated for ACh2 cells at rest (——) or stimulated with 100 nM (– · –), 10 μM (. . . . .) or 100 μM (– · · –) ACh. (Reprinted from Ransom *et al.* (2) with permission.)

8.2 Isolation of subclones using an on-line stimulus device

Protocol 5. Continuous sorting cloning of cells at a single time point in the response.

1. Prepare the flow cytometer as described in *Protocol 4*, steps 1a–c. Also, configure the instrument with a 'time-window' device (reference 7 or Cytek Development, Fremont, CA).

Protocol 5. *Continued*

2. Balance the tubing length from the sample input and stimulus input 'T'-junction and the relative air pressures so that the cells intercept the laser at the desired point in time after they have been mixed with the stimulus. If the two tubes leading into the T-junction are of approximately equal length and inner diameter, then equal pressure will drive fluid forward through the system towards the flow cell. Strongly biased pressure can force stimulus into the cell sample or cells into the stimulus vial. The pressure bias is controlled by the time-window biasing relay, and should be set at near zero bias to initiate set-up. The total pressure on the system is controlled by the sample pressure control which is part of the standard flow cytometer fluidics control panel. This pressure is the easiest way to control the delay between stimulus and laser intercept and should be set at the middle of its normal use range to initiate set-up. The delay is also determined by the length of tubing downstream from the T-junction, with tubing of 10 cm being convenient for delays of around 10 sec. These initial conditions are very approximate and subject to several aspects of each flow cytometer. Nevertheless the operating principles are general and performance as described can be expected with a little patience. With the device configured as described above, calibrate the sample and stimulus flow ratio and the delay using two bead suspensions as mock cells and stimulus.
   (a) With the two bead suspensions running under the conditions described above, clamp the stimulus line just upstream from the T-junction. Wait for the stimulus beads to clear from the common line, then open the clamp. The time taken for the stimulus beads to reappear at the laser is the delay. Small adjustments in the sample pressure will inversely increase or decrease the delay over a limited dynamic range. For longer or shorter delays, longer or shorter tubing can be used downstream from the T-junction.
   (b) Note that the ratio of stimulus and cells are controlled by the time-window biasing relay. Small adjustments in the relative pressure at the two vials will give coordinate flow rate changes, effectively controlling the stimulus concentration. If the concentration of the two bead suspensions is known, then the effective concentration after mixing can be easily derived from the ratio of the beads at the laser. It may prove convenient to use relative pressures giving a 1:1 mixture of beads, and adjusting the cell suspension and stimulus solution to 2× stocks.
   (c) Replace the two bead suspensions with cells and stimulus. It is useful to include a suspension of beads in the stimulus solution to monitor the stimulus delivery continuously during a long sort. When the cells and stimulus are running, and the appropriate delay is established, a constant fraction of cells should show elevated intracellular free Ca^{2+}. As

the delay is adjusted with small changes in sample pressure, the fraction of responding cells will change since the peak of the response will be apparent at only a given delay setting. The absolute number of beads and cells per second will also change if more or less pressure is applied to the combined system, but the ratio of stimulus to cells should not change over a relevant dynamic range.

(d) When the desired ratio and delay are established, the system can be sterilized with 70% ethanol by loading the ethanol as a cell sample with the stimulus chamber vented to air to allow the ethanol to backfill the stimulus line. Then the stimulus line can be clamped to force the ethanol forward to the flow cell to sterilize the T-junction and the common line.

(e) Finally a sterile cell suspension and stimulus solution (with reporter beads) is mounted and sort cloning proceeds by standard methods.

As an example of this technique, NG108-15 cells were analysed by flow cytometry on a FACStar Plus to determine the response heterogeneity within the population and the rate of onset of the response to AII (AII, *Figure 5A*).

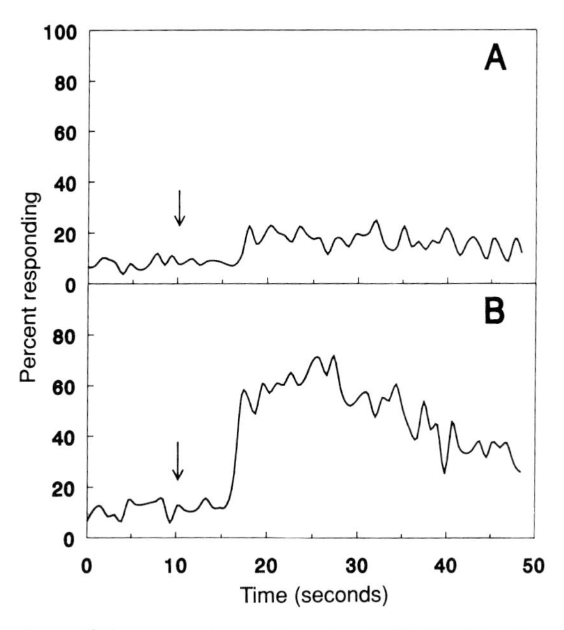

Figure 5. Comparison of the percentage of uncloned NG108-15 cells versus subcloned NG108-15.C16 cells that respond to AII. (**A**) Indo-1-loaded parental NG108-15 cells were analysed by flow cytometry as described in *Figure 2*. (**B**) NG108-15.C16 subclone cells that had been selected based on a response to AII at 6 sec were analysed for the percentage of cells responsive to AII.

The earliest response time window that yielded a reasonable percentage of detectable responding cells was 6 sec after addition of 1 μM AII. Approximately 15–20% of the cells exhibited a Ca^{2+} response at this time point. The time-window module was used to stimulate the cells with 1 μM AII so that the cells were continuously analysed 6 sec after mixing with AII. After the subclones were expanded, the characteristics of their responses to AII were analysed by bulk spectrofluorimetry and compared with the parent NG108-15 line. Clones that exhibited substantially faster or greater average responses than the parental line were further analysed by flow cytometry. The response of one typical subclone (NG108-15.C16) is shown (*Figure 5B*).

9. Characterization of receptor subtypes by non-parametric flow cytometry

The increased homogeneity of the resting intracellular Ca^{2+} level ($Ca^{2+}{}_i$) and the greater synchronicity and homogeneity of the Ca^{2+} response in the subclones has yielded many analytical advantages. One advantage is the ability to quantify responses to specific ligands in terms of the percentage of cells that respond instead of in terms of the average response magnitude throughout the entire population. Upon performing such an analysis with BK and the BK1 subclone we found that at low concentrations of agonist, only a few cells responded at the time point of analysis (*Figures 3E* and *6*). With greater (but submaximal) concentrations of BK the majority of cells responded. However, even in this case some cells did not respond. Thus, it is not the case that with low concentrations of agonist all of the cells respond to a lesser extent. Rather, with low concentration of agonist some of the cells respond but many do not appear to respond at all. A similar conclusion could be drawn from the response of the ACh2 cells to varying concentrations of ACh (*Figures 4E* and *7*). The molecular basis for this concentration-dependent variability remains unclear but may be related to factors such as the amount of receptor expressed on the surface or the efficiency of coupling of the receptors to the second messenger generating machinery within the cell.

When the response by BK1 cells to a series of BK receptor subtype-specific agonists was analysed in this manner, it became clear that the BK receptor which is coupled to Ca^{2+} mobilization in these cells is the B_2 receptor subtype (*Figure 6*). The EC_{50} value derived from the study agreed closely with one obtained by Appel and Barefoot (10) who used parental PC-12 cells and the bulk spectrofluorimetric measurement of the peak magnitude of the BK mediated Ca^{2+} response. The EC_{50} value also correlated well with values obtained for the B_2 receptor using different functional assays and tissues (1). When the technique was used to study the effect of receptor subtype specific antagonists this method was also proven to be accurate in the identification of BK receptor subtypes (*Figure 8*). The IC_{50} values for B_2 receptor antagonists

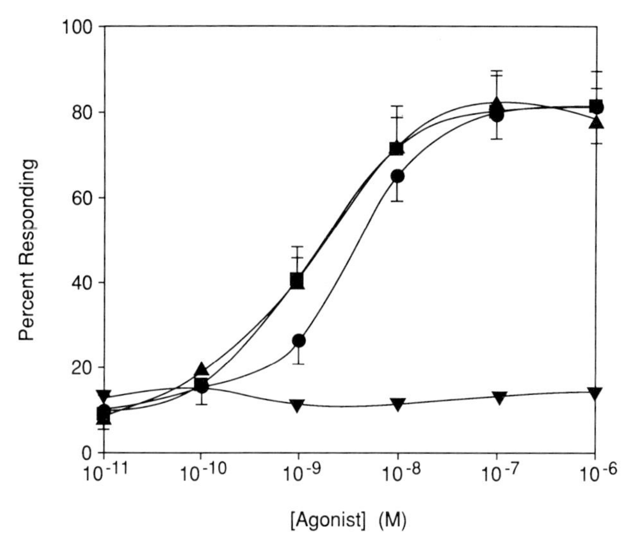

Figure 6. BK1 cells respond to B_2 but not B_1 receptor agonists. BK1 cells were stimulated with the indicated concentrations of BK (▲), Met-Lys-BK (●), or Lys-BK (■). The percentage of cells with elevated intracellular Ca^{2+} levels was determined after 25 sec (± SEM, $n = 4$). Des-Arg^9-BK (▼) data are from one experiment and similar results were seen in a duplicate experiment. The EC_{50} values (nM) were 2.0 ± 6.7, 1.9 ± 0.38, and 4.8 ± 0.48 for BK, Lys-BK, and Met-Lys-BK respectively. (Reprinted from Ransom *et al.* (1) with permission.)

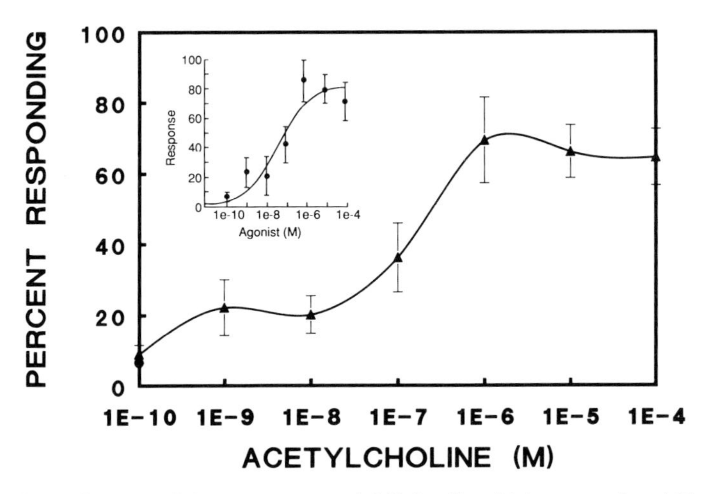

Figure 7. Quantification of the percentage of ACh2 cells which respond to ACh. Indo-1-loaded ACH2 cells were stimulated with ACh and the percentage of responding cells (▲) were determined by flow cytometry after 25 sec ($n = 5$, ± SEM). Resting cells (●). Inset: Data within each experiment were corrected for background and normalized as the percentage of the maximum response. Means (± SEM) were plotted as a function of the agonist concentration. The curve used for the EC_{50} calculation was determined by an iterative fitting procedure (3). The EC_{50} was determined to be 38.7 nM ($n = 5$). (Reprinted from Ransom *et al.* (2) with permission.)

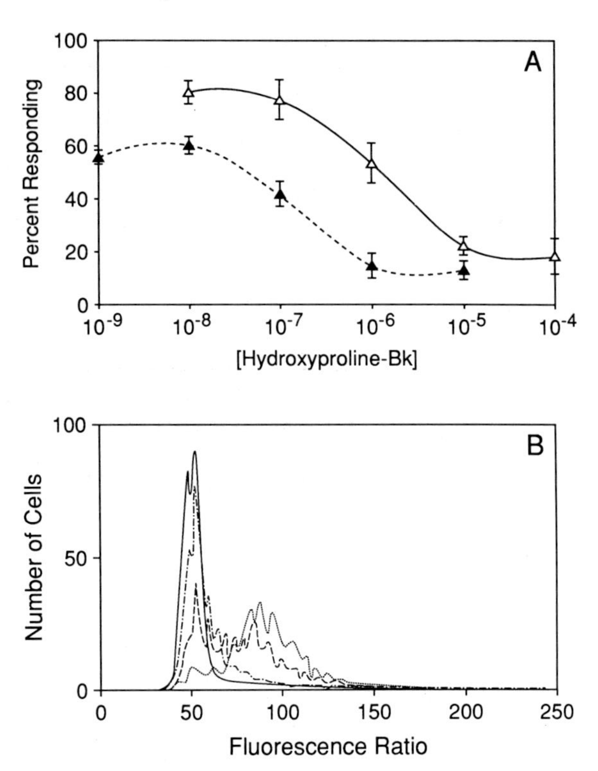

Figure 8. Inhibition of the BK-induced Ca^{2+} response by the B2 antagonist D-Arg^0-Hyp^3-D-Phe^7-BK. (**A**) BK1 cells were preincubated with the indicated concentrations of D-Arg^0-Hyp^3-D-Phe^7-BK then stimulated with 10 nM (▲) or 100 nM (△) BK (n = 4). (**B**) Fluorescence ratio histograms for a representative experiment with Hyp^3-BK are shown as described in *Figure 4B*. Histograms are of resting cells (——), or after stimulation with 100 nM BK only (. . . . .), or stimulated in the presence of 1 μM (– – –) or 10 μM (– · –) D-Arg^0-Hyp^3-D-Phe^7-BK. The calculated EC_{50} values against 10 nM and 100 nM BK were 114.3 ± 13.3 nM and 1.2 μM respectively. (Reprinted from Ransom *et al.* (1) with permission.)

correlated well with values obtained for the same antagonists using various functional assays in different tissues (1). Thus, the utility and reproducibility of flow cytometric characterization of receptor subtypes and receptor specific agonists and antagonists was validated in the study on BK1 cells.

Further utility of the technique was demonstrated when it was used to identify the M_4 muscarinic receptor expressed on the PC-12.ACh2 subclone. In this case the muscarinic response of the parental cells were so subtle that pharmacological characterization would have been impossible. The presence of a muscarinic receptor on the subclone was verified by studying the effect of increasing concentrations of the antagonists atropine or pirenzipine (*Figure 9*) on the ACh2 response to ACh. In combination with binding assays with

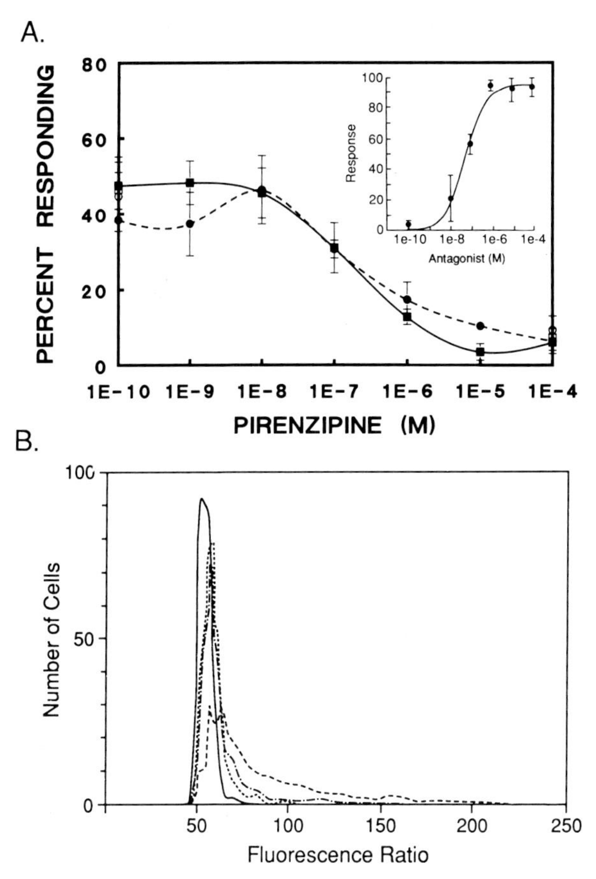

Figure 9. Pirenzipine inhibits the ACh-mediated Ca^{2+} response in all ACh2 cells. (**A**) Cells were pre-incubated with (■, ●) or without (□, ○) pirenzipine for 30–60 min and then stimulated with 10 μM (——) or 100 μM (– – –) ACh (n = 3–4, ± SEM). Unstimulated cells, right axis (□, ○). Insert: Data obtained with 100 μM ACh stimulation were normalized and plotted against the antagonist concentration. The calculated IC_{50} value was 55 nM against 100 μM ACh. (**B**) Fluorescence ratio histograms from one experiment calculated for cells at rest (——) or stimulated by 10 μM ACh in the presence of (– · –) 100 μM pirenzipine, (. . .) 1 μM pirenzipine, or (– – –) no pirenzipine (Reprinted from Ransom *et al.* (2) with permission.)

several muscarinic receptor antagonists it was determined that the muscarinic receptor was the M_4 subtype. The mechanism of coupling of the receptor to Ca^{2+} mobilization was found to be independent of extracellular Ca^{2+} and sensitive to inhibition by pertussis toxin (*Figure 10*). Our ability to characterize the M_4 receptor by use of a Ca^{2+} assay was surprising since the M_4 receptor has been shown to be more weakly coupled to phosphatidylinositol hydrolysis and Ca^{2+} mobilization than the M_1 and M_3 receptors (11).

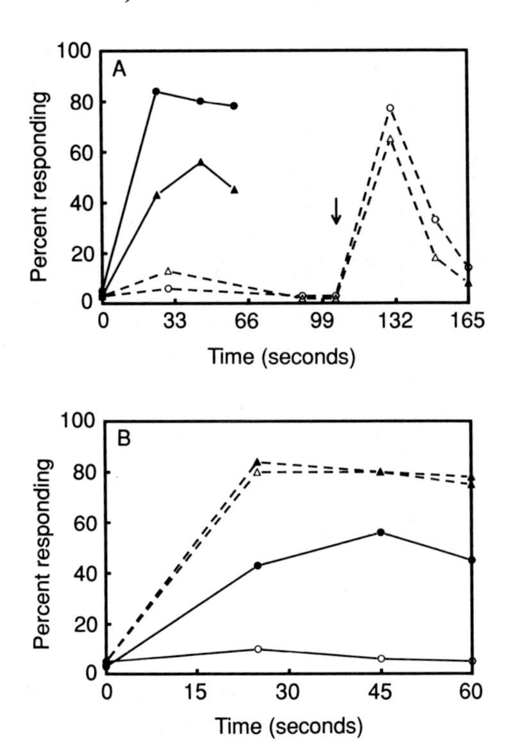

Figure 10. The muscarinic Ca^{2+} response is independent of Ca^{2+} and inhibited by pertussis toxin. (**A**) ACh2 cells were stimulated by 10 μM ACh (▲) or 0.1 μM BK (●) and the percentage of responding cells determined at the indicated time points. Alternatively, EGTA (4 mM, final pH 7.3) was added to the samples and analysed at 30 and 90 sec (△, ○). The cells were then stimulated (arrow) with BK (○) or ACh (△) and the response was determined. (**B**) The cells were incubated in culture medium (37 °C) in the presence (△, ○) or absence (▲, ●) of 50 ng/ml of PTX for 2 h. Then the cells were loaded with Indo-1, in the presence of PTX if appropriate, and stimulated with 10 μM Ach (○, ●) or 0.1 μM (△, ▲) BK. Results are typical of three separate experiments. (Reprinted from Ransom *et al.* (2) with permission.)

However, by analysing the percentage of responding cells rather than the response magnitude we obtained a very low EC_{50} value for the effect of ACh on the M_4 receptor (*Figure 9*). This indicated that occupany of only 2% of the M_4 receptors was required to observe a response in 50% of the cells. Thus, the M_4 receptor actually is efficiently coupled to Ca^{2+} mobilization, but the absolute magnitude of the response is relatively small.

10. Characterization of receptor subtypes by bulk spectrofluorimetry using sort-selected subclones

There are many reasons why it would be desirable or necessary to perform Ca^{2+} mobilization analyses in a cuvette-based bulk spectrofluorimeter. For

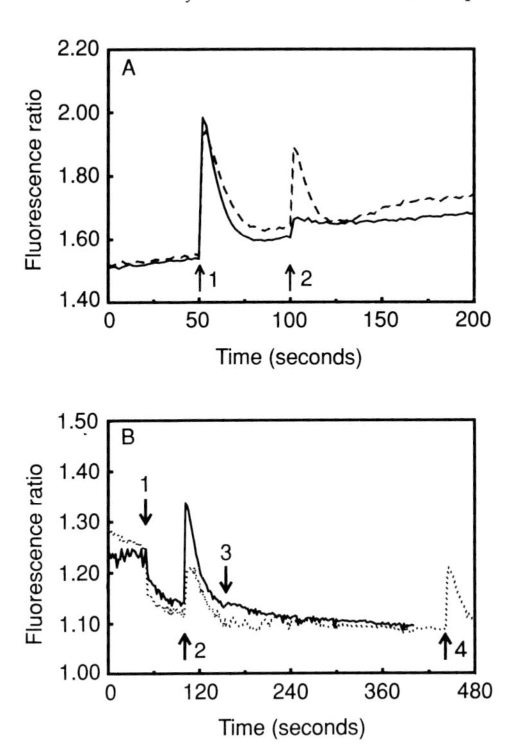

Figure 11. BK pre-stimulation reduces the mean magnitude of the ACh- evoked response. (**A**) Indo-1-loaded ACh2 cells were prestimulated at 50 sec (first arrow) with 1 μM BK (——) or 10 μM ACh (– – –) and the ratio of the dual fluorescence emissions was determined. At 100 sec (second arrow) the samples were stimulated with the identical concentrations of the converse agonist (BK (– – –) or ACh (——)). (**B**) EGTA (4mM) was added to ACH2 cell samples at 50 sec (first arrow). At 100 sec (second arrow) 1 μM BK (——) or 10 μM ACh (. . . . .) was added. At 150 sec (third arrow) 10 μM ACh was added to the sample prestimulated by BK (——). At 440 sec (fourth arrow) 1 μM BK was added to the sample prestimulated by ACh (. . . . .). (Reprinted from Ransom *et al.* (2) with permission.)

example, flow cytometers can be expensive to purchase and operate, it is difficult to regulate the temperature of the samples in a flow cytometer, and it may not be possible to obtain a subclone with a morphology that is compatible with flow cytometry. Many labs may find it more practical to perform repetitive assays on a conventional spectrofluorimeter after suitable subclones have been isolated by flow cytometry. The greater Ca^{2+} response synchronicity and homogeneity of the subclones yield improved signal-to-noise ratios when the average $Ca^{2+}{}_i$ levels are monitored in a spectrofluorimeter. For example, with the ACh2 subclone we determined by spectrofluorimetry that prestimulation of the cells with BK decreased the magnitude of a subsequent response to ACh (*Figure 11*). However, the prestimulation by BK did not decrease the percentage of cells that responded to ACh (2). This suggests that the BK receptor signal transduction pathway communicates negatively with the M_4 receptor signalling mechanisms at some point in the transduction pathway.

Another example of this benefit was the characterization of the angiotensin receptor expressed on NG108-15 cells as the AT_1 subtype. The analysis was performed by first isolating an AII responsive subclone (3). The subclone, C1, was isolated using the time-window module (see *Protocol 5*). All of the clones were initially screened on a spectrofluorimeter by comparing the magnitude and kinetics of the response to AII with that of the parental NG108-15 line. Once the panel of clones was obtained the response homogeneity was studied by flow cytometry to confirm that the increased response magnitude was due to increased homogeneity and synchronicity. The parental NG108-15 line did not exhibit the high degree of response heterogeneity that was observed with the parental PC-12 cells. None the less, there was an increase in response homogeneity by 3- to 4-fold in many of the subclones. Characterization of the AII receptor was performed on the spectrofluorimeter using a panel of agonists and antagonists specific for the AT_1 or AT_2 angiotensin receptors.

Protocol 6. Spectrofluorimetric analysis of Ca^{2+} mobilization

1. Harvest cells and load with fluorescent Ca^{2+} indicator dye as described in *Protocols 1* and *2*.
2. Wash the cells twice in Ringer's solution after dye loading. This is to remove excess unhydrolysed dye from the suspension. Resuspend the cells at 1×10^6 cells/ml in normal Ringer's.
3. Select the appropriate excitation and emission wavelengths for the fluorophore being used. For example, with Indo-1 we use 335 nm for excitation and emissions are read with two photomultipliers using 400 and 500 nm emissions. Other wavelength selections are unique to the dye being used and should be selected based on values cited in the literature.
4. The most consistent results are obtained if the cells are aliquoted (2.1 ml) into individual tubes prior to analysis. This ensures that all samples contain the same number of cells. Subtle differences in the resting $Ca^{2+}{}_i$ levels and response characteristics become apparent if successive samples are drawn from one common tube. This may be due to factors such as cell settling or sticking of the cells to each other in the tube.
5. If possible the samples should be analysed at 37°C by using a temperature-regulated water circulator to warm the cuvette holder. However, the individual aliquots of cells should be maintained at room temperature until 3–5 min prior to use. Prolonged warming of the cells causes more rapid rundown of the response and enhances dye leakage from the cells. Our system involves prewarming an aliquot of cells in the

water bath for 3–5 min before analysis. Then, when the cells are transferred to the cuvette the amount of time required to equilibrate the cell, buffer, and cuvette temperatures is minimal.

6. It is critical to allow the cells to equilibrate fully in the cuvette while the stirrer is running since the K_D of the dye for Ca^{2+} is very sensitive to temperature, pH, and ionic strength. If the cells are not equilibrated a sloping baseline value for resting $Ca^{2+}{}_i$ will be obtained. Without a stable horizontal baseline it is difficult to determine the effect of an agonist or antagonist on $Ca^{2+}{}_i$.

7. With each sample, once the baseline is stable the agonist can be added to the cells and the analysis quickly resumed.

8. If the effects of antagonists are to be tested, then the antagonists can be added to the dye-loaded, aliquoted cells 30–45 min before the analysis. Again, the cells should not be maintained at 37 °C during the preincubation. Each aliquot of cells and antagonist can then be transferred to the cuvette for analysis.

9. Since the dye reports $Ca^{2+}{}_i$ on a nonlinear scale it is desirable to linearize the data by determining the maximal (F_{max}) and minimal (F_{min}) dye dependent fluorescence values.
   (a) To determine the F_{max} add Triton X-100 to a final concentration of 0.03%. This will lyse the cells and allow the dye to become saturated with Ca^{2+}. A flat, stable signal indicates that the cells have fully lysed.
   (b) Determine the F_{min} value by adding 6 mM EGTA followed by 100 mM Tris base to chelate Ca^{2+} from the dye. Again a flat line indicates complete chelation of the Ca^{2+}.
   (c) These values can be used in the equations described by Grinkiewicz *et al.* (4) to obtain $Ca^{2+}{}_i$ values based on a linear scale.

10. The data can be analysed in several ways to gain information about the effects of increasing concentrations of agonist or antagonist on the response magnitude:
   (a) The baseline value just prior to addition of the agonist can be subtracted from the peak value obtained after addition of agonist (10).
   (b) If it is difficult to define a peak response consistently then it is possible to select a point in time in the ascending part of the response (3). Then, data for all samples can be derived from the same time point without operator bias or error in selecting the peak.
   (c) Some software makes it possible to subtract out the background and integrate the area under the response curve. This has proven satisfactory as long as the signals return to the original baseline.

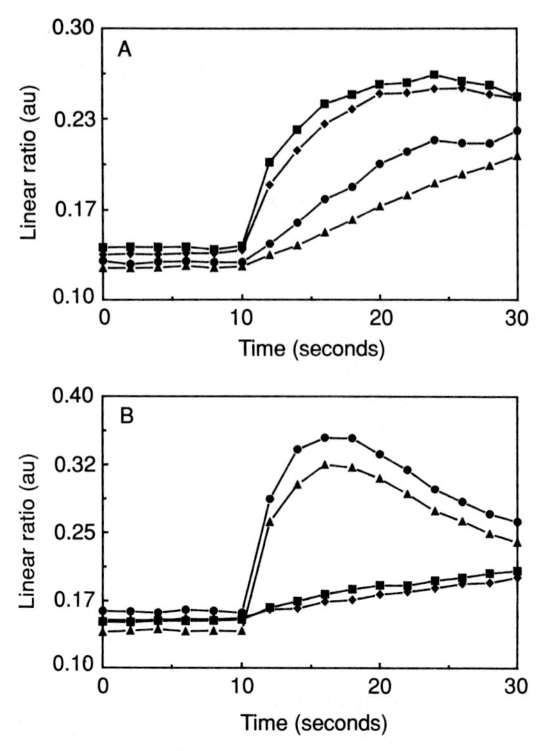

Figure 12. Comparison of the Ca^{2+} mobilization response kinetics of C1 cells to AI, AII, and AIII. (**A**) Fura-2-loaded C1 cells were stimulated by 0.1 μM (▲), 0.3 μM (●), 1 μM (◆), or 3 μM (■) of AII at t = 10 sec. (**B**) C1 cells were stimulated with 1 μM (◆) or 3 μM (■) of AI, or with 1 μM (▲) or 3 μM (●) of AIII at t = 10 sec. The Ca^{2+} mobilization data are in arbitrary linear units. (Reprinted from Ransom *et al.* (2) with permission.)

An example of the utility of the technique described in *Protocol 6*, step 10(b) is shown in *Figure 12*, where the responses of the NG108-15.C1 subclone to AII (A) or angiotensin I (AI) and angiotensin III (AIII, B) were studied. With high concentrations of AII it was difficult to identify a peak response value consistently and reliably. With lower concentrations a response was apparent but there was no detectable peak even when the assay was continued for a full minute. Thus, we chose to analyse all AII response data 6 sec after the addition of ligand. The response of the cells was still in the ascending phase for all concentrations of agonist at this time point. Similarly, we chose 4 and 10 sec as the analysis time point for AI and AIII respectively. When the data from a single experiment were normalized so that the maximal value was scored as 100% stimulation a dose-response curve could be constructed (*Figure 13*). The concentration of agonist that yielded 50% of the maximal response was estimated from several such experiments. A mean EC_{50} (57 ± 8 nM, n = 12) was then calculated for each agonist from the

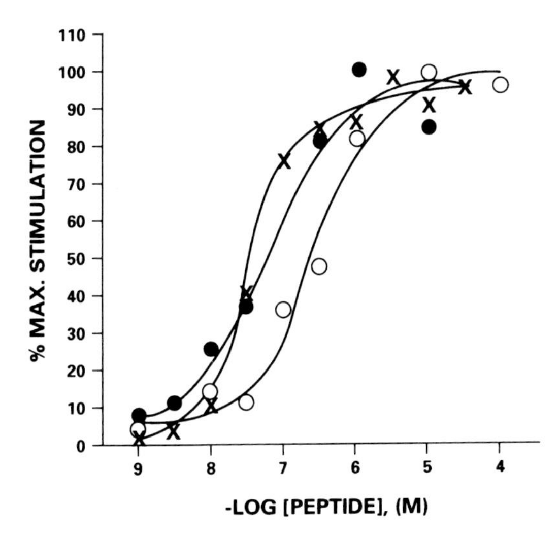

Figure 13. Concentration-response curves for AI (○), AII (●) and AIII (×) for mobilization of intracellular Ca^{2+} in C1 cells. Data are from one sample experiment. The mean EC_{50} values (nM) from several experiments like this were 437 ± 80, 57 ± 8, and 36 ± 5 for AI, AII, and AIII respectively. (Reprinted from Ransom *et al.* (3) with permission.)

experiments. The AT_1 angiotensin receptor antagonist DUP-753 and the AT_2 receptor-specific antagonist EXP-655 (also known as PD123177) were used to characterize the receptor subtype. While EXP-655 had no effect on the AII response, DUP-753 was very effective at inhibiting Ca^{2+} mobilization. The AII dose-response assay was repeated in the presence of increasing concentrations of DUP-753 (*Figure 14A*). Dextral shifts of the AII response curves, without a decrease in the maximal response, indicated that DUP-753 was a competitive antagonist, thereby identifying the AII receptor as belonging to the AT_1 subtype. The data were further analysed to yield an affinity estimate of the antagonist (DUP-753) for the receptor ($pA_2 = 8.5 \pm 0.2$, *Figure 14B*), this being very close to the affinity value obtained from tissue contraction bioassays (12).

11. Discussion and conclusions

In addition to voltage-gated Ca^{2+} channels, there are several common biochemical pathways that couple receptors to Ca^{2+} mobilization. These include receptor and ligand gated Ca^{2+} channels, GTP-binding proteins that activate phospholipase C, and tyrosine kinases that activate phospholipase C. Downstream of the activation of phospholipase C is the formation of inositol 1,4,5-trisphosphate which can release Ca^{2+} from an intracellular store (e.g. endoplasmic reticulum). The rate of development of Ca^{2+} responses may lie along

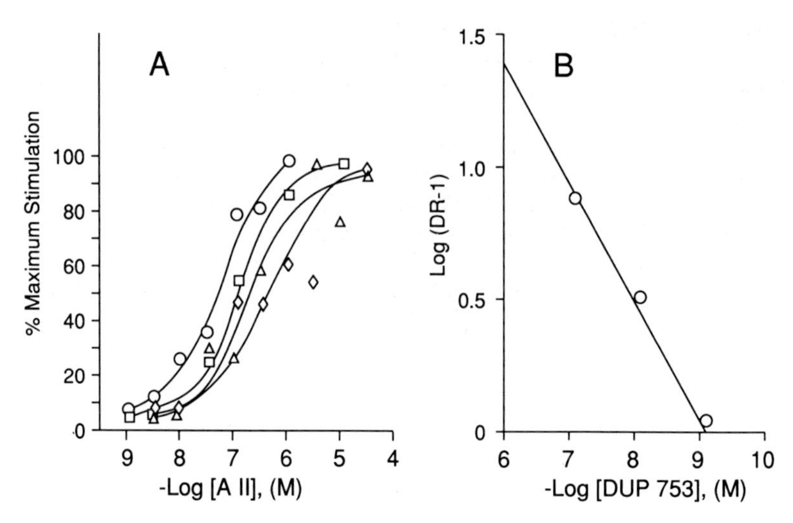

Figure 14. Antagonism of the AII-induced Ca^{2+} mobilization in C1 cells by DUP-753. (**A**) Concentration-response curves for AII determined on different days are shown in the absence (○) and presence of increasing concentrations (□, 1 nM; △, 10 nM; ◇, 100 nM) of DUP-753. Results from a typical experiment for each condition are shown. (**B**) Representative Schild analysis of the data shown in panel A. The affinity value (pA_2) estimated from this and several other experiments was 8.5. (Reprinted from Ransom *et al.* (3) with permission.) (DUP-753 is also now known as Losartan.)

a spectrum from relatively rapid, as in the case of voltage-gated Ca^{2+} channel activation, to relatively slow, as in the case of mobilization of Ca^{2+} in mitogen-stimulated lymphocytes. The Ca^{2+} response may also involve sustained oscillations in intracellular Ca^{2+} levels, and the periodicity and magnitude of oscillations are under regulatory control by mechanisms that are poorly understood. It is important to study the regulatory mechanisms of all aspects of a Ca^{2+} mobilization response. The ability to select clones based upon their selective expression of a particular Ca^{2+} response phenotype should prove useful in such endeavours.

References

1. Ransom, J. T., Cherwinski, H. M., Dunne, J. F., and Sharif, N. A. (1991). *J. Neurochem.*, **56,** 983.
2. Ransom, J. T., Cherwinski, H. M., Delmendo, R. E., Sharif, N. A., and Eglen, R. E. (1991). *J. Biol. Chem.*, **266,** 11738.
3. Ransom, J. T., Sharif, N. A., Dunne, J. F., Momiyama, M., and Melching, G. (1992). *J. Neurochem.*, **58,** 1883.
4. Grynkiewicz, G., Poenie, M., and Tsien, R. (1985). *J. Biol. Chem.*, **260,** 3440.
5. Van Dilla, M. A., Dean, P. N., Laerum, O. D., and Melamed, M. R. (1985). *Flow Cytometry: Instrumentation and Analyses*. Academic Press, Orlando, FL.

6. Kelly, K. A. (1989). *Cytometry*, **10,** 796.
7. Dunne, J. F. (1991). *Cytometry*, **12,** 597.
8. Rabinovitch, P. S., June, C. H., Grossmann, A, and Ledbetter, J. A. (1986). *J. Immunol.*, **137,** 952.
9. Ransom, J. T., Harris, L. D., and Cambier, J. C. (1986). *J. Immunol.*, **137,** 708.
10. Appel, K. C. and Barefoot, D. S. (1989). *Biochem. J.*, **263,** 11.
11. Ashkenazi, A., Peralta, E. G., Winslow, J. W., Ramachandran, J., and Capon, D. J. (1989). *Trends Pharmacol. Sci.*, **10** (Suppl. IV), 16.
12. Chiu, A. T., McCall, D. E., Price, W.-A., Wong, P. C., Carini, D. J., Duncia, J. V., Wexler, R. R., Yoo, S. E., Johnson, A. L., and Timmermans, P. B. M. W. M. (1990). *J. Pharmacol. Exp. Ther.*, **252,** 711.

Addresses of suppliers of materials and equipment

Aldrich Chemicals Co. Ltd., Gillingham, Dorset, UK. (For all chemicals.)

American Optical Co., P.O. Box 123, Buffalo, NY 14240, USA. (For sledge microtomes equipped with freezing stage.)

Amersham Corp., 2636 S. Clearbrook Drive, Arlington Heights, IL 60005, USA; **Amersham International plc,** Lincoln Place, Green End, Aylesbury, Buckinghamshire, HP20 2TP, UK. (For radioligands, Hyperfilm, radiation standards, and image analysis systems.)

Amicon Division, 4 Cherry Hill Drive, Danvers, MA 01923, USA. (For immunohistochemical reagents.)

ATCC, 123401 Parklawn Drive, Rockville, MD 20852–1776, USA. (For cell lines.)

Beckman Instruments Ltd., Stanford Industrial Park, P.O. Box 10015, Palo Alto, CA 94303, USA. (For centrifuges and robotics systems.)

Beckton-Dickinson, P.O. Box 7375, Mountain View, CA 94039, USA. (For flow cytometers such as FACStar Plus, and for automatic cell deposition unit.)

Bio-Rad Laboratories, 220 Maple Avenue, Rockville Center, NY 11571, USA. (For molecular biology reagents, protein reagents, ion exchange resins, and HPLC columns.)

Boehringer Mannheim Biochemicals, 9115 Hague Road, P.O. Box 50414, Indianapolis, IN 46250–0414, USA. (For enzymes and non-radioactive reagents for *in situ* histochemistry.)

Brandel Research & Development Labs Inc., 8561 Atlas Drive, Gaithersburg, MD 20877, USA. (For cell harvesters and other filtration equipment.)

Bright Equipment Co. Ltd., Clifton Road, Huntington, Cambridgeshire, UK. (For cryostat microtomes.)

Calbiochem, P.O. Box 12087, San Diego, CA 92112–4180, USA. (For peptides, reagents, and ion-sensitive dyes and stains.)

Cambridge Instruments Inc. (also known as **Leica**), 26011 Fallbrook, El Toro, CA 92630, USA: **Leica,** Cambridge CB3 8EL, UK. (For image analysis equipment, cryostat microtomes, and microscopes.)

Cappel Research Reagents, Organon Teknika Corp., Treyburn, 100 Akzo Avenue, Durham, NC 27704, USA; Veedijk 58, 2300 Turnhout, Belgium. (For immunohistochemical reagents.)

Carl Zeiss (Oberkochen) Ltd., P.O. Box 78, Woodfield Road, Welwyn Garden City, Hertfordshire, UK. (For microscopes and related equipment.)

Coster, 205 Broadway, Cambridge, MA 02139, USA. (For all cell culture materials and media.)

Creative Color Lab., 207 Orange Street, Henderson, NC 27536–4234, USA. For dark-room equipment and supplies.)

Cryoservices, Blackpole Trading Estate, Blackpole Road, Worcester WR3 8SG, UK. (For liquid nitrogen containers.)

Cytek Development, 1525 Fulton Place, Fremont, CA 94539, USA. (For Time-zero and Time-window devices needed in conjunction with the flow cytometer.)

Eastman Kodak Co., Laboratory Research Products Division, 343 State Street, Building 701, Rochester, NY 14652–32512, USA; P.O. Box 10, Dallimore Road, Manchester, M23 9NJ, UK. (For all photographic materials and equipment.)

Eutectic Electronics Inc., 6808 Jersey Court, Raleigh, NC 27612, USA. (For 3-D reconstruction computer programs.)

Fisher Scientific, 711 Forbes Avenue, Pittsburgh, PA 15219, USA. (For general laboratory equipment and supplies.)

Flow Labs. Ltd., P.O. Box 17, Second Avenue Industrial Estate, Irvine, Ayrshire KA12 8NB, UK; 7655 Old Spring Drive, McLean, VA 22102, USA. (For cell culture materials and media, cell lines, laminar flow hoods, and incubators.)

Fluka Chemika-BioChemika, 980 South Second Street, Ronkonkoma, NY 11779–728, USA. (For DPX mounting material, formamide, etc.)

GIBCO/BRL Life Technologies Inc., 3175 Staley Road, Grand Island, NY 14072, USA. (For all cell culture materials and media.)

Grant Instruments Ltd., Barrington, Cambridge CB2 5D1, UK. (For water baths.)

Hacker Instruments Inc., P.O. Box 657, Fairfield, NJ 07007, USA. (For cryostat microtomes.)

Hewlett-Packard Corp., 16399 West Bernado Drive, San Diego, CA 92127, USA. (For computers, plotters, and accessories.)

IBM AT Microcomputers, IBM Corp., Boca Raton, FL 33432, USA. (For computers and plotters.)

Imaging Technology Inc., Woburn, MA 01801, USA. (For image analysis equipment.)

Imaging Research Inc., Brock University, Ontario, L2S 3A1, Canada. (For the MCID image analysis system and accessories.)

Joyce-Loebl, Gateshead, NE11 OP2, UK. (For image analysis equipment.)

Labline Instruments, Labline Plaza, Melrose Park, IL 60160, USA. (For Environ shaker.)

Linscott Directory of Immunological & Biological Reagents, P.O. Box 5, East Grinstead, Sussex RH19 3YL, UK. (For antibodies, etc.)

Meridian Instruments Inc., 2310 Science Parkway, Okemos, MI 48864, USA; Meridian Instruments Europe Inc., Industriepark West 75, B–9100 St Niklaas, Belgium. (For confocal microscopes.)

Miles Labs., P.O. Box 37, Stoke Court, Stoke Poges, Slough S12 4LY. UK; Naperville, IL 60566, USA. (For OCT compound and antibodies.)

Molecular Probes, P.O. Box 22010, 4849 Pitchford Avenue, Eugene, OR 97402, USA. (For various molecular biological materials and dyes.)

New England Nuclear, 549 Albany Street, Boston, MA 02118, USA; **NEN Du Pont (UK) Ltd.,** Wedgewood Way, Stevenage, Hertfordshire SG1 4QN, UK. (For radioligands.)

Nikon Inc., Garden City, NY 11530, USA (For cameras and lenses.)

Nunc, 2000 North Aurora Road, Napeville, IL 60653, USA. (For all cell culture materials.)

Olympus Optical Co. (UK) Ltd., London EC1Y 0TX, UK. (For cameras and accessories.)

Paines & Byrne Ltd., Pabyrn Labs, 155/179 Bilton Road, Perivale, Greenford, Middlesex, UB6 7HG, UK. (For electrophoresis equipment.)

Perkin Elmer Cetus, 761 Main Avenue, Norwalk, CT 06859, USA. (For analytical equipment and spectrofluorimeters.)

Pharmacia LKB Biotechnology Inc., 800 Cenntennial Avenue, P.O. Box 1327, Piscataway, NJ 08855–1327, USA. (For HPLC columns, filtration systems coupled to β-counters, radiation-sensitive films, spectrophotometers, and immunostaining apparatus.)

Polaroid Inc., Cambridge, MA 02139, USA. (For all photographic materials and equipment.)

Seescan Ltd., Unit 9, 25 Gwyder Street, Cambridge CB1 2LG, UK. (For image analysis systems.)

Sigma Chemical Company Ltd., P.O. Box 14508, St Louis, MO 63178, USA; Fancy Road, Poole, Dorset BH17 7BR, UK. (For all laboratory chemicals and drugs, and cell culture and molecular biological materials.)

Stratagene, 11099 North Torrey Pines Road, La Jolla, CA 94912, USA. (For molecular biology reagents and equipment.)

United Dessicants, 6845 Westfield Avenue, Pennsauken, NJ 08110, USA. (For desiccants.)

Universal Imaging Corporation, Media, PA 19063, USA. (For image analysis equipment.)

Vector Labs Inc., 30 Ingold Road, Burlingame, CA 94010, USA. (For Elite Kit.)

Wild Leitz USA, Inc., 1123 Grandview Drive, South San Francisco, CA 94080, USA. (For macroscope and microscopes and accessories.)

Index

Index

ORDER OTHER TITLES OF INTEREST TODAY

Price list for: UK, Europe, Rest of World (excluding US and Canada)

138. Plasmids (2/e) Hardy, K.G. (Ed)
.... Spiralbound hardback 0-19-963445-9 **£30.00**
.... Paperback 0-19-963444-0 **£19.50**

136. RNA Processing: Vol. II Higgins, S.J. & Hames, B.D. (Eds)
.... Spiralbound hardback 0-19-963471-8 **£30.00**
.... Paperback 0-19-963470-X **£19.50**

135. RNA Processing: Vol. I Higgins, S.J. & Hames, B.D. (Eds)
.... Spiralbound hardback 0-19-963344-4 **£30.00**
.... Paperback 0-19-963343-6 **£19.50**

134. NMR of Macromolecules Roberts, G.C.K. (Ed)
.... Spiralbound hardback 0-19-963225-1 **£32.50**
.... Paperback 0-19-963224-3 **£22.50**

133. Gas Chromatography Baugh, P. (Ed)
.... Spiralbound hardback 0-19-963272-3 **£40.00**
.... Paperback 0-19-963271-5 **£27.50**

132. Essential Developmental Biology Stern, C.D. & Holland, P.W.H. (Eds)
.... Spiralbound hardback 0-19-963423-8 **£30.00**
.... Paperback 0-19-963422-X **£19.50**

131. Cellular Interactions in Development Hartley, D.A. (Ed)
.... Spiralbound hardback 0-19-963391-6 **£30.00**
.... Paperback 0-19-963390-8 **£18.50**

129 Behavioural Neuroscience: Volume II Sahgal, A. (Ed)
.... Spiralbound hardback 0-19-963458-0 **£32.50**
.... Paperback 0-19-963457-2 **£22.50**

128 Behavioural Neuroscience: Volume I Sahgal, A. (Ed)
.... Spiralbound hardback 0-19-963368-1 **£32.50**
.... Paperback 0-19-963367-3 **£22.50**

127. Molecular Virology Davison, A.J. & Elliott, R.M. (Eds)
.... Spiralbound hardback 0-19-963358-4 **£35.00**
.... Paperback 0-19-963357-6 **£25.00**

126. Gene Targeting Joyner, A.L. (Ed)
.... Spiralbound hardback 0-19-963407-6 **£30.00**
.... Paperback 0-19-9634036-8 **19.50**

125. Glycobiology Fukuda, M. & Kobata, A. (Eds)
.... Spiralbound hardback 0-19-963372-X **£32.50**
.... Paperback 0-19-963371-1 **£22.50**

124. Human Genetic Disease Analysis (2/e) Davies, K.E. (Ed)
.... Spiralbound hardback 0-19-963309-6 **£30.00**
.... Paperback 0-19-963308-8 **£18.50**

122. Immunocytochemistry Beesley, J. (Ed)
.... Spiralbound hardback 0-19-963270-7 **£35.00**
.... Paperback 0-19-963269-3 **£22.50**

123. Protein Phosphorylation Hardie, D.G. (Ed)
.... Spiralbound hardback 0-19-963306-1 **£32.50**
.... Paperback 0-19-963305-3 **£22.50**

121. Tumour Immunobiology Gallagher, G., Rees, R.C. & others (Eds)
.... Spiralbound hardback 0-19-963370-3 **£40.00**
.... Paperback 0-19-963369-X **£27.50**

120. Transcription Factors Latchman, D.S. (Ed)
.... Spiralbound hardback 0-19-963342-8 **£30.00**
.... Paperback 0-19-963341-X **£19.50**

119. Growth Factors McKay, I. & Leigh, I. (Eds)
.... Spiralbound hardback 0-19-963360-6 **£30.00**
.... Paperback 0-19-963359-2 **£19.50**

118. Histocompatibility Testing Dyer, P. & Middleton, D. (Eds)
.... Spiralbound hardback 0-19-963364-9 **£32.50**
.... Paperback 0-19-963363-0 **£22.50**

117. Gene Transcription Hames, B.D. & Higgins, S.J. (Eds)
.... Spiralbound hardback 0-19-963292-8 **£35.00**
.... Paperback 0-19-963291-X **£25.00**

116. Electrophysiology Wallis, D.I. (Ed)
.... Spiralbound hardback 0-19-963348-7 **£32.50**
.... Paperback 0-19-963347-9 **£22.50**

115. Biological Data Analysis Fry, J.C. (Ed)
.... Spiralbound hardback 0-19-963340-1 **£50.00**
.... Paperback 0-19-963339-8 **£27.50**

114. Experimental Neuroanatomy Bolam, J.P. (Ed)
.... Spiralbound hardback 0-19-963326-6 **£32.50**
.... Paperback 0-19-963325-8 **£22.50**

113. Preparative Centrifugation Rickwood, D. (Ed)
.... Spiralbound hardback 0-19-963208-1 **£45.00**
.... Paperback 0-19-963211-1 **£25.00**
.... Paperback 0-19-963099-2 **£25.00**

112. Lipid Analysis Hamilton, R.J. & Hamilton, Shiela (Eds)
.... Spiralbound hardback 0-19-963098-4 **£35.00**
.... Paperback 0-19-963099-2 **£25.00**

111. Haemopoiesis Testa, N.G. & Molineux, G. (Eds)
.... Spiralbound hardback 0-19-963366-5 **£32.50**
.... Paperback 0-19-963365-7 **£22.50**

110. Pollination Ecology Dafni, A.
.... Spiralbound hardback 0-19-963299-5 **£32.50**
.... Paperback 0-19-963298-7 **£22.50**

109. In Situ Hybridization Wilkinson, D.G. (Ed)
.... Spiralbound hardback 0-19-963328-2 **£30.00**
.... Paperback 0-19-963327-4 **£18.50**

108. Protein Engineering Rees, A.R., Sternberg, M.J.E. & others (Eds)
.... Spiralbound hardback 0-19-963139-5 **£35.00**
.... Paperback 0-19-963138-7 **£25.00**

107. Cell-Cell Interactions Stevenson, B.R., Gallin, W.J. & others (Eds)
.... Spiralbound hardback 0-19-963319-3 **£32.50**
.... Paperback 0-19-963318-5 **£22.50**

106. Diagnostic Molecular Pathology: Volume I Herrington, C.S. & McGee, J. O'D. (Eds)
.... Spiralbound hardback 0-19-963237-5 **£30.00**
.... Paperback 0-19-963236-7 **£19.50**

105. Biomechanics-Materials Vincent, J.F.V. (Ed)
.... Spiralbound hardback 0-19-963223-5 **£35.00**
.... Paperback 0-19-963222-7 **£25.00**

104. Animal Cell Culture (2/e) Freshney, R.I. (Ed)
.... Spiralbound hardback 0-19-963212-X **£30.00**
.... Paperback 0-19-963213-8 **£19.50**

103. Molecular Plant Pathology: Volume II Gurr, S.J., McPherson, M.J. & others (Eds)
.... Spiralbound hardback 0-19-963352-5 **£32.50**
.... Paperback 0-19-963351-7 **£22.50**

102 Signal Transduction Milligan, G. (Ed)
.... Spiralbound hardback 0-19-963296-0 **£30.00**
.... Paperback 0-19-963295-2 **£18.50**

101. Protein Targeting Magee, A.I. & Wileman, T. (Eds)
.... Spiralbound hardback 0-19-963206-5 **£32.50**
.... Paperback 0-19-963210-3 **£22.50**

100. Diagnostic Molecular Pathology: Volume II: Cell and Tissue Genotyping Herrington, C.S. & McGee, J.O'D. (Eds)
.... Spiralbound hardback 0-19-963239-1 **£30.00**
.... Paperback 0-19-963238-3 **£19.50**

99. Neuronal Cell Lines Wood, J.N. (Ed)
.... Spiralbound hardback 0-19-963346-0 **£32.50**
.... Paperback 0-19-963345-2 **£22.50**

98. Neural Transplantation Dunnett, S.B. & Björklund, A. (Eds)
.... Spiralbound hardback 0-19-963286-3 **£30.00**
.... Paperback 0-19-963285-5 **£19.50**

97. Human Cytogenetics: Volume II: Malignancy and Acquired Abnormalities (2/e) Rooney, D.E. & Czepulkowski, B.H. (Eds)
.... Spiralbound hardback 0-19-963290-1 **£30.00**
.... Paperback 0-19-963289-8 **£22.50**

96. Human Cytogenetics: Volume I: Constitutional Analysis (2/e) Rooney, D.E. & Czepulkowski, B.H. (Eds)
.... Spiralbound hardback 0-19-963288-X **£30.00**
.... Paperback 0-19-963287-1 **£22.50**

95. Lipid Modification of Proteins Hooper, N.M. & Turner, A.J. (Eds)
.... Spiralbound hardback 0-19-963274-X **£32.50**
.... Paperback 0-19-963273-1 **£22.50**

94. Biomechanics-Structures and Systems Biewener, A.A. (Ed)
.... Spiralbound hardback 0-19-963268-5 **£42.50**
.... Paperback 0-19-963267-7 **£25.00**

93. Lipoprotein Analysis Converse, C.A. & Skinner, E.R. (Eds)
.... Spiralbound hardback 0-19-963192-1 **£30.00**
.... Paperback 0-19-963231-6 **£19.50**

92. Receptor-Ligand Interactions Hulme, E.C. (Ed)
.... Spiralbound hardback 0-19-963090-9 **£35.00**
.... Paperback 0-19-963091-7 **£27.50**

91. Molecular Genetic Analysis of Populations Hoelzel, A.R. (Ed)
.... Spiralbound hardback 0-19-963278-2 **£32.50**
.... Paperback 0-19-963277-4 **£22.50**

90. Enzyme Assays Eisenthal, R. & Danson, M.J. (Eds)
.... Spiralbound hardback 0-19-963142-5 **£35.00**
.... Paperback 0-19-963143-3 **£25.00**

89. Microcomputers in Biochemistry Bryce, C.F.A. (Ed)
.... Spiralbound hardback 0-19-963253-7 **£30.00**
.... Paperback 0-19-963252-9 **£19.50**

88. The Cytoskeleton Carraway, K.L. & Carraway, C.A.C. (Eds)
.... Spiralbound hardback 0-19-963257-X **£30.00**
.... Paperback 0-19-963256-1 **£19.50**

87. Monitoring Neuronal Activity Stamford, J.A. (Ed)
.... Spiralbound hardback 0-19-963244-8 **£30.00**
.... Paperback 0-19-963243-X **£19.50**

86. Crystallization of Nucleic Acids and Proteins Ducruix, A. & Giegé, R. (Eds)
.... Spiralbound hardback 0-19-963245-6 **£35.00**
.... Paperback 0-19-963246-4 **£25.00**

85. Molecular Plant Pathology: Volume I Gurr, S.J., McPherson, M.J. & others (Eds)
.... Spiralbound hardback 0-19-963103-4 **£30.00**
.... Paperback 0-19-963102-6 **£19.50**

84. Anaerobic Microbiology Levett, P.N. (Ed)
.... Spiralbound hardback 0-19-963204-9 **£32.50**
.... Paperback 0-19-963262-6 **£22.50**

83. Oligonucleotides and Analogues Eckstein, F. (Ed)
.... Spiralbound hardback 0-19-963280-4 **£32.50**
.... Paperback 0-19-963279-0 **£22.50**

82. Electron Microscopy in Biology Harris, R. (Ed)
.... Spiralbound hardback 0-19-963219-7 **£32.50**
.... Paperback 0-19-963215-4 **£22.50**

81. Essential Molecular Biology: Volume II Brown, T.A. (Ed)
.... Spiralbound hardback 0-19-963112-3 **£32.50**
.... Paperback 0-19-963113-1 **£22.50**

80. Cellular Calcium McCormack, J.G. & Cobbold, P.H. (Eds)
.... Spiralbound hardback 0-19-963131-X **£35.00**
.... Paperback 0-19-963130-1 **£25.00**

79. Protein Architecture Lesk, A.M.
.... Spiralbound hardback 0-19-963054-2 **£32.50**
.... Paperback 0-19-963055-0 **£22.50**

78. Cellular Neurobiology Chad, J. & Wheal, H. (Eds)
.... Spiralbound hardback 0-19-963106-9 **£32.50**
.... Paperback 0-19-963107-7 **£22.50**

77. PCR McPherson, M.J., Quirke, P. & others (Eds)
.... Spiralbound hardback 0-19-963226-X **£30.00**
.... Paperback 0-19-963196-4 **£19.50**

76. Mammalian Cell Biotechnology Butler, M. (Ed)
.... Spiralbound hardback 0-19-963207-3 **£30.00**
.... Paperback 0-19-963209-X **£19.50**

75. Cytokines Balkwill, F.R. (Ed)
.... Spiralbound hardback 0-19-963218-9 **£35.00**
.... Paperback 0-19-963214-6 **£25.00**

74. Molecular Neurobiology Chad, J. & Wheal, H. (Eds)
.... Spiralbound hardback 0-19-963108-5 **£30.00**
.... Paperback 0-19-963109-3 **£19.50**

73. Directed Mutagenesis McPherson, M.J. (Ed)
.... Spiralbound hardback 0-19-963141-7 **£30.00**
.... Paperback 0-19-963140-9 **£19.50**

72. Essential Molecular Biology: Volume I Brown, T.A. (Ed)
.... Spiralbound hardback 0-19-963110-7 **£32.50**
.... Paperback 0-19-963111-5 **£22.50**

71. Peptide Hormone Action Siddle, K. & Hutton, J.C.
.... Spiralbound hardback 0-19-963070-4 **£32.50**
.... Paperback 0-19-963071-2 **£22.50**

70. Peptide Hormone Secretion Hutton, J.C. & Siddle, K. (Eds)
.... Spiralbound hardback 0-19-963068-2 **£35.00**
.... Paperback 0-19-963069-0 **£25.00**

69. Postimplantation Mammalian Embryos Copp, A.J. & Cockroft, D.L. (Eds)
.... Spiralbound hardback 0-19-963088-7 **£15.00**
.... Paperback 0-19-963089-5 **£12.50**

68. Receptor-Effector Coupling Hulme, E.C. (Ed)
.... Spiralbound hardback 0-19-963094-1 **£30.00**
.... Paperback 0-19-963095-X **£19.50**

67. Gel Electrophoresis of Proteins (2/e) Hames, B.D. & Rickwood, D. (Eds)
.... Spiralbound hardback 0-19-963074-7 **£35.00**
.... Paperback 0-19-963075-5 **£25.00**

66. Clinical Immunology Gooi, H.C. & Chapel, H. (Eds)
.... Spiralbound hardback 0-19-963086-0 **£32.50**
.... Paperback 0-19-963087-9 **£22.50**

65. Receptor Biochemistry Hulme, E.C. (Ed)
.... Paperback 0-19-963093-3 **£25.00**

64. Gel Electrophoresis of Nucleic Acids (2/e) Rickwood, D. & Hames, B.D. (Eds)
.... Spiralbound hardback 0-19-963082-8 **£32.50**
.... Paperback 0-19-963083-6 **£22.50**

63. Animal Virus Pathogenesis Oldstone, M.B.A. (Ed)
.... Spiralbound hardback 0-19-963100-X **£15.00**
.... Paperback 0-19-963101-8 **£12.50**

62. Flow Cytometry Ormerod, M.G. (Ed)
.... Paperback 0-19-963053-4 **£22.50**

61. Radioisotopes in Biology Slater, R.J. (Ed)
.... Spiralbound hardback 0-19-963080-1 **£32.50**
.... Paperback 0-19-963081-X **£22.50**

60. Biosensors Cass, A.E.C. (Ed)
.... Spiralbound hardback 0-19-963046-1 **£30.00**
.... Paperback 0-19-963047-X **£19.50**

59. Ribosomes and Protein Synthesis Spedding, G. (Ed)
.... Spiralbound hardback 0-19-963104-2 **£15.00**
.... Paperback 0-19-963105-0 **£12.50**

58. Liposomes New, R.R.C. (Ed)
.... Spiralbound hardback 0-19-963076-3 **£35.00**
.... Paperback 0-19-963077-1 **£22.50**

57. Fermentation McNeil, B. & Harvey, L.M. (Eds)
.... Spiralbound hardback 0-19-963044-5 **£30.00**
.... Paperback 0-19-963045-3 **£19.50**

56. Protein Purification Applications Harris, E.L.V. & Angal, S. (Eds)
.... Spiralbound hardback 0-19-963022-4 **£30.00**
.... Paperback 0-19-963023-2 **£18.50**

55. Nucleic Acids Sequencing Howe, C.J. & Ward, E.S. (Eds)
.... Spiralbound hardback 0-19-963056-9 **£30.00**
.... Paperback 0-19-963057-7 **£19.50**

54. Protein Purification Methods Harris, E.L.V. & Angal, S. (Eds)
.... Spiralbound hardback 0-19-963002-X **£30.00**
.... Paperback 0-19-963003-8 **£22.50**

53. Solid Phase Peptide Synthesis Atherton, E. & Sheppard, R.C.
.... Spiralbound hardback 0-19-963066-6 **£15.00**
.... Paperback 0-19-963067-4 **£12.50**

52. Medical Bacteriology Hawkey, P.M. & Lewis, D.A. (Eds)
.... Paperback 0-19-963009-7 **£25.00**

51. Proteolytic Enzymes Beynon, R.J. & Bond, J.S. (Eds)
.... Spiralbound hardback 0-19-963058-5 **£30.00**
.... Paperback 0-19-963059-3 **£19.50**

50. Medical Mycology Evans, E.G.V. & Richardson, M.D. (Eds)
.... Spiralbound hardback 0-19-963010-0 **£37.50**
.... Paperback 0-19-963011-9 **£25.00**

49. Computers in Microbiology Bryant, T.N. & Wimpenny, J.W.T. (Eds)
.... Paperback 0-19-963015-1 **£12.50**

48. Protein Sequencing Findlay, J.B.C. & Geisow, M.J. (Eds)
.... Spiralbound hardback 0-19-963012-7 **£15.00**
.... Paperback 0-19-963013-5 **£12.50**
47. Cell Growth and Division Baserga, R. (Ed)
.... Spiralbound hardback 0-19-963026-7 **£15.00**
.... Paperback 0-19-963027-5 **£12.50**
46. Protein Function Creighton, T.E. (Ed)
.... Spiralbound hardback 0-19-963006-2 **£32.50**
.... Paperback 0-19-963007-0 **£22.50**
45. Protein Structure Creighton, T.E. (Ed)
.... Spiralbound hardback 0-19-963000-3 **£32.50**
.... Paperback 0-19-963001-1 **£22.50**
44. Antibodies: Volume II Catty, D. (Ed)
.... Spiralbound hardback 0-19-963018-6 **£30.00**
.... Paperback 0-19-963019-4 **£19.50**
43. HPLC of Macromolecules Oliver, R.W.A. (Ed)
.... Spiralbound hardback 0-19-963020-8 **£30.00**
.... Paperback 0-19-963021-6 **£19.50**
42. Light Microscopy in Biology Lacey, A.J. (Ed)
.... Spiralbound hardback 0-19-963036-4 **£30.00**
.... Paperback 0-19-963037-2 **£19.50**
41. Plant Molecular Biology Shaw, C.H. (Ed)
.... Paperback 1-85221-056-7 **£12.50**
40. Microcomputers in Physiology Fraser, P.J. (Ed)
.... Spiralbound hardback 1-85221-129-6 **£15.00**
.... Paperback 1-85221-130-X **£12.50**
39. Genome Analysis Davies, K.E. (Ed)
.... Spiralbound hardback 1-85221-109-1 **£30.00**
.... Paperback 1-85221-110-5 **£18.50**
38. Antibodies: Volume I Catty, D. (Ed)
.... Paperback 0-947946-85-3 **£19.50**
37. Yeast Campbell, I. & Duffus, J.H. (Eds)
.... Paperback 0-947946-79-9 **£12.50**
36. Mammalian Development Monk, M. (Ed)
.... Hardback 1-85221-030-3 **£15.00**
.... Paperback 1-85221-029-X **£12.50**
35. Lymphocytes Klaus, G.G.B. (Ed)
.... Hardback 1-85221-018-4 **£30.00**
34. Lymphokines and Interferons Clemens, M.J., Morris, A.G. & others (Eds)
.... Paperback 1-85221-035-4 **£12.50**
33. Mitochondria Darley-Usmar, V.M., Rickwood, D. & others (Eds)
.... Hardback 1-85221-034-6 **£32.50**
.... Paperback 1-85221-033-8 **£22.50**
32. Prostaglandins and Related Substances Benedetto, C., McDonald-Gibson, R.G. & others (Eds)
.... Hardback 1-85221-032-X **£15.00**
.... Paperback 1-85221-031-1 **£12.50**
31. DNA Cloning: Volume III Glover, D.M. (Ed)
.... Hardback 1-85221-049-4 **£15.00**
.... Paperback 1-85221-048-6 **£12.50**
30. Steroid Hormones Green, B. & Leake, R.E. (Eds)
.... Paperback 0-947946-53-5 **£19.50**
29. Neurochemistry Turner, A.J. & Bachelard, H.S. (Eds)
.... Hardback 1-85221-028-1 **£15.00**
.... Paperback 1-85221-027-3 **£12.50**
28. Biological Membranes Findlay, J.B.C. & Evans, W.H. (Eds)
.... Hardback 0-947946-84-5 **£15.00**
.... Paperback 0-947946-83-7 **£12.50**
27. Nucleic Acid and Protein Sequence Analysis Bishop, M.J. & Rawlings, C.J. (Eds)
.... Hardback 1-85221-007-9 **£35.00**
.... Paperback 1-85221-006-0 **£25.00**
26. Electron Microscopy in Molecular Biology Sommerville, J. & Scheer, U. (Eds)
.... Hardback 0-947946-64-0 **£15.00**
.... Paperback 0-947946-54-3 **£12.50**
25. Teratocarcinomas and Embryonic Stem Cells Robertson, E.J. (Ed)
.... Paperback 1-85221-004-4 **£19.50**
24. Spectrophotometry and Spectrofluorimetry Harris, D.A. & Bashford, C.L. (Eds)
.... Hardback 0-947946-69-1 **£15.00**
.... Paperback 0-947946-46-2 **£12.50**
23. Plasmids Hardy, K.G. (Ed)
.... Paperback 0-947946-81-0 **£12.50**
22. Biochemical Toxicology Snell, K. & Mullock, B. (Eds)
.... Paperback 0-947946-52-7 **£12.50**
19. Drosophila Roberts, D.B. (Ed)
.... Hardback 0-947946-66-7 **£32.50**
.... Paperback 0-947946-45-4 **£22.50**
17. Photosynthesis: Energy Transduction Hipkins, M.F. & Baker, N.R. (Eds)
.... Hardback 0-947946-63-2 **£15.00**
.... Paperback 0-947946-51-9 **£12.50**
16. Human Genetic Diseases Davies, K.E. (Ed)
.... Hardback 0-947946-76-4 **£15.00**
.... Paperback 0-947946-75-6 **£12.50**
14. Nucleic Acid Hybridisation Hames, B.D. & Higgins, S.J. (Eds)
.... Hardback 0-947946-61-6 **£15.00**
.... Paperback 0-947946-23-3 **£12.50**
13. Immobilised Cells and Enzymes Woodward, J. (Ed)
.... Hardback 0-947946-60-8 **£15.00**
12. Plant Cell Culture Dixon, R.A. (Ed)
.... Paperback 0-947946-22-5 **£19.50**
11a. DNA Cloning: Volume I Glover, D.M. (Ed)
.... Paperback 0-947946-18-7 **£12.50**
11b. DNA Cloning: Volume II Glover, D.M. (Ed)
.... Paperback 0-947946-19-5 **£12.50**
10. Virology Mahy, B.W.J. (Ed)
.... Paperback 0-904147-78-9 **£19.50**
9. Affinity Chromatography Dean, P.D.G., Johnson, W.S. & others (Eds)
.... Paperback 0-904147-71-1 **£19.50**
7. Microcomputers in Biology Ireland, C.R. & Long, S.P. (Eds)
.... Paperback 0-904147-57-6 **£18.00**
6. Oligonucleotide Synthesis Gait, M.J. (Ed)
.... Paperback 0-904147-74-6 **£18.50**
5. Transcription and Translation Hames, B.D. & Higgins, S.J. (Eds)
.... Paperback 0-904147-52-5 **£12.50**
3. Iodinated Density Gradient Media Rickwood, D. (Ed)
.... Paperback 0-904147-51-7 **£12.50**

Sets

Essential Molecular Biology: 2 vol set Brown, T.A. (Ed)
.... Spiralbound hardback 0-19-963114-X **£58.00**
.... Paperback 0-19-963115-8 **£40.00**
Antibodies: 2 vol set Catty, D. (Ed)
.... Paperback 0-19-963063-1 **£33.00**
Cellular and Molecular Neurobiology: 2 vol set Chad, J. & Wheal, H. (Eds)
.... Spiralbound hardback 0-19-963255-3 **£56.00**
.... Paperback 0-19-963254-5 **£38.00**
Protein Structure and Protein Function: 2 vol set Creighton, T.E. (Ed)
.... Spiralbound hardback 0-19-963064-X **£55.00**
.... Paperback 0-19-963065-8 **£38.00**
DNA Cloning: 2 vol set Glover, D.M. (Ed)
.... Paperback 1-85221-069-9 **£30.00**
Molecular Plant Pathology: 2 vol set Gurr, S.J., McPherson, M.J. & others (Eds)
.... Spiralbound hardback 0-19-963354-1 **£56.00**
.... Paperback 0-19-963353-3 **£37.00**
Protein Purification Methods, and Protein Purification Applications: 2 vol set Harris, E.L.V. & Angal, S. (Eds)
.... Spiralbound hardback 0-19-963048-8 **£48.00**
.... Paperback 0-19-963049-6 **£32.00**
Diagnostic Molecular Pathology: 2 vol set Herrington, C.S. & McGee, J. O'D. (Eds)
.... Spiralbound hardback 0-19-963241-3 **£54.00**
.... Paperback 0-19-963240-5 **£35.00**
RNA Processing: 2 vol set Higgins, S.J. & Hames, B.D. (Eds)
.... Spiralbound hardback 0-19-963473-4 **£54.00**
.... Paperback 0-19-963472-6 **£35.00**
Receptor Biochemistry; Receptor-Effector Coupling; Receptor-Ligand Interactions: 3 vol set Hulme, E.C. (Ed)
.... Paperback 0-19-963097-6 **£62.50**
Human Cytogenetics: 2 vol set (2/e) Rooney, D.E. & Czepulkowski, B.H. (Eds)
.... Hardback 0-19-963314-2 **£58.50**
.... Paperback 0-19-963313-4 **£40.50**
Behavioural Neuroscience: 2 vol set Sahgal, A. (Ed)
.... Spiralbound hardback 0-19-963460-2 **£58.00**
.... Paperback 0-19-963459-9 **£40.00**
Peptide Hormone Secretion/Peptide Hormone Action: 2 vol set Siddle, K. & Hutton, J.C. (Eds)
.... Spiralbound hardback 0-19-963072-0 **£55.00**
.... Paperback 0-19-963073-9 **£38.00**

ORDER FORM for UK, Europe and Rest of World

(Excluding USA and Canada)

Qty	ISBN	Author	Title	Amount
			P&P	
			*VAT	
			TOTAL	

Please add postage and packing: £1.75 for UK orders under £20; £2.75 for UK orders over £20; overseas orders add 10% of total.
* EC customers please note that VAT must be added (excludes UK customers)

Name ....

Address ....

....

.... Post code ....

[ ] Please charge £ .... to my credit card

Access/VISA/Eurocard/AMEX/Diners Club (circle appropriate card)

Card No .... Expiry date ....

Signature ....

Credit card account address if different from above:

....

.... Postcode ....

[ ] I enclose a cheque for £....................

Please return this form to: OUP Distribution Services, Saxon Way West, Corby, Northants NN18 9ES, UK

OR ORDER BY CREDIT CARD HOTLINE: Tel +44-(0)536-741519 or
Fax +44-(0)536-746337

ORDER OTHER TITLES OF INTEREST TODAY

Price list for: USA and Canada

128. Behavioural Neuroscience: Volume I Sahgal, A. (Ed)
.... Spiralbound hardback 0-19-963368-1 **$57.00**
.... Paperback 0-19-963367-3 **$37.00**

127. Molecular Virology Davison, A.J. & Elliott, R.M. (Eds)
.... Spiralbound hardback 0-19-963358-4 **$49.00**
.... Paperback 0-19-963357-6 **$32.00**

126. Gene Targeting Joyner, A.L. (Ed)
.... Spiralbound hardback 0-19-963407-6 **$49.00**
.... Paperback 0-19-9634036-8 **$34.00**

124. Human Genetic Disease Analysis (2/e) Davies, K.E. (Ed)
.... Spiralbound hardback 0-19-963309-6 **$54.00**
.... Paperback 0-19-963308-8 **$33.00**

123. Protein Phosphorylation Hardie, D.G. (Ed)
.... Spiralbound hardback 0-19-963306-1 **$65.00**
.... Paperback 0-19-963305-3 **$45.00**

122. Immunocytochemistry Beesley, J. (Ed)
.... Spiralbound hardback 0-19-963270-7 **$62.00**
.... Paperback 0-19-963269-3 **$42.00**

121. Tumour Immunobiology Gallagher, G., Rees, R.C. & others (Eds)
.... Spiralbound hardback 0-19-963370-3 **$72.00**
.... Paperback 0-19-963369-X **$50.00**

120. Transcription Factors Latchman, D.S. (Ed)
.... Spiralbound hardback 0-19-963342-8 **$48.00**
.... Paperback 0-19-963341-X **$31.00**

119. Growth Factors McKay, I. & Leigh, I. (Eds)
.... Spiralbound hardback 0-19-963360-6 **$48.00**
.... Paperback 0-19-963359-2 **$31.00**

118. Histocompatibility Testing Dyer, P. & Middleton, D. (Eds)
.... Spiralbound hardback 0-19-963364-9 **$60.00**
.... Paperback 0-19-963363-0 **$41.00**

117. Gene Transcription Hames, B.D. & Higgins, S.J. (Eds)
.... Spiralbound hardback 0-19-963292-8 **$72.00**
.... Paperback 0-19-963291-X **$50.00**

116. Electrophysiology Wallis, D.I. (Ed)
.... Spiralbound hardback 0-19-963348-7 **$56.00**
.... Paperback 0-19-963347-9 **$39.00**

115. Biological Data Analysis Fry, J.C. (Ed)
.... Spiralbound hardback 0-19-963340-1 **$80.00**
.... Paperback 0-19-963339-8 **$60.00**

114. Experimental Neuroanatomy Bolam, J.P. (Ed)
.... Spiralbound hardback 0-19-963326-6 **$59.00**
.... Paperback 0-19-963325-8 **$39.00**

113. Preparative Centrifugation Rickwood, D. (Ed)
.... Spiralbound hardback 0-19-963208-1 **$78.00**
.... Paperback 0-19-963211-1 **$44.00**

111. Haemopoiesis Testa, N.G. & Molineux, G. (Eds)
.... Spiralbound hardback 0-19-963366-5 **$59.00**
.... Paperback 0-19-963365-7 **$39.00**

110. Pollination Ecology Dafni, A.
.... Spiralbound hardback 0-19-963299-5 **$56.95**
.... Paperback 0-19-963298-7 **$39.95**

109. In Situ Hybridization Wilkinson, D.G. (Ed)
.... Spiralbound hardback 0-19-963328-2 **$58.00**
.... Paperback 0-19-963327-4 **$36.00**

108. Protein Engineering Rees, A.R., Sternberg, M.J.E. & others (Eds)
.... Spiralbound hardback 0-19-963139-5 **$64.00**
.... Paperback 0-19-963138-7 **$44.00**

107. Cell-Cell Interactions Stevenson, B.R., Gallin, W.J. & others (Eds)
.... Spiralbound hardback 0-19-963319-3 **$55.00**
.... Paperback 0-19-963318-5 **$38.00**

106. Diagnostic Molecular Pathology: Volume I Herrington, C.S. & McGee, J. O'D. (Eds)
.... Spiralbound hardback 0-19-963237-5 **$50.00**
.... Paperback 0-19-963236-7 **$33.00**

105. Biomechanics-Materials Vincent, J.F.V. (Ed)
.... Spiralbound hardback 0-19-963223-5 **$70.00**
.... Paperback 0-19-963222-7 **$50.00**

104. Animal Cell Culture (2/e) Freshney, R.I. (Ed)
.... Spiralbound hardback 0-19-963212-X **$55.00**
.... Paperback 0-19-963213-8 **$35.00**

103. Molecular Plant Pathology: Volume II Gurr, S.J., McPherson, M.J. & others (Eds)
.... Spiralbound hardback 0-19-963352-5 **$65.00**
.... Paperback 0-19-963351-7 **$45.00**

102. Signal Transduction Milligan, G. (Ed)
.... Spiralbound hardback 0-19-963296-0 **$60.00**
.... Paperback 0-19-963295-2 **$38.00**

101. Protein Targeting Magee, A.I. & Wileman, T. (Eds)
.... Spiralbound hardback 0-19-963206-5 **$75.00**
.... Paperback 0-19-963210-3 **$50.00**

100. Diagnostic Molecular Pathology: Volume II: Cell and Tissue Genotyping Herrington, C.S. & McGee, J.O'D. (Eds)
.... Spiralbound hardback 0-19-963239-1 **$60.00**
.... Paperback 0-19-963238-3 **$39.00**

99. Neuronal Cell Lines Wood, J.N. (Ed)
.... Spiralbound hardback 0-19-963346-0 **$68.00**
.... Paperback 0-19-963345-2 **$48.00**

98. Neural Transplantation Dunnett, S.B. & Bj‹148›rklund, A. (Eds)
.... Spiralbound hardback 0-19-963286-3 **$69.00**
.... Paperback 0-19-963285-5 **$42.00**

97. Human Cytogenetics: Volume II: Malignancy and Acquired Abnormalities (2/e) Rooney, D.E. & Czepulkowski, B.H. (Eds)
.... Spiralbound hardback 0-19-963290-1 **$75.00**
.... Paperback 0-19-963289-8 **$50.00**

96. Human Cytogenetics: Volume I: Constitutional Analysis (2/e) Rooney, D.E. & Czepulkowski, B.H. (Eds)
.... Spiralbound hardback 0-19-963288-X **$75.00**
.... Paperback 0-19-963287-1 **$50.00**

95. Lipid Modification of Proteins Hooper, N.M. & Turner, A.J. (Eds)
.... Spiralbound hardback 0-19-963274-X **$75.00**
.... Paperback 0-19-963273-1 **$50.00**

94. Biomechanics-Structures and Systems Biewener, A.A. (Ed)
.... Spiralbound hardback 0-19-963268-5 **$85.00**
.... Paperback 0-19-963267-7 **$50.00**

93. Lipoprotein Analysis Converse, C.A. & Skinner, E.R. (Eds)
.... Spiralbound hardback 0-19-963192-1 **$65.00**
.... Paperback 0-19-963231-6 **$42.00**

92. Receptor-Ligand Interactions Hulme, E.C. (Ed)
.... Spiralbound hardback 0-19-963090-9 **$75.00**
.... Paperback 0-19-963091-7 **$50.00**

91. Molecular Genetic Analysis of Populations Hoelzel, A.R. (Ed)
.... Spiralbound hardback 0-19-963278-2 **$65.00**
.... Paperback 0-19-963277-4 **$45.00**

90. **Enzyme Assays** Eisenthal, R. & Danson, M.J. (Eds)
.... Spiralbound hardback 0-19-963142-5 **$68.00**
.... Paperback 0-19-963143-3 **$48.00**
89. **Microcomputers in Biochemistry** Bryce, C.F.A. (Ed)
.... Spiralbound hardback 0-19-963253-7 **$60.00**
.... Paperback 0-19-963252-9 **$40.00**
88. **The Cytoskeleton** Carraway, K.L. & Carraway, C.A.C. (Eds)
.... Spiralbound hardback 0-19-963257-X **$60.00**
.... Paperback 0-19-963256-1 **$40.00**
87. **Monitoring Neuronal Activity** Stamford, J.A. (Ed)
.... Spiralbound hardback 0-19-963244-8 **$60.00**
.... Paperback 0-19-963243-X **$40.00**
86. **Crystallization of Nucleic Acids and Proteins** Ducruix, A. & Giegé, R. (Eds)
.... Spiralbound hardback 0-19-963245-6 **$60.00**
.... Paperback 0-19-963246-4 **$50.00**
85. **Molecular Plant Pathology: Volume I** Gurr, S.J., McPherson, M.J. & others (Eds)
.... Spiralbound hardback 0-19-963103-4 **$60.00**
.... Paperback 0-19-963102-6 **$40.00**
84. **Anaerobic Microbiology** Levett, P.N. (Ed)
.... Spiralbound hardback 0-19-963204-9 **$75.00**
.... Paperback 0-19-963262-6 **$45.00**
83. **Oligonucleotides and Analogues** Eckstein, F. (Ed)
.... Spiralbound hardback 0-19-963280-4 **$65.00**
.... Paperback 0-19-963279-0 **$45.00**
82. **Electron Microscopy in Biology** Harris, R. (Ed)
.... Spiralbound hardback 0-19-963219-7 **$65.00**
.... Paperback 0-19-963215-4 **$45.00**
81. **Essential Molecular Biology: Volume II** Brown, T.A. (Ed)
.... Spiralbound hardback 0-19-963112-3 **$65.00**
.... Paperback 0-19-963113-1 **$45.00**
80. **Cellular Calcium** McCormack, J.G. & Cobbold, P.H. (Eds)
.... Spiralbound hardback 0-19-963131-X **$75.00**
.... Paperback 0-19-963130-1 **$50.00**
79. **Protein Architecture** Lesk, A.M.
.... Spiralbound hardback 0-19-963054-2 **$65.00**
.... Paperback 0-19-963055-0 **$45.00**
78. **Cellular Neurobiology** Chad, J. & Wheal, H. (Eds)
.... Spiralbound hardback 0-19-963106-9 **$73.00**
.... Paperback 0-19-963107-7 **$43.00**
77. **PCR** McPherson, M.J., Quirke, P. & others (Eds)
.... Spiralbound hardback 0-19-963226-X **$55.00**
.... Paperback 0-19-963196-4 **$40.00**
76. **Mammalian Cell Biotechnology** Butler, M. (Ed)
.... Spiralbound hardback 0-19-963207-3 **$60.00**
.... Paperback 0-19-963209-X **$40.00**
75. **Cytokines** Balkwill, F.R. (Ed)
.... Spiralbound hardback 0-19-963218-9 **$64.00**
.... Paperback 0-19-963214-6 **$44.00**
74. **Molecular Neurobiology** Chad, J. & Wheal, H. (Eds)
.... Spiralbound hardback 0-19-963108-5 **$56.00**
.... Paperback 0-19-963109-3 **$36.00**
73. **Directed Mutagenesis** McPherson, M.J. (Ed)
.... Spiralbound hardback 0-19-963141-7 **$55.00**
.... Paperback 0-19-963140-9 **$35.00**
72. **Essential Molecular Biology: Volume I** Brown, T.A. (Ed)
.... Spiralbound hardback 0-19-963110-7 **$65.00**
.... Paperback 0-19-963111-5 **$45.00**
71. **Peptide Hormone Action** Siddle, K. & Hutton, J.C.
.... Spiralbound hardback 0-19-963070-4 **$70.00**
.... Paperback 0-19-963071-2 **$50.00**
70. **Peptide Hormone Secretion** Hutton, J.C. & Siddle, K. (Eds)
.... Spiralbound hardback 0-19-963068-2 **$70.00**
.... Paperback 0-19-963069-0 **$50.00**
69. **Postimplantation Mammalian Embryos** Copp, A.J. & Cockroft, D.L. (Eds)
.... Spiralbound hardback 0-19-963088-7 **$70.00**
.... Paperback 0-19-963089-5 **$50.00**
68. **Receptor-Effector Coupling** Hulme, E.C. (Ed)
.... Spiralbound hardback 0-19-963094-1 **$70.00**
.... Paperback 0-19-963095-X **$45.00**
67. **Gel Electrophoresis of Proteins (2/e)** Hames, B.D. & Rickwood, D. (Eds)
.... Spiralbound hardback 0-19-963074-7 **$75.00**
.... Paperback 0-19-963075-5 **$50.00**
66. **Clinical Immunology** Gooi, H.C. & Chapel, H. (Eds)
.... Spiralbound hardback 0-19-963086-0 **$69.95**
.... Paperback 0-19-963087-9 **$50.00**
65. **Receptor Biochemistry** Hulme, E.C. (Ed)
.... Paperback 0-19-963093-3 **$50.00**
64. **Gel Electrophoresis of Nucleic Acids (2/e)** Rickwood, D. & Hames, B.D. (Eds)
.... Spiralbound hardback 0-19-963082-8 **$75.00**
.... Paperback 0-19-963083-6 **$50.00**
63. **Animal Virus Pathogenesis** Oldstone, M.B.A. (Ed)
.... Spiralbound hardback 0-19-963100-X **$68.00**
.... Paperback 0-19-963101-8 **$40.00**
62. **Flow Cytometry** Ormerod, M.G. (Ed)
.... Paperback 0-19-963053-4 **$50.00**
61. **Radioisotopes in Biology** Slater, R.J. (Ed)
.... Spiralbound hardback 0-19-963080-1 **$75.00**
.... Paperback 0-19-963081-X **$45.00**
60. **Biosensors** Cass, A.E.G. (Ed)
.... Spiralbound hardback 0-19-963046-1 **$65.00**
.... Paperback 0-19-963047-X **$43.00**
59. **Ribosomes and Protein Synthesis** Spedding, G. (Ed)
.... Spiralbound hardback 0-19-963104-2 **$75.00**
.... Paperback 0-19-963105-0 **$45.00**
58. **Liposomes** New, R.R.C. (Ed)
.... Spiralbound hardback 0-19-963076-3 **$70.00**
.... Paperback 0-19-963077-1 **$45.00**
57. **Fermentation** McNeil, B. & Harvey, L.M. (Eds)
.... Spiralbound hardback 0-19-963044-5 **$65.00**
.... Paperback 0-19-963045-3 **$39.00**
56. **Protein Purification Applications** Harris, E.L.V. & Angal, S. (Eds)
.... Spiralbound hardback 0-19-963022-4 **$54.00**
.... Paperback 0-19-963023-2 **$36.00**
55. **Nucleic Acids Sequencing** Howe, C.J. & Ward, E.S. (Eds)
.... Spiralbound hardback 0-19-963056-9 **$59.00**
.... Paperback 0-19-963057-7 **$38.00**
54. **Protein Purification Methods** Harris, E.L.V. & Angal, S. (Eds)
.... Spiralbound hardback 0-19-963002-X **$60.00**
.... Paperback 0-19-963003-8 **$40.00**
53. **Solid Phase Peptide Synthesis** Atherton, E. & Sheppard, R.C.
.... Spiralbound hardback 0-19-963066-6 **$58.00**
.... Paperback 0-19-963067-4 **$39.95**
52. **Medical Bacteriology** Hawkey, P.M. & Lewis, D.A. (Eds)
.... Paperback 0-19-963009-7 **$50.00**
51. **Proteolytic Enzymes** Beynon, R.J. & Bond, J.S. (Eds)
.... Spiralbound hardback 0-19-963058-5 **$60.00**
.... Paperback 0-19-963059-3 **$39.00**
50. **Medical Mycology** Evans, E.G.V. & Richardson, M.D. (Eds)
.... Spiralbound hardback 0-19-963010-0 **$69.95**
.... Paperback 0-19-963011-9 **$50.00**
49. **Computers in Microbiology** Bryant, T.N. & Wimpenny, J.W.T. (Eds)
.... Paperback 0-19-963015-1 **$40.00**
48. **Protein Sequencing** Findlay, J.B.C. & Geisow, M.J. (Eds)
.... Spiralbound hardback 0-19-963012-7 **$56.00**
.... Paperback 0-19-963013-5 **$38.00**
47. **Cell Growth and Division** Baserga, R. (Ed)
.... Spiralbound hardback 0-19-963026-7 **$62.00**
.... Paperback 0-19-963027-5 **$38.00**
46. **Protein Function** Creighton, T.E. (Ed)
.... Spiralbound hardback 0-19-963006-2 **$65.00**
.... Paperback 0-19-963007-0 **$45.00**
45. **Protein Structure** Creighton, T.E. (Ed)
.... Spiralbound hardback 0-19-963000-3 **$65.00**
.... Paperback 0-19-963001-1 **$45.00**
44. **Antibodies: Volume II** Catty, D. (Ed)
.... Spiralbound hardback 0-19-963018-6 **$58.00**
.... Paperback 0-19-963019-4 **$39.00**
43. **HPLC of Macromolecules** Oliver, R.W.A. (Ed)
.... Spiralbound hardback 0-19-963020-8 **$54.00**
.... Paperback 0-19-963021-6 **$45.00**
42. **Light Microscopy in Biology** Lacey, A.J. (Ed)
.... Spiralbound hardback 0-19-963036-4 **$62.00**
.... Paperback 0-19-963037-2 **$38.00**
41. **Plant Molecular Biology** Shaw, C.H. (Ed)
.... Paperback 1-85221-056-7 **$38.00**
40. **Microcomputers in Physiology** Fraser, P.J. (Ed)
.... Spiralbound hardback 1-85221-129-6 **$54.00**
.... Paperback 1-85221-130-X **$36.00**
39. **Genome Analysis** Davies, K.E. (Ed)
.... Spiralbound hardback 1-85221-109-1 **$54.00**
.... Paperback 1-85221-110-5 **$36.00**
38. **Antibodies: Volume I** Catty, D. (Ed)
.... Paperback 0-947946-85-3 **$38.00**
37. **Yeast** Campbell, I. & Duffus, J.H. (Eds)
.... Paperback 0-947946-79-9 **$36.00**

36. Mammalian Development Monk, M. (Ed)
.... Hardback 1-85221-030-3 **$60.00**
.... Paperback 1-85221-029-X **$45.00**
35. Lymphocytes Klaus, G.G.B. (Ed)
.... Hardback 1-85221-018-4 **$54.00**
34. Lymphokines and Interferons Clemens, M.J., Morris, A.G. & others (Eds)
.... Paperback 1-85221-035-4 **$44.00**
33. Mitochondria Darley-Usmar, V.M., Rickwood, D. & others (Eds)
.... Hardback 1-85221-034-6 **$65.00**
.... Paperback 1-85221-033-8 **$45.00**
32. Prostaglandins and Related Substances Benedetto, C., McDonald-Gibson, R.G. & others (Eds)
.... Hardback 1-85221-032-X **$58.00**
.... Paperback 1-85221-031-1 **$38.00**
31. DNA Cloning: Volume III Glover, D.M. (Ed)
.... Hardback 1-85221-049-4 **$56.00**
.... Paperback 1-85221-048-6 **$36.00**
30. Steroid Hormones Green, B. & Leake, R.E. (Eds)
.... Paperback 0-947946-53-5 **$40.00**
29. Neurochemistry Turner, A.J. & Bachelard, H.S. (Eds)
.... Hardback 1-85221-028-1 **$56.00**
.... Paperback 1-85221-027-3 **$36.00**
28. Biological Membranes Findlay, J.B.C. & Evans, W.H. (Eds)
.... Hardback 0-947946-84-5 **$54.00**
.... Paperback 0-947946-83-7 **$36.00**
27. Nucleic Acid and Protein Sequence Analysis Bishop, M.J. & Rawlings, C.J. (Eds)
.... Hardback 1-85221-007-9 **$66.00**
.... Paperback 1-85221-006-0 **$44.00**
26. Electron Microscopy in Molecular Biology Sommerville, J. & Scheer, U. (Eds)
.... Hardback 0-947946-64-0 **$54.00**
.... Paperback 0-947946-54-3 **$40.00**
24. Spectrophotometry and Spectrofluorimetry Harris, D.A. & Bashford, C.L. (Eds)
.... Hardback 0-947946-69-1 **$56.00**
.... Paperback 0-947946-46-2 **$39.95**
23. Plasmids Hardy, K.G. (Ed)
.... Paperback 0-947946-81-0 **$36.00**
22. Biochemical Toxicology Snell, K. & Mullock, B. (Eds)
.... Paperback 0-947946-52-7 **$40.00**
19. Drosophila Roberts, D.B. (Ed)
.... Hardback 0-947946-66-7 **$67.50**
.... Paperback 0-947946-45-4 **$46.00**
17. Photosynthesis: Energy Transduction Hipkins, M.F. & Baker, N.R. (Eds)
.... Hardback 0-947946-63-2 **$54.00**
.... Paperback 0-947946-51-9 **$36.00**
16. Human Genetic Diseases Davies, K.E. (Ed)
.... Hardback 0-947946-76-4 **$60.00**
.... Paperback 0-947946-75-6 **$34.00**
14. Nucleic Acid Hybridisation Hames, B.D. & Higgins, S.J. (Eds)
.... Hardback 0-947946-61-6 **$60.00**
.... Paperback 0-947946-23-3 **$36.00**
12. Plant Cell Culture Dixon, R.A. (Ed)
.... Paperback 0-947946-22-5 **$36.00**
11a. DNA Cloning: Volume I Glover, D.M. (Ed)
.... Paperback 0-947946-18-7 **$36.00**
11b. DNA Cloning: Volume II Glover, D.M. (Ed)
.... Paperback 0-947946-19-5 **$36.00**
10. Virology Mahy, B.W.J. (Ed)
.... Paperback 0-904147-78-9 **$40.00**
9. Affinity Chromatography Dean, P.D.G., Johnson, W.S. & others (Eds)
.... Paperback 0-904147-71-1 **$36.00**
7. Microcomputers in Biology Ireland, C.R. & Long, S.P. (Eds)
.... Paperback 0-904147-57-6 **$36.00**
6. Oligonucleotide Synthesis Gait, M.J. (Ed)
.... Paperback 0-904147-74-6 **$38.00**
5. Transcription and Translation Hames, B.D. & Higgins, S.J. (Eds)
.... Paperback 0-904147-52-5 **$38.00**
3. Iodinated Density Gradient Media Rickwood, D. (Ed)
.... Paperback 0-904147-51-7 **$36.00**

Sets

Essential Molecular Biology: 2 vol set Brown, T.A. (Ed)
.... Spiralbound hardback 0-19-963114-X **$118.00**
.... Paperback 0-19-963115-8 **$78.00**
Antibodies: 2 vol set Catty, D. (Ed)
.... Paperback 0-19-963063-1 **$70.00**
Cellular and Molecular Neurobiology: 2 vol set Chad, J. & Wheal, H. (Eds)
.... Spiralbound hardback 0-19-963255-3 **$133.00**
.... Paperback 0-19-963254-5 **$79.00**
Protein Structure and Protein Function: 2 vol set Creighton, T.E. (Ed)
.... Spiralbound hardback 0-19-963064-X **$114.00**
.... Paperback 0-19-963065-8 **$80.00**
DNA Cloning: 2 vol set Glover, D.M. (Ed)
.... Paperback 1-85221-069-9 **$92.00**
Molecular Plant Pathology: 2 vol set Gurr, S.J., McPherson, M.J. & others (Eds)
.... Spiralbound hardback 0-19-963354-1 **$110.00**
.... Paperback 0-19-963353-3 **$75.00**
Protein Purification Methods, and Protein Purification Applications: 2 vol set Harris, E.L.V. & Angal, S. (Eds)
.... Spiralbound hardback 0-19-963048-8 **$98.00**
.... Paperback 0-19-963049-6 **$68.00**
Diagnostic Molecular Pathology: 2 vol set Herrington, C.S. & McGee, J. O'D. (Eds)
.... Spiralbound hardback 0-19-963241-3 **$105.00**
.... Paperback 0-19-963240-5 **$69.00**
Receptor Biochemistry; Receptor-Effector Coupling; Receptor-Ligand Interactions: 3 vol set Hulme, E.C. (Ed)
.... Paperback 0-19-963097-6 **$130.00**
Human Cytogenetics: (2/e): 2 vol set Rooney, D.E. & Czepulkowski, B.H. (Eds)
.... Hardback 0-19-963314-2 **$130.00**
.... Paperback 0-19-963313-4 **$90.00**
Peptide Hormone Secretion/Peptide Hormone Action: 2 vol set Siddle, K. & Hutton, J.C. (Eds)
.... Spiralbound hardback 0-19-963072-0 **$135.00**
.... Paperback 0-19-963073-9 **$90.00**

ORDER FORM for USA and Canada

Qty	ISBN	Author	Title	Amount
			S&H	
			CA and NC residents add appropriate sales tax	
			TOTAL	

Please add shipping and handling: US $2.50 for first book, (US $1.00 each book thereafter)

Name ....

Address ....

....

.... Zip ....

[ ] Please charge $ .... to my credit card
Mastercard/VISA/American Express (circle appropriate card)

Acct. .... Expiry date ....

Signature ....

Credit card account address if different from above:

....

.... Zip ....

[ ] I enclose a cheque for US $...........

Mail orders to: Order Dept. Oxford University Press, 2001 Evans Road, Cary, NC 27513